T0396581

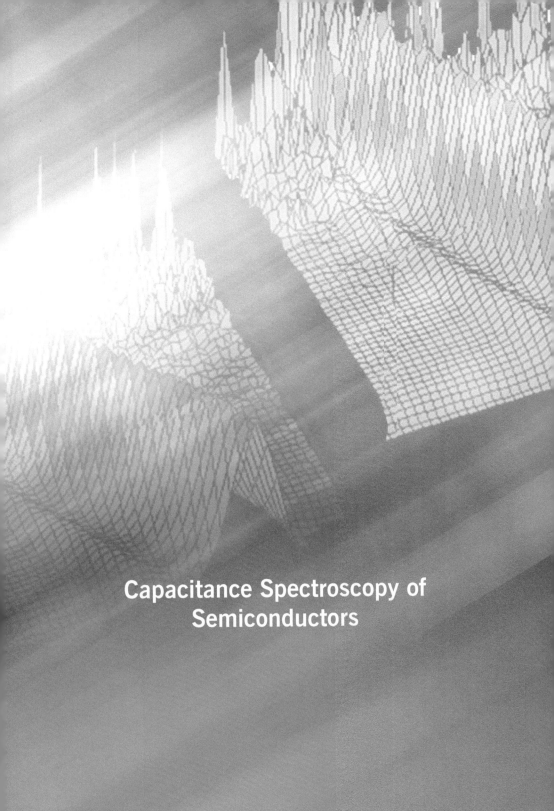

Capacitance Spectroscopy of Semiconductors

Capacitance Spectroscopy of Semiconductors

edited by

Jian V. Li
Giorgio Ferrari

PAN STANFORD PUBLISHING

Published by

Pan Stanford Publishing Pte. Ltd.
Penthouse Level, Suntec Tower 3
8 Temasek Boulevard
Singapore 038988

Email: editorial@panstanford.com
Web: www.panstanford.com

British Library Cataloguing-in-Publication Data
A catalogue record for this book is available from the British Library.

Capacitance Spectroscopy of Semiconductors

ISBN 978-981-4774-54-3 (Hardcover)
ISBN 978-1-315-15013-0 (eBook)

Contents

SECTION III: APPLICATIONS

8. Comparison of Capacitance Spectroscopy for PV Semiconductors 221

Adam Halverson

Preface

Capacitance spectroscopy techniques refer to a loosely coupled collection of experimental methods for inspecting the alternating-current responses of materials and devices with respect to bias voltage, modulation amplitude, frequency, and time. Since their invention starting in the 1950s, they have made increasingly important contributions to modern semiconductor research. Earlier developments were thoroughly captured by Blood and Orton's book *The Electrical Characterization of Semiconductors: Majority Carriers and Electron States* (1992). Although the fundamental principles involved have not changed, the subject matter embraced by this book has since grown in importance and in complexity. New techniques have been invented, new analysis have been developed, and the instrumentation has undergone vast improvement. The applications of capacitance spectroscopy have also proliferated to previously unexplored materials and devices, especially in this age of nanotechnology. With this in mind, the editors and the chapter authors came together and decided to work on this volume.

We have attempted to arrange the sections and chapters in a logical sequence to assist the reader in finding a particular subject. The book is organized into four sections. Section I focuses on the physical principles. Chapter 1 presents an introduction to the basic principles of the capacitance spectroscopy technique. Chapters 2, 3, and 4 provide detailed accounts as well as last-minute progress report on the conventional capacitance spectroscopy techniques of admittance spectroscopy, deep-level transient spectroscopy, capacitance-voltage profiling, and its more advanced cousin of drive-level capacitance profiling. Section II focuses on the electronic instrumentation. Chapter 5 reviews the main techniques and methods used to perform capacitance and impedance spectroscopy. Chapter 6 examines advanced circuits and techniques aimed at high-sensitivity capacitance measurements. The time-domain approach for the measurement of impedance is comprehensively covered in Chapter 7. Section III provides reviews on the applications of capacitance spectroscopy to major fields of semiconductor research. Chapter 8 compares the applications to mainstream photovoltaic materials. Chapter 9 discusses how to apply capacitance spectroscopy techniques in the field of organic semiconductors. Chapter 10 summarizes the applications to the technologically important metal–oxide–semiconductor system. Capacitance-based techniques for the characterization of quantum dots and single-charge devices are introduced and reviewed in Chapter 11. Section IV focuses on emerging technologies and pays special attention to those that are based on the scanning probe microscope and enable the study of properties and phenomena at nanometer scales. Chapter 12 describes the high-frequency scanning capacitance microscopy that is commercially available. Chapter 13 summarizes the recent achievements in the quantitative measurement of dielectric constant using scanning probe techniques. Chapter 14 describes the development of a low-to-medium-frequency operation of the conventional capacitance spectroscopies toward nanoscale resolution. Chapter 15 reviews the principle and progress of scanning microwave microscopy.

We are indebted to all those who helped in making this book possible. First, we would like to thank all our colleagues who accepted the invitation to write the chapters for the book. Their work will help both students and established researchers learn basic and advanced techniques and applications of capacitance spectroscopy. J. V. L. would like to thank Steve W. Johnston, Richard S. Crandall, Jennifer T. Heath, David L. Young, Adam Halverson, and D. Westley Miller for helpful discussions on the subject. G. F. would like to thank Marco Sampietro and all the past and present members of the I3N Laboratory for the inspiring environment. We are grateful to Stanford Chong and the editorial team at Pan Stanford Publishing for their helpful cooperation and prompt actions. Finally, we wish to thank our families for their constant support.

Jian V. Li
Giorgio Ferrari

SECTION I: PHYSICS

Chapter 1

An Introduction to Capacitance Spectroscopy in Semiconductors

Jian V. Li[a] and Jennifer T. Heath[b]

[a]*Department of Aeronautics and Astronautics,*
National Cheng Kung University, Tainan, Taiwan
[b]*Department of Physics,*
Linfield College, McMinnville, Oregon 97128, USA

jianvli@mail.ncku.edu.tw, jheath@linfield.edu

The term "capacitance spectroscopy" encompasses a suite of measurements that inspect the charge responses of a device to various experimental controls such as temperature, frequency, time, and electric potential. These measurements yield useful information regarding the electrical properties of free carriers as well as bulk and interfacial electronic states energetically located in the bandgap. An incomplete list of extractable properties include carrier density, its dependence on the depletion width, the built-in voltage, interface charge, the energy level of in-gap states, their capture cross section, and density. In this chapter, we describe the concept of capacitance (Section 1.1), the related

Capacitance Spectroscopy of Semiconductors
Edited by Jian V. Li and Giorgio Ferrari
Copyright © 2018 Pan Stanford Publishing Pte. Ltd.
ISBN 978-981-4774-54-3 (Hardcover), 978-1-315-15013-0 (eBook)
www.panstanford.com

electric quantities, the associated physical phenomena, major types of capacitance in semiconductors (Section 1.2), and a survey of capacitance spectroscopy techniques in semiconductors (Sections 1.3 and 1.4).

1.1 Capacitance

This section provides the definition of capacitance, its types, and its extraction from equivalent circuits.

1.1.1 The Definitions of Capacitance

The capacitance characterizes the relationship between charge storage capacity and electric potential. At the same time, the capacitance is also a measure of the electric energy storage capacity. In its most basic form, the capacitance is defined as the ratio between the charge Q and electric potential or voltage V of the system, taking on the mathematical relationship of $C = Q/V$. This definition implies that the capacitance C is constant with respective to voltage.

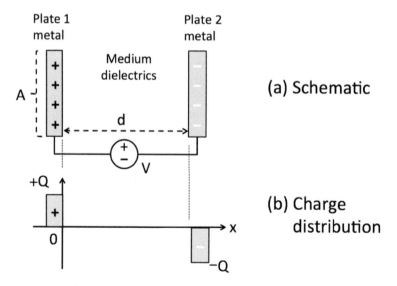

Figure 1.1 The schematic of a parallel-plate capacitor (a) and the spatial distribution of charge response (b).

The parallel-plate capacitor (Fig. 1.1a) consisting of two metal plates separated by an insulating (or dielectric) medium is the simplest form of capacitance. Upon the application of a voltage, electrons rearrange. For satisfaction of charge neutrality, the resulting charge is zero everywhere except on the two plates where there are charges of the same amount but opposing polarities (Fig. 1.1b). This charge distribution leads to an electric field in the dielectric to satisfy the self-consistent charge-field-potential relationship required by Maxwell's equations. The value of the parallel-plate capacitance is

$$C = \varepsilon\varepsilon_0 A/d, \tag{1.1}$$

where ε is the relative permittivity or dielectric constant of the dielectric, ε_0 the permittivity of vacuum, A the area of metal plates, and d their separation.

In general, the charge Q need not vary linearly with voltage V, in which case, the capacitance C is a function of V. In particular, the mechanisms of charge storage and separation in semiconductors do not typically lead to a constant capacitance, but one that is voltage-dependent. In a very basic 1D model of a semiconductor diode with depletion region width W, the voltage-dependence of C comes about because the separation W changes with V (Fig. 1.2a), and because Q is spatially distributed (Fig. 1.2b). Section 1.2 describes further the physics and properties of some capacitance responses encountered in semiconductors.

For non-linear capacitances, such as are commonly encountered in semiconductors, the definition of $C = Q/V$ is mathematically cumbersome and not very useful. A more useful definition is the differential capacitance $C = \delta Q/\delta V$, where δQ is a small variation of charge and δV is a small voltage. These two definitions of capacitance converge for a linear and constant capacitance such as that described in Fig. 1.1. In typical analyses of the ideal 1D semiconductor diode, the distributed nature of the depletion charge throughout the depletion region does not affect the differential capacitance, as the incremental change of charge occurs only at the edge of the depletion region. This makes its capacitance response equivalent to a parallel plate capacitor with distance W between the plates. From here forward, the capacitance C refers to the differential capacitance unless otherwise noted.

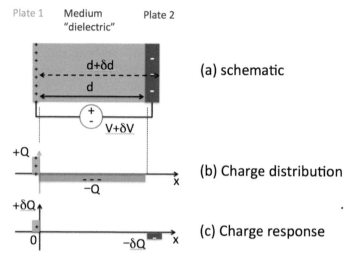

Figure 1.2 The schematic of a parallel-plate capacitor where the separation between plates changes with voltage (a), its static charge distribution (b), and its dynamic charge response (c).

The differential capacitance is thus a measure of the *small-signal* response of charge to a varying electric potential, e.g., those occurring at discrete spatial locations, as opposed to the *total* charge distributed over the entire volume of electric potential distribution. The differential capacitance reflects incremental changes in the quantity and location of stored charge, which can vary with voltage, time, temperature, light, or other experimental variables. In this way, the differential capacitance provides insight into distinct physical characteristics of semiconductor materials and devices, forming the foundation of the various capacitance spectroscopy techniques discussed in this book.

1.1.2 Extraction of Capacitance from Equivalent Circuits

The capacitance, resistance, and inductance are the three fundamental two-terminal passive elements of electrical circuits. Inductance will not be discussed further in this book due to the scarcity of its appearance in capacitance spectroscopy.

The resistive elements dissipate energy, i.e., they are lossy. In a time-varying AC circuit, the phase difference between the current I flowing through a pure resistance and the voltage V across it is 0°. On the other hand, the capacitive elements store energy in an electric field, i.e., they are lossless. In a time-varying AC circuit, the phase difference between the current I flowing through a pure capacitance and the voltage V across it is 90°.

A two-terminal circuit element generally exhibits a combination of both resistive and capacitive characteristics. In a time-varying AC circuit, the phase angle θ between the current I flowing through such a general element and the voltage V across it is between 0° and 90°, assuming the equivalent circuit consists of any combination of positive capacitance and resistance values. The admittance Y is defined to measure how well, in a generalized sense, the circuit element *conducts (admits)* current under a given voltage. That is, $I = YV$. The admittance Y is generally a complex number $Y = |Y| \exp(j\omega\theta) = Y' + jY''$, where $|Y|$ is the magnitude of Y and θ the phase angle, an example of which is shown in Fig. 1.3a. Experimentally, the current response to an applied bias is measured, with Y' corresponding to the in-phase response and Y'' corresponding to the part that is 90° out of phase.

The reciprocal quantity of admittance is impedance Z, which measures how well in a generalized sense the circuit element *resists (impedes)* current under a given voltage. That is, $I = Z^{-1} V$, where Z is also generally a complex number with $Z = |Z| \exp(-j\omega\theta_z) = Z' + jZ''$, and $Z = Y^{-1}$, an example of which is shown in Fig. 1.3b.

The capacitances of real semiconductor devices are extracted assuming certain equivalent circuit models. Equivalent circuit models containing two elements, namely the parallel model (Fig. 1.3a) and the series model (Fig. 1.3b), are the simplest in concept. There can also be numerous other equivalent circuit models containing more than two elements. One of them (Fig. 1.3c), which adds a series resistance to the simple two-element parallel model, is often encountered in practice and will be discussed below.

The parallel equivalent circuit model, as shown in Fig. 1.3a, is the most commonly used model in the context of capacitance or admittance spectroscopy. Using this model, the real and imaginary parts of admittance neatly separate the conductance and capacitance, with $Y' = 1/R_p$ and $Y'' = \omega C_p$. The phase angle θ of the admittance measurement is defined as $\arctan(Y''/Y')$, which is an indicator of how capacitive (θ approaching 90°) or how conductive (θ approaching 0°) the circuit is.

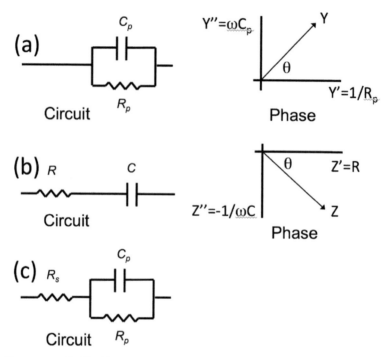

Figure 1.3 (Left) Equivalent circuits and (right) phase of (a) the two-element parallel admittance model, (b) the two-element serial impedance, and (c) the three-element serial-parallel model.

The parallel equivalent circuit is a common, although not necessary, assumption when measuring semiconductor devices. For example, in a typical reverse-biased p-n or Schottky diode, the capacitance originates largely from the depletion region, while the parallel conductance ideally has a low value, corresponding to a combination of leakage current that bypasses the depletion region and recombination current within the depletion region. In this approach, any series resistance, for example from carriers moving through the undepleted part of the semiconductor, is considered negligible. This model is returned to repeatedly throughout this text.

The parallel circuit model is also closely related to collection and initial analysis of AC experimental data, as Y' and Y'' are directly proportional to the real and imaginary parts (i.e., 0° and 90° components) of the current response to an applied ac bias. Experimental techniques and assumptions are further detailed in Chapters 5–7. Even when a device is ultimately analyzed using a more sophisticated equivalent circuit, Y' and Y'' are typically translated into "measured" values of capacitance C_m and resistance R_m (or conductance G_m) using the parallel circuit model. These measured values are then used as inputs to calculate the elements of the appropriate equivalent circuit model.

The other two-element model is a series equivalent circuit model, as shown in Fig. 1.3b. In this model, the real and imaginary (0° and 90°) impedances separate cleanly, with $Z' = R$ and $Z'' = 1/\omega C$. For this reason, impedance measurements and the series resistance model are typically used to extract "measured" values of capacitance and resistance in systems with significant series resistance, and where any parallel resistance can be neglected. This approach is normally used in impedance spectroscopy, described in Section 1.4.

Of course, a more complex circuit model is likely to be more accurate. The next most common circuit includes both series and parallel resistances, as pictured in Fig. 1.3c. In the presence of a series resistance (Fig. 1.3c), the values of Z' and Z'' depend on a combination of the R_s, R_p, and C_p values; similarly for Y' and Y''. In the case of a reverse-biased diode, where the depletion region response (i.e., from the parallel part of the circuit) is of primary concern, then the voltage applied to the parallel circuit, i.e., the junction, is reduced by the voltage drop across R_s. Note that the reduced voltage is also shifted in phase from the voltage at the contacts, as it depends on the phase of the total current, which is determined by all of the circuit elements. It is still possible to write $Y' = 1/R_m$ and $Y'' = \omega C_m$. However, the values R_m and C_m do not directly correspond simply to resistance and capacitances in the lumped circuit model (R_p and C_p), and their frequency response reflects the influence of the series resistance.

1.2 Capacitances in Semiconductors

This section describes the basic forms of capacitances encountered in the context of semiconductor materials and devices.

1.2.1 Capacitance of an Insulator

An ideal insulator is absent of mobile charges. When sandwiched between two metal plates, its dielectric response, characterized by the dielectric constant ε, modifies the electric field, leading to the two-plate capacitance of Eq. 1.1. While the free charge is entirely stored on the plates, the dielectric response (polarizability) of the insulator affects the relationship between charge and voltage, and hence the capacitance, as well as the total energy stored in the device.

Under certain conditions, a semiconductor may act as an insulating dielectric. This occurs when the measurement occurs too rapidly for free carriers to respond, as discussed in Section 1.2.2, and when the free carriers have been swept away by an electric field resulting in a depletion region, as discussed in Section 1.2.3.

1.2.2 Capacitance of a Semiconductor at Equilibrium

A semiconductor at thermal equilibrium can possess the dual characteristics of a conductor and of an insulating dielectric depending on measurement conditions. For a DC or low-frequency varying voltage, the semiconductor exhibits a conductance $\sigma = qn\mu$, where n and μ are the density and mobility of the majority carriers, respectively. This conductance determines the drift current due to majority carriers responding to the bias voltage. For a voltage varying at a sufficiently high frequency, the majority carriers cannot respond to the voltage (or electric field) therefore producing zero drift current. The semiconductor then appears purely insulating. The demarcation frequency of these two regimes is called the dielectric relaxation frequency $\omega_{dr} = \sigma/(\varepsilon\varepsilon_0)$. The reciprocal quantity of ω_{dr}, the dielectric relaxation time, describes the characteristic time required for the majority carriers to respond to an electric-field induced perturbation of their density via the drift mechanism in order to return to equilibrium.

The equivalent circuit of a charge-neutral semiconductor at equilibrium therefore consists of the parallel connection of the geometrical conductivity $G = \sigma A/t$ and the geometrical capacitance $C = (\varepsilon\varepsilon_0 A)/t$, where t is the thickness of the semiconductor. Although not of primary significance in all cases, the geometrical capacitance can dominate the capacitance response under certain experimental conditions.

1.2.3 Capacitance of a Semiconductor Depleted of Carriers

A semiconductor is under the condition of depletion when the free and mobile carriers are depleted, usually under the influence of an electric field. The field is screened by the remaining fixed charges, such that the width W of the depletion region depends on the built-in potential V_{bi} at the junction, any applied dc bias V_{dc}, as well as the average doping density N_B. The resulting charge density, field, and potential distributions are illustrated in Fig. 1.4.

In a simple 1D semiconductor *p-n* junction diode,

$$W = \sqrt{\frac{2\varepsilon\varepsilon_0(V_{bi} - V_{dc} - 2kT/e)}{eN_B}} \tag{1.2}$$

where, for a one-sided p$^+$-n or n$^+$-p junction, N_B is approximately the doping density on the more lightly doped side, either N_D or N_A respectively; similarly for a Schottky barrier. For a two-sided p-n abrupt junction, i.e., with similar doping densities on each side such that neither can be neglected, $N_B = N_A N_D/(N_A + N_D)$.

When the applied potential changes by δV, i.e., due to an applied small signal ac bias, the depletion width changes slightly. This requires a charge response δQ, which is due to the majority carriers and occurs at the two boundaries of the depletion region. This response results in a depletion capacitance identical in form to that of the parallel plate capacitor (Eq. 1.1). Observation of the depletion capacitance only occurs at frequencies lower than ω_{dr}, such that the majority charge is able to move in and out of the depletion region edge in time with the applied ac bias.

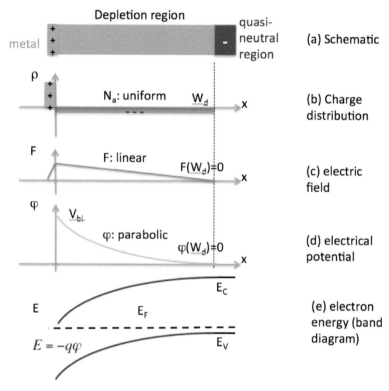

Figure 1.4 The schematic (a), charge distribution (b), electric field distribution (c), electric potential distribution (d), and band diagram of the capacitance of a depleted semiconductor (e).

The depletion capacitance of a semiconductor is distinctively different from the capacitance of an insulator in that the depleted semiconductor is present with immobile charges continuously distributed in space, namely the space charge. The distributions of the space charge, the electric field, and the electric potential in a depleted semiconductor are self-consistently governed by Poisson's equation and Fermi statistics. A change in the electric potential (voltage) requires the depletion width to change towards a new self-consistency of charge-field-potential distributions, which in turn changes the depletion capacitance. The depletion capacitance is the cornerstone of many capacitance spectroscopy techniques described in this book.

1.2.4 Capacitance of a Semiconductor with Excess Carriers

A semiconductor is under a non-equilibrium condition with excess carriers when carriers are induced by electrostatics, electrically injected from neighboring materials, or generated by physical stimulation such as optical illumination. The charge response in such a semiconductor to varying electric potential, hence its capacitance, may be dominated by the excess carriers.

For a p-n junction under a forward bias, injection of minority carriers leads to a separation of charge, resulting in the diffusion capacitance. The minority carrier density depends exponentially on the bias voltage, and hence the diffusion capacitance also increases exponentially with the bias voltage. In forward bias, the diffusion capacitance may dominate the depletion capacitance, whose dependence on the bias voltage, from Eqs. 1.1 and 1.2, is typically about $V^{-1/2}$. The diffusion capacitance, though less often the focus of study, is important to junctions under a forward bias or optical illumination.

Under certain electrostatic conditions, excess majority or minority carriers may concentrate in a semiconductor adjacent to an insulator, because current is not able to flow freely. When the density of excess majority carriers is greater than that of the majority carriers at equilibrium, this condition is known as accumulation, while a concentration of minority carriers greater than that of the majority carriers at equilibrium is known as inversion. The excess carriers, governed by the Fermi statistics, may be of so high a density and so concentrated in a thin region that they effectively screen the semiconductor beyond, preventing the charges there from responding to varying electric potential and contributing to the capacitance. The accumulation and inversion capacitances are important in the metal-insulator-semiconductor structures and certain hetero-structures.

1.2.5 Capacitance of a Semiconductor with Carrier Traps

The electron occupation probability of the electronic states in the bandgap of a semiconductor, also called carrier traps, is governed by Fermi statistics. When the Fermi level or the quasi-Fermi level is modified by a varying electric potential (Fig. 1.5), the electron occupation of the traps change accordingly by exchanging carriers with the conduction or valence bands. The carrier exchange to and from the traps, in response to a varying electric potential, underlies the capacitive behavior of a semiconductor with carrier traps. The interfacial traps are by definition spatially localized but their energetic distribution tends to be continuous. The traps distributed in the bulk of semiconductor may be either discrete or extended energetically. The energetic and spatial distributions of the traps determine the location and magnitude of the charge response and hence the capacitance.

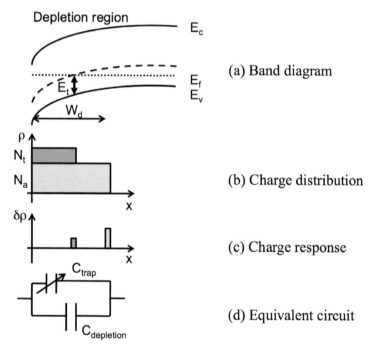

Figure 1.5 The band diagram (a), charge distribution (b), charge response (c), and equivalent circuit (d) of the capacitance due to a trap in the semiconductors.

The trap-band carrier exchange takes on four possible processes—electron capture, electron emission, hole capture, and hole emission. Together these four processes satisfy the detailed balance requirements at thermal equilibrium. Because the charge carriers exchange energy with phonons during these processes, their dynamic characteristics are governed by exchange rates that are thermally activated. The electrostatic and dynamic behaviors of traps are the subject of study for many capacitance spectroscopy techniques described in this book.

1.3 Capacitance Spectroscopy

This section briefly describes the basic capacitance spectroscopic techniques in terms of their measurements, physics, and applications. We first describe the techniques that use the admittance-type equivalent circuit model to for data extraction and physical interpretation—capacitance–voltage profiling, admittance spectroscopy, and deep-level transient spectroscopy. The impedance spectroscopy techniques that use the impedance-type equivalent circuit model are described next.

1.3.1 Capacitance-Voltage Profiling

The band bending within the depletion region, and hence the capacitance, varies with applied bias. This yields an opportunity to spatially profile device properties through the lightly doped side of a one-sided diode, just by varying the dc bias V_{dc} and measuring the capacitance response. When the lightly doped layer has only shallow densities of states, i.e., donors and acceptors, the change in capacitance reflects only a change in depletion width, and CV data can be analyzed in a straightforward way. The CV data provide a depth profile of the doping density through the semiconductor film, plus an estimate of the built-in potential at the junction. Chapter 4 provides further discussion of CV and depth profiles for both this simple situation and for less ideal circumstances.

1.3.2 Drive-Level Capacitance Profiling

The drive-level capacitance profiling (DLCP) technique, also discussed in detail in Chapter 4, addresses some of the shortcomings of CV by adding energy resolution to the spatial profile. In this way, contributions to the capacitance by shallow and deeper trap states can be distinguished, and individual densities and widths of the trap distributions can be found. While in CV, densities are calculated based on the variation of capacitance with DC bias, the DLCP technique is purely dynamic, and densities are calculated based on the variation of capacitance with the amplitude of the ac bias signal. The DLCP technique, like the CV technique, does depend on the depletion approximation though and is only valid in reverse bias or for small values of forward bias. The two techniques also have similar spatial resolution.

1.3.3 Admittance Spectroscopy

As discussed in Section 1.2.5, the traps located inside the depletion region of a semiconductor junction (Fig. 1.5) make a capacitive contribution to the depletion capacitance by exchanging carriers with the band edges. The rate of this trap-band carrier exchange is determined by the energetic separation of the trap from the band edge and the temperature due to its thermally activated nature. If the AC modulation rate of electric potential is slower than this rate (i.e., lower modulation frequency), then sufficient time is allowed to complete the trap-band carrier exchange. In this scenario, the capacitance due to the trap-band carrier exchange is observed in addition to the depletion capacitance. If the AC modulation rate of electric potential is faster than the trap-band exchange rate (i.e., higher modulation frequency), then insufficient time is allowed to complete the trap-band carrier exchange. In this scenario, the capacitance due to the trap-band carrier exchange is not observed in addition to the depletion capacitance. One can survey the frequency response of the overall capacitance and detect the demarcation frequency which is proportional to the trap-band carrier exchange rate. The trap properties—energetic location, concentration, density of states, and capture cross section—can be extracted by analyzing the admittance spectroscopy data, as discussed in detail in Chapter 2.

1.3.4   Deep-Level Transient Spectroscopy

Due to the frequency-time duality, it is also feasible to detect the trap-band charge exchange rate and its thermal activation in the time domain. The deep-level transient spectroscopy (DLTS) measurement, which is discussed in detail in Chapter 3, records the time-varying capacitance of a semiconductor junction after a section of the depletion region and the traps located there are temporarily filled with carriers. The time dependence of that capacitive contribution due to the emission of carriers from the traps contains the information on the trap properties obtainable from admittance spectroscopy—energetic location, concentration, density of states, and capture cross section. The DLTS technique is a large perturbation method whereas the admittance spectroscopy is usually a small perturbation method. The DLTS technique is inherently capable of detecting both majority-carrier trapping and minority carrier trapping traps because the filling mechanism can supply both types of carriers to the traps. The admittance spectroscopy technique is more commonly used to detect majority-carrier trapping traps.

1.4   Impedance Spectroscopy

While the prior sections of this chapter serve as introduction to the capacitance measurements discussed throughout the text, here we take a brief detour to introduce the closely related technique of impedance spectroscopy (IS). We would be remiss to ignore this technique entirely, as impedance and admittance spectroscopies are two approaches to analyzing the same data; once you have collected the data for one you have it for the other. Despite their strong overlaps in experimental approach and theory, these techniques have different histories and communities. Although it is not common to see data in the literature which has been analyzed with both approaches, each technique brings out distinct physical information about the device. For this reason, a basic introduction to IS is included here, and the reader is encouraged to become familiar with the relevant literature.

For the researcher focused primarily on admittance measurements, as is the focus of this book, impedance spectroscopy provides a validity check of the lumped circuit model being employed to analyze the data. While admittance spectroscopy studies nearly always assume a simple parallel model (Fig. 1.3a), the IS approach has historically been applied to compare data to more complicated circuit models, at the minimum including an additional series resistance (Fig. 1.3c).

Note that in general, a particular lumped circuit model is not unique, and components can be moved around while maintaining the same total impedance. This becomes particularly true for more complex circuit models. In addition, the dielectric response of a device is, in general, distributed—with spatial and temporal distributions, for example, of potential barriers, trap energies, and distances. Thus, a truly general analysis does not rely on a lumped circuit model at all, but investigates the total ac current response or the sample in a more general way (see, e.g., Ref. 1). Variation of the response with temperature, bias, or other parameters provides necessary corroboration of the analysis [2].

As will be described below, the complex impedance diagram of Z'' vs. Z', often called the Nyquist diagram, gives strong visual confirmation of the applicability of different circuit models, and directly yields three circuit elements. Here, it is analyzed using the series/parallel model of Fig. 1.3c, and hence indicates the range of experimental parameters within which the series resistance can be neglected, i.e., a parallel model is appropriate. Since a stray series resistance shifts the phase and, if neglected, distorts the extracted R_p and C_p values, a quick analysis of the impedance can provide peace of mind or, if series resistance is discovered, can allow the data to be re-analyzed to remove its influence. The IS analysis can also give more in-depth information about conduction mechanisms and other physical characteristics. The introduction to IS described here was already largely understood in the 1940s (see, e.g. refs. 3, 4); much theoretical effort since then has succeeded in explaining the physical mechanisms behind different types of IS response. For an excellent introduction to these methods, see Refs. [4, 5].

When using a series model (Fig. 1.3b), Z' and Z'' separate cleanly into measured values of R and C, as discussed above. For this reason IS is naturally suited to analyzing results from devices with significant series resistances. Not surprisingly, the IS technique is most often applied in materials systems which have relatively poor conduction—often through diffusive, ionic, or hopping mechanisms. Electrochemical impedance spectroscopy (EIS) is an entire subset of the IS technique, and has its own active research community. In EIS, contacts are made using electrochemical means, through liquid electrolytes; the materials under study can be liquid or solid, including batteries, biological materials, magnetic materials, and semiconductors. This approach allows materials to be measured without application of metallic contact layers, either to speed the study of a series of novel materials or because stable, good quality contacting mechanisms have not yet been discovered or are not well understood. Electrolytic contacts are also valuable for semiconductor devices being developed for biological applications, consistent with their operating conditions (see, e.g., 6, 7). And, recent work in graphene and other 2D materials have also employed electrochemical contacting schemes as a means of controlling surface conditions in highly surface sensitive films and of creating novel sensors [8–10]. Although EIS will not be further discussed here, the resources cited above also discuss the electrolyte/electrode interface, electrode design, diffusion models, double-layer capacitance, and other specific issues related to EIS.

1.4.1 Experiment

The experimental approach to impedance spectroscopy is identical to that of admittance spectroscopy, in that the dielectric response of the sample to an applied AC signal is measured as a function of frequency. Temperature and DC bias are also important experimental parameters.

Typically currents are measured in response to applied AC voltages. A variety of different experimental apparatus can be employed; experimental setup is discussed more extensively in Chapters 5–7. Common setups directly measure the sample response with an LCR meter, network analyzer, or lock-in amplifier. Experimental apparatus should ideally allow the AC amplitude and frequency to be adjusted over an appropriate range, as well as allow for the addition of a DC bias. Capacitance bridges are sometimes employed for very high sensitivity measurements.

1.4.2   Graphical Analysis

The complex impedance diagram, a graph of Z'' vs. Z' (implicitly as a function of f), allows R_s and R_p to be directly extracted from IS data. For the simple series/parallel circuit (Fig. 1.3c), this diagram is a semicircle centered at $R_s + R_p/2$ with diameter $R_p/2$. Although this analysis is straightforward, we outline the major steps here.

The impedance of the series/parallel circuit can be found as

$$Z = Z' + jZ'' = R_s + (R_p^{-1} + j\omega C)^{-1}. \tag{1.3}$$

Yielding

$$Z' = R_s + \frac{R_p}{1 + (\omega R_p C)^2} \tag{1.4}$$

$$Z'' = -\frac{\omega C R_p^2}{1 + (\omega R_p C)^2} \tag{1.5}$$

Assuming that the lumped circuit elements are frequency independent, this analysis describes any system with a single relaxation time τ, which appears in this model as $\tau = R_p C$. Analysis of such a system, with a single relaxation time, is known as Debye analysis. When the impedance is analyzed as a function of frequency, Z' will show an inflection at $\omega = 1/R_p C$, while Z'' reaches a peak value at that same frequency. This is consistent with the Kramers–Kronig relationship between the real and imaginary parts of any AC response; a step in one parameter will yield a peak in the other [11]. The value of Z' reaches an asymptotic value of $Z' \approx R_s + R_p$ when $\omega \ll R_p C$, and of $Z' \approx R_s$ when $\omega \gg R_p C$.

The complex impedance diagram traces out a semicircle, as can be seen by using Eq. 1.4 to eliminate ω from Eq. 1.5. With some algebra, one obtains

$$\omega = \frac{1}{R_p C} \sqrt{\frac{R_p}{Z' - R_s} - 1}$$

and

$$(Z'')^2 = R_p(Z' - R_s) - (Z' - R_s)^2,$$

which can be written in the equation of a circle:

$$(Z'')^2 + [(Z' - R_s) - {R_p}/{2}]^2 = {R_p^2}/{4} \qquad (1.6)$$

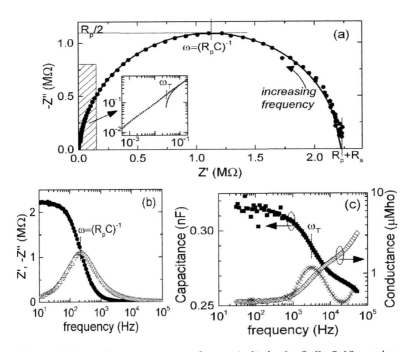

Figure 1.6 Impedance response of a p-n⁺ diode (a Cu(In,Ga)Se₂ solar cell). (a) Nyquist plot fits well to a circle, using R_s = 24 kΩ, R_p = 2.2 MΩ, and C = 0.32 nF. Inset: close inspection of the high frequency data shows a shift from the fit line, likely corresponding to the frequency at which the components themselves become frequency dependent. (b) Impedance spectroscopy plot of Z' (solid circles) and Z'' (open triangles). The transition frequency occurs at the peak of Z''. (c) Capacitance C_m (solid squares), conductance G_m (open diamonds) and derivative data $-\omega dC/d\omega$ (X). The admittance emphasizes a different transition frequency. Note, this transition is barely discernable in the inset of (a). Admittance spectroscopy (AS) is further discussed in Chapter 2.

Thus, agreement of the experimental data with the model can be quickly and easily determined from the semicircular fit, as illustrated in Fig. 1.6a. Since this data extends over many orders of magnitude, it is important to appropriately weight the fitting algorithm, to obtain a reasonable fit. Weights could be proportional to the data, or, better, to the value of the theoretical curve itself [12]. Note also that when $R_s \ll R_p$, as in Fig. 1.6a, the uncertainty in the value of R_s extracted from the curve fit is significant.

The IS literature, of course, goes far beyond simple Debye analysis to treat more sophisticated responses. We will just briefly note a couple of the most common and relevant results here. Physical quantities that are distributed in space, time, or energy may require distributed, rather than lumped, circuit elements. These are often modeled using constant phase elements, which rotate the current signal by a constant phase and cannot be represented simply by capacitors or resistors. Such behavior is described by the Havriliak–Negami (HN) response [13]:

$$Z_{HN} = R_{HN}[1 + (i\omega\tau)^{\alpha}]^{-\beta} \qquad (1.7)$$

When $\beta = 1$, this yields a symmetric arc centered below the axis. Materials or electrodes exhibiting diffusion could yield HN response, as can materials with Gaussian or exponential distributions of states at which charges can be trapped and/ or recombine. The HN response in the frequency domain is closely related to stretched exponential behavior in the time domain [14].

1.5 Summary

In summary, capacitance is a basic quantity describing various physical mechanisms and their manifestations in measurement circuits. The scope of semiconductor properties investigable by the family of capacitance spectroscopy techniques covers many aspects—carrier distribution, doping, transport, defects, and interfaces, to name a few. Readers are encouraged to proceed to the following chapters and sections for a dedicated description of specific techniques, their measurement principles, their applications, and emerging innovations.

References

1. Laux, S. F., and Hess, K. (1999) Revisiting the analytic theory of p-n junction impedance: improvements guided by computer simulation leading to a new equivalent circuit. *IEEE Trans. Electr. Devices* **46**, pp. 396–412.

2. Macdonald, J. R., and Johnson, W. B. (2005) Fundamentals of Impedance Spectroscopy, in E. Barsoukov and J. R. Macdonald (eds.), *Impedance Spectroscopy: Theory, Experiment, and Applications*, 2nd ed., Chapter 1 (Wiley, New Jersey) pp. 1–26.

3. Cole, K. S. and Cole, R. H. (1941) Dispersion and absorption in dielectrics I. Alternating current characteristics, *J. Chem. Phys.* **9**, pp. 98–105.

4. Macdonald, D. D. (2006) Reflections on the history of electrochemical impedance spectroscopy, *Electrochim. Acta* **41**, pp. 1376–1388.

5. Jonscher, A. K. (1983) *Dielectric Relaxation in Solids* (Chelsea Dielectrics Press, London).

6. Randviir, E. P., and Banks, C. E. (2013) Electrochemical impedance spectroscopy: an overview of bioanalytical applications, *Anal. Methods* **5**, pp. 1098–1115.

7. Lisdat, F., and Schafer, D. (2008) The use of electrochemical impedance spectroscopy for biosensing, *Anal. Bioanal. Chem.* **391**, pp. 1555–1567.

8. Das, A., Pisana, S., Chakraborty, B., Piscanec, S., Saha, S. K., Waghmare, U. V., Novoselov, K. S., Krishnamurthy, H. R., Geim, A. K., Ferrari, A. C., and Sood, A. K. (2008) Monitoring dopants by Raman scattering in an electrochemically top-gated graphene transistor, *Nat. Nanotechnol.* **3**, pp. 210–215.

9. Lin, Z. Y., Liu, Y., Yao, Y. G., Hildreth, O. J., Li, Z., Moon, K., and Wong, C. P. (2011) Superior capacitance of functionalized graphene. *J. Phys. Chem. C* **115**, pp. 7120–7125.

10. Shao, Y. Y., Wang, J., Wu, H., Liu, J., Aksay, I. A., and Lin, Y. H. (2010) Graphene based electrochemical sensors and biosensors: a review. *Electroanalysis* **22**, pp. 1027–1036.

11. Macdonald, J. R., and Brachman, M. D. (1956) Linear system integral transform relations, *Rev. Mod. Phys.* **28**, pp. 393–422.

12. Macdonald, J. R. (1992) Impedance spectroscopy, *Ann. Biomed. Eng.* **20**, pp. 289–305.

13. Havrilia, S., and Negami, S. (1967) A complex plane representation of dielectric and mechanical relaxation processes in some semi-conductors, *Polymer* **8**, pp. 161–210.

14. Alvarez, F., Alegría, J., and Colmenero, J. (1991) Relationship between the time-domain Kohlrausch-Williams-Watts and frequency-domain Havriliak-Negami relaxation functions, *Phys. Rev. B* **44**, pp. 7306–7312.

Chapter 2

Admittance Spectroscopy

Thomas Walter

Institute for Medical Engineering and Mechatronics,
University of Applied Sciences Ulm, Ulm, Germany

walter.th@hs-ulm.de

Admittance spectroscopy is one of the most commonly applied characterization techniques for semiconductor junctions, in particular solar cells. Basically, the amplitude and phase of the current response to a small AC signal are determined as a function of the AC frequency. Usually these measurements are undertaken at different temperatures in order to derive activation energies. In literature, related characterization techniques such as differential capacitance spectra, e.g., (Li, et al., 2010), have been published. Nevertheless, this chapter focuses on the "classical" admittance spectroscopy (AS).

The thermodynamic limit for the conversion efficiency of solar cells is given by the Shockley–Queisser Limit (Shockley and Queisser, 1961) containing radiative recombination as the limiting recombination mechanism. However, even in highly efficient thin-film solar cells recombination is dominated by

Capacitance Spectroscopy of Semiconductors
Edited by Jian V. Li and Giorgio Ferrari
Copyright © 2018 Pan Stanford Publishing Pte. Ltd.
ISBN 978-981-4774-54-3 (Hardcover), 978-1-315-15013-0 (eBook)
www.panstanford.com

non-radiative recombination processes due to defects in the semiconductor material. In (Rau, 2007) it is demonstrated that for a 16% efficient Cu(In,Ga)Se$_2$ CIGS-based thin-film solar cell the ratio of non-radiative to radiative recombination is about 10^3, pointing out clearly the importance of defect characterization for the optimization of such optoelectronic devices. Furthermore, with respect to the mass production of thin-film solar modules a quality gate for the control of defect densities is a requirement regarding high production yields.

In this chapter, the principles and possible interpretations of admittance spectra for polycrystalline solar cells are introduced and discussed. A key conclusion is the necessity for co-simulation of other device characteristics in order to reduce ambiguities in the interpretation of AS as not only traps/defects can lead to certain admittance signatures. This chapter mainly deals with the application of AS on CIGS-based thin-film solar cells. However, publications on CdTe are also considered and cited in this work. The principles and interpretations are applicable to semiconductors in general.

This chapter contains the following sections:

- Principles of admittance spectroscopy
- Interpretation of admittance spectra
- Simulation
- Defect spectroscopy
- Carrier freeze-out
- Back-contact barrier
- Phototransistor model
- Discussion
- Conclusion

2.1   Principles of Admittance Spectroscopy

Admittance spectroscopy implies the determination of the current response to a small AC bias applied to a device and the deduction of the frequency and temperature dependence of the capacitance and conductance. Usually, the admittance is interpreted as a parallel circuit of a capacitance and a conductance. Primarily the capacitance value (and its frequency and temperature dependence) are used and analyzed. Alternatively, the conductance could be used and analyzed. However, capacitance and conductance are connected by the Kramers–Kronig relations, meaning that they can be analyzed without losing information. Nevertheless, in (Leon, et al., 1996) it was pointed out that conductance and capacitance can both be used in order to eliminate the influence of parasitic resistances on the admittance data.

The main advantage of this characterization technique is its (relative) simplicity for implementation. The frequency range up to 1–10 MHz does not exhibit principle technical problems and a temperature range from room temperature down to 80 K can be accomplished in a liquid-nitrogen cooled cryostat. A typical junction capacitance ranges around 20 nF/cm^2 in a Cu(In,Ga)Se$_2$ (CIGS)-based thin-film solar cell implying that capacitance values (depending on the area) of some hundreds of pF have to be measured, which again does not represent a major challenge. Initially this technique was applied for the determination of defect distributions in solar cell junctions (Walter T., Herberholz, Müller, and Schock, 1996), (Losee, 1975). However, this interpretation of admittance data is under certain circumstances ambiguous. This ambiguity in the interpretation represents the major challenge of this measurement technique leading to discussions on the interpretation of certain features (N1 defect in CIGS) which started more than two decades ago and still continue. It should be pointed out that the application of admittance spectroscopy to thin-film solar cells has a history of more than two decades. Nevertheless, this characterization technique is still in use and under discussion as can be deduced from a pretty high number of recent journal or conference publications, e.g., (Mansfield, et al., 2015).

2.2 Interpretation of Admittance Spectroscopy

2.2.1 Series Resistance

As indicated above usually the complex admittance of a solar cell is interpreted as a parallel circuit consisting of a capacitance and a conductance. In the majority of the publications, only the capacitance value is taken as a basis for the interpretation (which contains according to the Kramers–Kronig relations all information). In the presence of a series resistance, such an equivalent circuit can lead to the interpretation of a frequency dependent capacitance value despite the fact that the value of the capacitance does not depend at all on the frequency. This shall be illustrated using a simple series connection of a capacitor (C_S) and an ohmic resistor (R_S) (Fig. 2.1). Transforming this serial circuit into a parallel equivalent circuit yields the following capacitance value (C_P):

$$C_P = \frac{C_S}{1 + \omega^2 (R_S C_S)^2} \qquad (2.1)$$

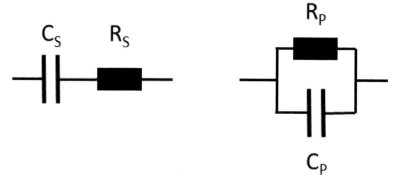

Figure 2.1 Serial and equivalent parallel circuit.

Thus, apparently the so measured capacitance value exhibits a step between low and high frequencies despite the fact that the assumed capacitor has a constant capacitance value.

2.2.2 Defect Spectroscopy

Traps, i.e., defects in a semiconductor which can attract and re-emit carriers, are commonly observed especially in polycrystalline semiconductors. With respect to admittance spectroscopy where a small AC-signal is applied to the junction, it is important to determine whether the trapping and especially re-emission rates are fast enough to follow the applied AC frequency or not. Thus, if the frequency of the external AC signal is low enough, the defects are charged and discharged adding an additional capacitance. As the thermal reemission time depends on the depth of the trap in the bandgap and on the temperature, a measurement of the device capacitance as a function of temperature and frequency can determine the density, energetic depth and capture cross-sections of defects in the bandgap of a semiconductor. It should be noted that a temperature dependence of the characteristic time constant is inherent to this mechanism as capture and re-emission of trapped carriers depend on temperature. The underlying theory and essential equations will be discussed below.

2.2.3 Carrier Freeze-Out

In a semiconductor, a local disturbance of the majority carrier concentration leads to an accumulation or depletion of charges and in consequence to an electric field, which homogenizes the charge distribution with a characteristic time constant called the dielectric relaxation time. This time constant mainly depends on the resistivity of the semiconductor. So, only for low AC signal frequencies the majorities at the edge of the space charge region can follow, leading to a charging and discharging of the junction capacitance. For high bias frequencies, the junction capacitance cannot be charged and discharged resulting in a lower capacitance, which corresponds to the geometrical capacitance of the solar cell, i.e., to an apparent space charge width stretching to the back contact. Below, the governing equations and equivalent electrical circuits will be discussed. A temperature dependence of this characteristic time constant may arise from a temperature dependence of the resistivity.

2.2.4 Back Contact Diode

In case of a non-ohmic back contact, a second junction with the corresponding junction capacitance has to be considered. Figure 2.2 represents the corresponding equivalent circuit in a small signal approach. From this equivalent circuit, it is quite evident that, similar to the series resistance discussed above, a frequency dependence of the measured capacitance could be expected despite the fact that the single components do not depend on frequency. It should be noted that a temperature dependence of the characteristic time constant is inherent to this interpretation as the small signal conductance of a junction exhibits a thermal activation as a consequence of the thermal current transport across a barrier.

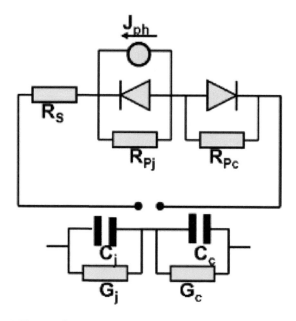

Figure 2.2 Electrical equivalent circuits of a back barrier diode from (Eisenbarth, Unold, Caballero, Kaufmann, and Schock, 2010).

2.2.5 Simulation

For the discussion and evaluation of underlying models and mechanisms the availability of simulation tools is essential. Especially, if the interpretation of admittance data is ambiguous these simulators can also co-simulate other device characteristics, which then can be compared to experimental results reducing the possible ambiguity of the pure admittance measurements.

For the simulation of thin-film solar cells, SCAPS1D is a commonly used simulator developed at the University of Gent (Niemegeers and Burgelman, 1996). In order to demonstrate the ambiguity for the interpretation of a capacitance step in the admittance spectra, the four presented possible mechanisms were simulated using SCAPS1D. The parameters were adjusted in such a way that this capacitance step occurs at a frequency of approximately 1–10 kHz. The basic doping in the CIGS was kept constant resulting in a junction capacitance (C_j) of approximately 20 nF/cm^2. In Figs. 2.3–2.6, the admittance results for a series resistance (Fig. 2.3), a defect (hole trap) (Fig. 2.4), carrier freeze-out (Fig. 2.5) and a back contact barrier (Fig. 2.6) simulated using SCAPS1D are shown (on the same capacitance scale for comparison). At a first glance, the shape of the capacitance step looks similar. However, looking closer at these diagrams some differences are revealed: Only the interpretation of admittance due to defects results at low frequencies in a capacitance value which is much higher than the junction capacitance (due to an additional capacitance related to the charging and discharging of defects). In case that the space charge width can be estimated from other measurements, defects as origin have a certain probability if the low frequency capacitance is significantly higher than the estimated junction capacitance. In case that for high frequencies the capacitance approaches the geometrical capacitance (which can be determined rather accurately) a carrier freeze-out appears to be plausible. If the high-frequency limit of the measured capacitance approaches zero a series resistance can be considered as underlying mechanism. In Table 2.1 the low- and high-frequency capacitances are listed together with the corresponding transition frequencies ω_0. These values and the underlying theory will be discussed in the following. One key statement of this article is the necessity for co-simulation of other device properties based on the assumed model for the interpretation of admittance data. As an example, the impact of a back contact diode on the admittance spectra and on the temperature dependent *IV*-characteristics will be discussed and evaluated. The second key argument concerns quantitative parameters as predicted by the assumed models.

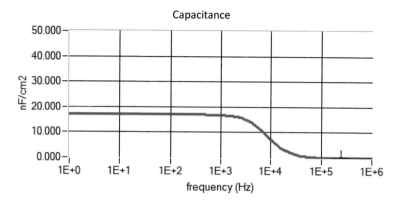

Figure 2.3 SCAPS1D simulation of series resistance.

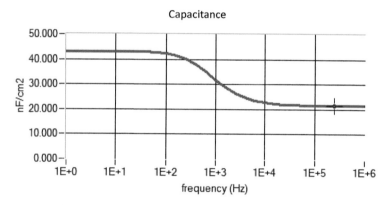

Figure 2.4 SCAPS1D simulation of traps/defects.

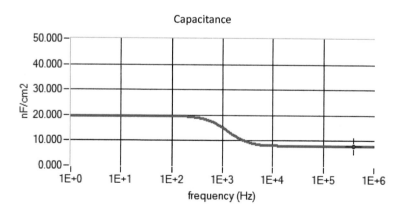

Figure 2.5 SCAPS1D simulation of carrier freeze-out.

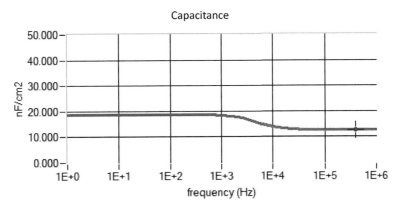

Figure 2.6 SCAPS1D simulation of back contact diode.

Table 2.1 Mechanisms and signatures

Mechanism	$C\,(\omega = 0)$	$C\,(\omega = \infty)$	ω_0
Series resistance	C_j	0	$\dfrac{1}{R_s \times C_s}$
Defect (trap)	$C_j + C_{tr}$	C_j	$2\beta_p N_v\, e^{-\frac{E_t - E_v}{kT}}$
Carrier freeze-out	C_j	C_{geo}	$\dfrac{w}{t}\dfrac{\sigma}{\varepsilon}$
Back contact diode	C_j	$\dfrac{C_j \times C_c}{C_j + C_c}$	$\dfrac{G_c}{C_c + C_c}$

Parameter	Description
C_j	Junction capacitance
C_s, R_s	Capacitance and resistance in a serial equivalent circuit
β_p	Capture coefficient for holes
N_v, E_v	Effective density of states and energy level of the valence band
E_t	Energy level of the trap/defect
C_{geo}	Geometrical capacitance
G_c, C_c	Conductance and capacitance of the back contact
w, t, σ	Space charge width, width of the semiconductor, conductivity

Assuming that an equivalent electrical circuit can be extracted from an assumed model, a circuit simulator such as LTSPICE can be used for the simulation of AC characteristics. Such a simulator also allows for the simulation of 2D effects, which is essential for distributed components. Such distributed electrical components have to be taken into account in case of devices with lateral dimensions giving rise to distributed parasitic elements such as sheet resistances of the top or bottom electrode. Quite recently, 2D circuit simulators were developed and made accessible, e.g., PVMOS (Pieters, 2014), allowing even the simulation of large area modules with local inhomogeneities. This approach will become certainly more and more important as a quality control in a production line.

2.3 Series Resistance

As shown before, a series resistance transformed into a parallel equivalent electronic circuit can lead to an apparent frequency dependent capacitance value. However, it should be noted that for the high-frequency limit, the value of the capacitance approaches basically zero forming a major difference and signature in comparison to the other proposed mechanisms. In (Kneisel, Siemer, Luck, and Bräunig, 2000), the impact of a laterally distributed series resistance on the admittance spectra was considered and discussed (see Fig. 2.7). With respect to a quantitative discussion, a transmission line model was developed for the configuration shown in Fig. 2.7. One of the major arguments in this publication is the dependence of the admittance spectrum on the cell dimensions. A spread or distributed series resistance certainly contributes more to the admittance if the lateral device dimensions increase. This forms a signature for the confirmation of a distributed series resistance as the underlying mechanism.

A very interesting example for this dependency is illustrated in Fig. 2.8. For the larger area (a), the curves converge (as indicated in the figure) and the capacitance values approach basically zero whereas for the smaller cell area (b) this convergence does not occur in this frequency range. Regarding the temperature dependence which is a feature being observed quite frequently,

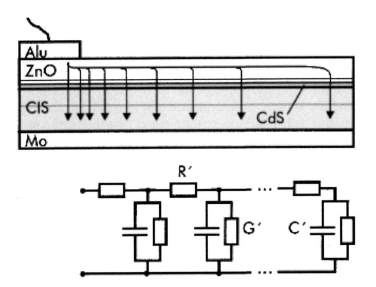

Figure 2.7 Distributed series resistance from (Kneisel, Siemer, Luck, and Bräunig, 2000).

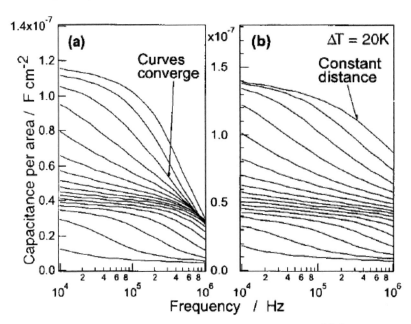

Figure 2.8 Impact of device area on admittance spectra (a) large area; (b) small area from (Kneisel, Siemer, Luck, and Bräunig, 2000).

the RC time constant should be temperature dependent implying that the resistivity is temperature dependent. Thus, a model that accounts for the observed admittance data should also fit the *IV*-characteristics and their temperature dependence. In consequence, a series resistance as origin of admittance data can be made more or less plausible by carefully simulating and comparing other device characteristics. Furthermore, in case of a series resistance, the characteristic frequency should be a real cut off frequency (and not only a transition frequency) exhibiting high-frequency capacitance values approaching zero.

2.4   Dielectric Relaxation

The relaxation of a disturbance of the majority carriers occurs with a characteristic time constant called dielectric relaxation time:

$$\tau_0 = \rho\varepsilon \tag{2.2}$$

This relaxation time primarily depends on the resistivity ρ of the semiconductor. For frequencies lower than the transition frequency the charging and discharging of the edge of the space charge region can occur leading to a capacitance value equal to the value of the pn-junction capacitance whereas for higher frequencies the junction capacitance cannot be charged and discharged. The semiconductor then behaves like a dielectric leading to a capacitance value which equals the geometrical capacitance of the device meaning that the complete semiconductor in the solar cell forms the dielectric (Jin Woo Lee, Cohen, and Shafarman, 2004 (online)). This model is illustrated in Fig. 2.9. The difference in the capacitance values (low- and high-frequency range) depends on the difference between the space charge width and the thickness of the semiconductor (Li, et al., 2010), (Li, Crandall, Repins, Nardes, and Levi, 2011), ((Jin Woo Lee, Cohen, and Shafarman, 2004 (online)). These considerations lead to the following equation which describes the measured capacitance in the context of dielectric relaxation:

$$C = C_d \frac{1 + \dfrac{C_d}{C_g}(\omega\varepsilon\rho)^2}{1 + \left(\dfrac{C_d}{C_g}\right)^2 (\omega\varepsilon\rho)^2} \tag{2.3}$$

with C_d being the depletion capacitance, and C_g the geometric capacitance.

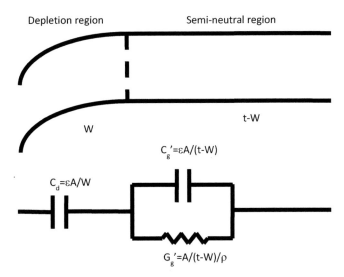

Figure 2.9 Equivalent electrical circuit for carrier freeze-out from (Li, Crandall, Repins, Nardes, and Levi, 2011).

Using this knowledge, a plausibility check can be made regarding this interpretation of admittance data. Looking quantitatively at the value of the dielectric relaxation time, a rather low majority mobility is required in order to see this capacitance step in the usual frequency window. Consequently, this effect should be seen for a "well behaved" device in a rather high-frequency range (Li, et al., 2010). Regarding the underlying model and the resulting equations the reader should refer to (Li, Crandall, Repins, Nardes, and Levi, 2011) and (Jin Woo Lee, Cohen, and Shafarman, 2004 (online)).

In (Li, Crandall, Repins, Nardes, and Levi, 2011) a plausible interpretation of a measured capacitance step due to carrier freeze-out is given (Fig. 2.10). For temperatures around 90 K, the space charge width "jumps" by a factor of more than three. The low temperature (or high frequency) limit of the space charge width agrees well with the geometrical dimension of the absorber layer. Again, this quantitative analysis of the capacitance values and their high and low frequency limits reduces ambiguities in the interpretation. A discussion on the observed temperature dependence and the underlying physics is published in (Li, Crandall, Repins, Nardes, and Levi, 2011).

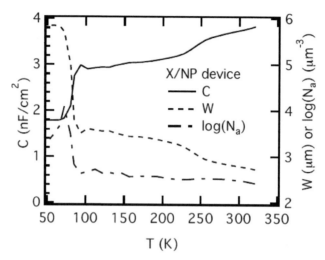

Figure 2.10 Space charge width for carrier freeze-out below 100 K from (Li, et al., 2010).

2.5 Defect Spectroscopy

As indicated above, knowledge about defects in semiconductors is both essential for the optimization and quality control of solar cells. An additional capacitance arises from the fact that defects are charged and discharged—if this process can follow the applied frequency of the AC bias.

The capture and reemission of an electron by an electron trap (equivalent equations are valid for the trapping and reemission of holes) are governed by the following equation:

$$\frac{dn_t}{dt} = \beta_n n \left(N_t - n_t\right) - \beta_n N_c n_t e^{-\frac{E_c - E_t}{kT}} \tag{2.4}$$

with n_t being the density of occupied traps, β_n the capture coefficient for electrons, N_c the effective density of states in the conduction band and E_t the energy level of the defect in the bandgap.

In a small signal approach, all parameters consist of a steady value and an AC component:

$$n_t = n_t^= + n_t^\approx \tag{2.5}$$

Taking into account the steady state conditions and introducing the Fermi function $f_n(E)$ the following equation for the capacitance arising from the charging and discharging of a defect can be deduced:

$$C = \frac{q}{u_{ext}^\approx} \frac{\omega_1 \omega_2}{\omega^2 + \omega_0^2} \tag{2.6}$$

with ω_1 and ω_2 being parameters which include the Fermi function, u_{ext} the e xternally applied small signal voltage, and q the elementary charge. ω_0 is the characteristic transition frequency which is defined by the following equation:

$$\omega_0 = 2\beta_n N_c e^{-\frac{E_c - E_t}{kT}} \tag{2.7}$$

The obtained expression for the additional capacitance may be interpreted in the following way. If the angular frequency of the applied AC voltage is smaller than the characteristic frequency ω_0 the capture and reemission can follow and the charging and discharging of the defects contributes to the capacitance. For high angular frequencies of the applied AC bias this additional capacitance approaches zero as capture and reemission are too slow with respect to the frequency of the applied bias. The term $\omega_1 \omega_2 / \omega_0^2$ indicates and determines which traps contribute in a junction to this capacitance. A closer look at this expression demonstrates that indeed only those trapswhich intersect with the Fermi level can be detected as a change in the occupation of the defect can only occur in the vicinity of the Fermi level. This immediately imposes a restriction on the interpretation of admittance spectroscopy. If (in case of a hole trap) the defect level is closer to the valence band than the Fermi level (that means a very shallow defect) a contradiction occurs as there should not be any intersect of the two levels (Walter T., Herberholz, Müller, and Schock, 1996). The amount of the contribution to the capacitance by a defect also depends on the density of the defect and on the "angle" between the defect level and the Fermi level, thus on the electric field and therefore on the parameters of the space charge region. How can now the properties of a defect level be determined from admittance spectra as shown in Fig. 2.11? One possibility arises from the inflection point of the capacitance sweep or (which is equivalent according to the Kramers–Kronig relation) taking the maxima of the conductance divided by the angular frequency. Taking these maxima and plotting them in an Arrhenius diagram both yield the depth of this defect and the capture cross-section or the attempt-to-escape frequency, respectively as shown in Fig. 2.12.

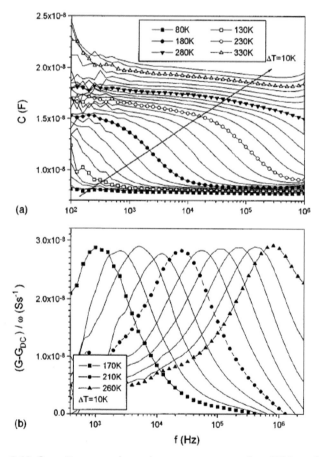

Figure 2.11 Capacitance and conductance spectra of a CIS-based solar cell from (Walter T., Herberholz, Müller, and Schock, 1996).

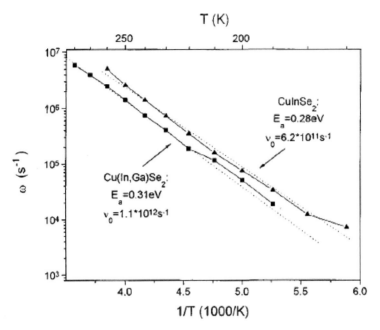

Figure 2.12 Arrhenius plot and activation energy from (Walter T., Herberholz, Müller, and Schock, 1996).

A complication arises if an energetic distribution of defects is present which is plausible for a moderate degree of disordering in a semiconductor. In (Walter T., Herberholz, Müller, and Schock, 1996) a formula was derived which yields for a parabolic band the defect spectrum in a pn⁺ junction:

$$N_t(E_\omega) = \frac{2U_d^{3/2}}{w\sqrt{q}\sqrt{qU_d - (E_g - E_\omega)}} \frac{dC}{d\omega} \frac{\omega}{kT} \qquad (2.8)$$

$$E_\omega = kT \ln \frac{2\beta_p N_V}{\omega}, \qquad (2.9)$$

with U_d being the diffusion voltage of the junction, w the space charge width, β_p the capture coefficient for holes and N_V the effective density of states in the valence band. The scaling of the energy axis has to be derived from a distinct defect and the procedure described above or by "choosing" the attempt to escape frequency in such a way that the density of states diagram results in a "smooth" and "continuous" plot without "echoes" for different temperatures. An example for such an interpretation is shown in Figs. 2.13 and 2.14. The density of states consists of a peak at about 300 meV followed by a sort of exponential tail stretching into the center of the bandgap. In principle, it is not possible to attribute this density of states to either hole or electron traps in the semiconductor. The fact that the defects start at around 300 meV at least does not contradict hole traps as the Fermi level in this n⁺p-junction is certainly not deeper into the bandgap than this 300 meV. The rather abrupt decrease of the density of states for lower energies might indicate the position of the Fermi level in the p-CIGS. However, additional measurements such as DLTS are required in order to increase the degree of plausibility with respect to the nature of such traps.

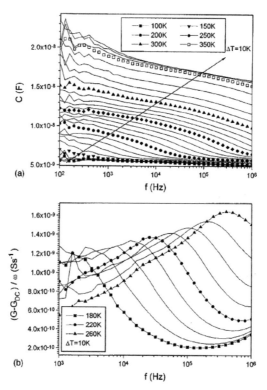

Figure 2.13 Capacitance and conductance spectra of a CIGS-based solar cell from (Walter T., Herberholz, Müller, and Schock, 1996).

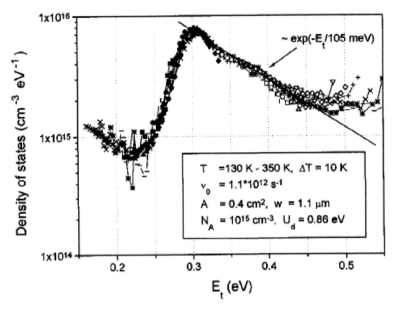

Figure 2.14 Corresponding defect spectrum (DOS) from (Walter T., Herberholz, Müller, and Schock, 1996).

The key question which was already addressed in the introduction of this chapter concerns the relevance of such measurements for the device performance. If the defects determined from admittance spectroscopy act as recombination centers a higher density of such recombination centers should coincide with a reduced device performance. In case of recombination in the space charge region such a distribution of defects should affect the diode ideality factor as only defects between the demarcation levels act as recombination centers. Thus, with increasing bias more recombination centers are between the demarcation levels and the dark current increases more steeply as compared to a single deep defect. As a consequence, the diode ideality factor decreases as compared to the single deep defect which corresponds to an ideality factor of approximately two. Especially for exponential tails of defects as deduced from Fig. 2.14 an analytical formula for the ideality factor can be given (Walter, Herberholz, and Schock, Distribution of defects in polycrystalline chalcopyrite thin films, 1996), (Walter, Menner, Köble, and Schock, 1994):

$$n = 2\frac{T^{*}}{T + T^{*}},\tag{2.10}$$

with kT^* being the characteristic slope energy of the tail-like defects. For the defect spectrum of Fig. 2.14 an ideality factor of approximately 1.5 can be expected which was a quite often observed value at the time of these measurements presented in Fig. 2.14.

However, such defects might also have a beneficial impact on the device performance if they add charge to the junction. In (Heath, Cohen, and Shafarman, 2004) considerations regarding the impact of defects on the overall charge distribution in a junction are discussed. Acceptor-like defects below the Fermi level increase the negative charge in the absorber leading to a narrowing of the space charge region. In case of recombination in the space charge region a smaller space charge width increases V_{oc} and possibly also reduces current collection, which should be visible in the EQE of the device. In (Seymour, Kaydanov, and Ohno, 2006) it is pointed out that a high density of defects can lead to a Fermi level pinning and an associated broadening of the admittance signatures. Again, co-simulation of other device parameters and a careful comparison with measurements should reduce ambiguities in the interpretation of admittance spectra. This aspect of admittance spectroscopy, i.e. the impact of a high trap density on the overall charge distribution in a junction is quite often neglected in publications.

In Figs. 2.15 and 2.16, admittance data and the corresponding density of states of a CIGS-based heterojunction are shown. The defect spectrum exhibits two features: a rather sharp peak at about 100 meV labeled N1 and a much broader feature labeled as N2. This N1 peak gave rise to a still ongoing discussion regarding its physical origin. This defect level is pretty shallow excluding plausibly the origin of a hole trap. Furthermore, this defect is affected by so called metastabilities of CIGS with respect to dark anneal and light soaking. Additionally the width of this peak is rather small especially compared to the N2 feature. In (Herberholz, et al., 1996) additional DLTS measurements were undertaken and the authors concluded that the N1 defect is located at the heterointerface and represents a minority carrier trap. The shift of this peak after different air anneals in the admittance spectra was consistent with the observed shift in the DLTS measurement. Thus, the observed shift might reflect the shift of the Fermi level at the heterointerface sampling the (certainly high) defect density at the heterointerface. Furthermore, the authors in (Herberholz, et al., 1996) observed that the energy of this N1 defect and the associated trapping parameters obey to a large degree the Meyer-Neldel (MN) rule. A model for the origin of the MN rule that is compatible with the widespread occurrence of MN behavior was given by (Yelon, Movaghar, and Branz, 1992) based on the thermodynamics of multi-phonon excitations.

In the course of research on admittance spectroscopy, after this rather early work the appearance of these two features depending on the deposition techniques and the device quality was discussed in numerous research papers, conferences, and workshops. One of the major outcomes was that the density of the N2 peak correlated with the device performance and sort of disappeared with improving device efficiency giving some evidence that the N2 peak is related to defects in the semiconductor which also govern the recombination in this device.

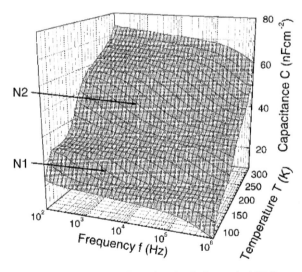

Figure 2.15 Admittance spectra from (Herberholz, et al., 1996).

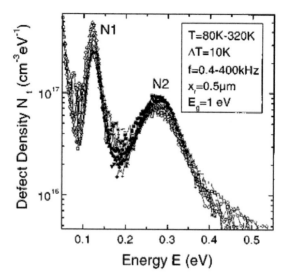

Figure 2.16 Corresponding defect spectrum (DOS) from (Herberholz, et al., 1996).

2.6 Back Contact Diode

Another possible interpretation arises from a Schottky barrier at the back contact as shown in the band diagram of Fig. 2.17. In a small signal equivalent circuit, a junction consists of a capacitance and of a conductance in parallel. For a solar cell with a back contact barrier, the corresponding equivalent electrical circuit is shown in Fig. 2.2.

This equivalent circuit principally consists of the capacitance and conductance of the main junction in series with the corresponding parallel circuit from the back contact diode. As under realistic circumstances the barrier of the main junction is much higher than the back barrier, the conductance of the main junction can be neglected. An analysis of such a circuit reveals that under low-frequency conditions the capacitance of the main junction (C_J) is measured whereas for high frequencies the series connection of the capacitance of the main junction with the back contact barrier (C_C)—thus a lower value—should be obtained. The low and high-frequency values together with the characteristic frequencies are given in Table 2.1. As the conductance of the back diode depends on the barrier of the Schottky contact and on the temperature due to thermal emission, this transition frequency will depend on temperature. Figure 2.18 shows a simulated capacitance spectrum for three different temperatures clearly demonstrating the increase of the transition frequency for higher temperatures. From the inflection point of the capacitance curve, an activation energy can be extracted which corresponds to the barrier height of the back contact barrier (Eisenbarth, Unold, Caballero, Kaufmann, and Schock, 2010).

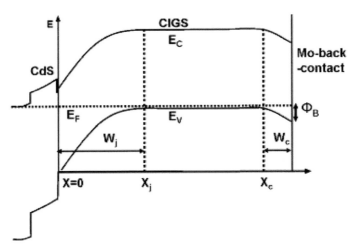

Figure 2.17 Band diagram of back contact diode from (Eisenbarth, Unold, Caballero, Kaufmann, and Schock, 2010).

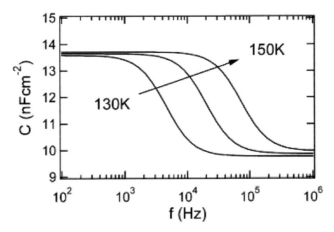

Figure 2.18 Simulated admittance spectrum for back contact diode from (Eisenbarth, Unold, Caballero, Kaufmann, and Schock, 2010).

In (Eisenbarth, Unold, Caballero, Kaufmann, and Schock, 2010) a new input for the interpretation of admittance spectra was published. Prior to this publication, the interpretation as traps/defects was quite common in most publications. Especially with respect to the interpretation of this N1 defect in CIGS (Herberholz, Igalson, and Schock, Distinction between bulk and interface states in $CuInSe_2/CdS/ZnO$ by space charge, 1998) as an interface related defect, the postulated presence of a Schottky back contact broadened the range of possible underlying mechanisms. However, as stated above for several times, co-simulation of other device characteristics based on the assumed model for the interpretation of admittance spectroscopy should be a valuable tool to verify or exclude certain models. Especially, the presence of a Schottky barrier at the back contact must have a significant impact on the DC device characteristics. Such an approach for the verification of assumed models will be discussed in detail below.

Of course, when analyzing admittance spectra the proposed mechanisms might appear simultaneously in one device. As an example for the interpretation of a rather complex admittance spectrum, a typical admittance spectrum of a $CuInS_2$ based solar cell is shown in Fig. 2.19. The authors (Kneisel, Siemer, Luck, and Bräunig, 2000) interpreted the features in the measured spectra as follows:

(1) is due to dielectric relaxation
(2) represents the depletion capacitance
(3) results from a trap peak
(4) is identified as continuous trap distribution
(5) is the cutoff due to a series resistance

These results can be discussed in view of Table 2.1, which contains quantitative capacitance values for the proposed underlying mechanisms.

A depletion capacitance of about 50 nF/cm^2 is plausible for this kind of semiconductor. Feature (3) and (4) imply higher capacitance values than the depletion capacitance making the interpretation of the authors (traps) highly probable as the defect interpretation is the only mechanism resulting in capacitance values much higher than the depletion capacitance for high temperatures/low frequencies. Regarding feature (1) the high-frequency capacitance value should approach the geometrical capacitance of the absorber. The authors show that the measured value coincides nicely with the expected geometrical capacitance. Furthermore, the activation energy of the capacitance step agrees well with a measured activation energy of the conductivity making the interpretation as a carrier freeze-out again plausible. Of course, this plausible interpretation critically depends on the knowledge of other device parameters such as the junction capacitance of this device. The assumption of defects adding an additional capacitance to the junction capacitance gains more plausibility due to the measured values of the capacitance at low frequencies/high temperatures which exceed—as mentioned before—the junction depletion capacitance by a significant amount. Nevertheless, it would be extremely helpful to see, e.g., the temperature dependent *IV*-characteristics of such a device in order to exclude the existence of a back contact diode.

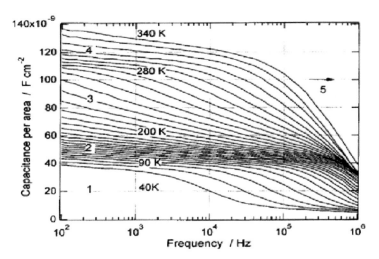

Figure 2.19 Capacitance spectra of CuInS$_2$-based solar cell from (Kneisel, Siemer, Luck, and Bräunig, 2000).

2.7 Phototransistor Model

As pointed out earlier, co-simulation of other device characteristics is considered as a viable method to reduce ambiguities. This will be explained for a back contact diode. In (Li, et al., 2010) and (Li, Crandall, Repins, Nardes, and Levi, 2011) it is explained that the barrier of the back contact can be also extracted from temperature dependent dark *IV*-measurements of a CdTe-based solar cell. The obtained barrier heights are in good agreement with activation energies obtained from admittance spectroscopy measurements indicating that a back contact barrier and not a deep defect is responsible for the observed admittance signature.

This method can also be applied to CIGS-based heterojunctions. However, the extraction of the back contact barrier height can be also achieved based on a phototransistor model which was published rather recently (Ott, Walter, and Unold, Phototransistor effects in Cu(In, Ga)Se$_2$ solar cells, 2013). Looking at the band diagram in Fig. 2.17 it is quite obvious that under certain conditions (e.g., diffusion length in the order of the base thickness) such a solar cell behaves like a bipolar transistor with an n-ZnO emitter, a p-CIGS base and a Schottky collector. This Schottky diode at the back contact limits the hole current into the solar cell. Therefore, the forward dark current of the main junction is limited by the reverse saturation current of the Schottky diode. This reverse saturation current depends on the barrier height and on the temperature. In (Ott, et al., 2015) it could be shown that at higher temperatures the device (depending on the barrier height at the back contact) behaves like a solar cell whereas for lower temperatures the device behaves like a phototransistor. This is illustrated in Fig. 2.20 containing simulations of a solar cell with a back contact diode at 300 K and at 150 K. For the high temperature, the expected solar cell characteristics can be observed whereas at the low temperature for voltages exceeding V_{oc} a transistor like behavior can be seen. This transition from a solar cell to a phototransistor also affects the temperature dependence of V_{oc} (Fig. 2.20). From solar cell device physics it can be deduced that V_{oc} extrapolates to the bandgap energy (divided by q) at 0 K. In case of a back barrier V_{oc} extrapolates (low temperature regime) to the bandgap energy reduced by the back barrier height. Basically, the back contact diode also produces a V_{oc} with opposite polarity as compared to the main junction. As V_{oc} of the main junction extrapolates to E_g at 0 K and V_{oc} of the back diode extrapolates to the back contact barrier height, the V_{oc} of the complete device extrapolates to $E_g - \Phi_B$. Consequently, extrapolating V_{oc} at high temperatures and at low temperatures can be used to determine the back contact barrier height Φ_B (Ott, et al., 2015). Thus, this saturation of V_{oc} at low temperatures is a signature for a back contact diode.

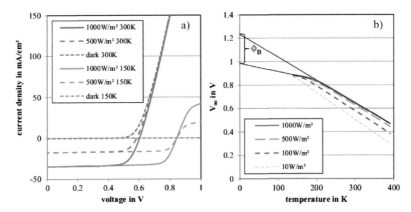

Figure 2.20 Simulated *IV*-characteristics at 300 and 150 K (a) and V_{oc} (*T*) (b) for a back contact diode from (Ott, et al., 2015).

In Table 2.2 this method for the extraction of the back contact barrier height is illustrated by SCAPS1D simulation results. A barrier of 250 meV was used as input parameter for the simulation. With AS 254 meV and with $V_{oc}(T)$ 266 meV could be extracted from the simulation proving that both methods yield the input parameter to this simulation.

Table 2.2 Comparison of AS and $V_{oc}(T)$ for determination of Φ_B (simulations) from (Ott, Walter, and Schäffler, On the Interpretation of Admittance and *IV*(*T*) Measurements of CIGS Thin Film Solar Cells, 2015)

	meV
Φ_B input	250
Φ_B AS	254
Φ_B $V_{oc}(T)$	266

In Fig. 2.21 measurements of CIGS-based solar cells are shown at two different temperatures (Ott, Lavrenko, Walter, and Schäffler, 2014). In the high-temperature regime, the expected solar cell behavior is obvious whereas in the low temperature regime the transistor characteristics become apparent. Furthermore, the associated saturation of V_{oc} in the transistor regime is clearly visible. From the latter measurement a back contact barrier height of approximately 320 meV could be extracted as explained above.

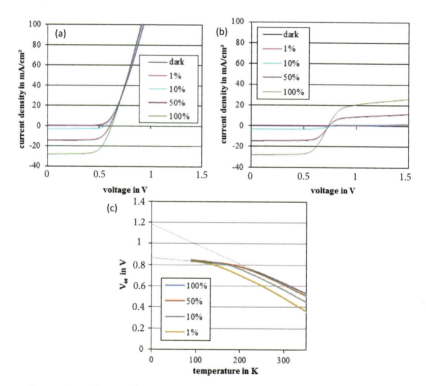

Figure 2.21 Measured *IV*-characteristics at 300 K (a), 180 K (b) and V_{oc} (*T*) (c) for a back contact diode from (Ott, Lavrenko, Walter, and Schäffler, 2014).

Coming back to admittance spectroscopy a back contact barrier leads to a temperature dependent step of the capacitance with the back contact barrier as activation energy. On the other hand, such a back contact barrier leads to a transistor like behavior in the low temperature regime being a signature which can be easily checked experimentally. Furthermore, the back contact barrier height can be extracted by the procedure explained above.

Table 2.3 contains experimentally extracted Φ_B values using AS and $V_{oc}(T)$ for two different CIGS devices in the initial state and after accelerated ageing (Ott, On the interpretation of admittance and $IV(T)$ measurements of CIGS thin film solar cells, 2015). In the initial state the obtained values from the two methods differ significantly whereas for the aged devices a very good correlation could be obtained. The origin of these rather low Φ_B values in the initial state are not clear. As these devices contain one interconnect (part of a module) and as the associated P1 shunt varies significantly after a rather short dark anneal, the assumed equivalent electrical model might be one reason for the observed differences in the initial state of the device.

Table 2.3 Comparison of AS and $V_{oc}(T)$ for the determination of Φ_B (experimental) from (Ott, On the interpretation of admittance and $IV(T)$ measurements of CIGS thin film solar cells, 2015)

	Φ_B, initial (meV)	Φ_B, aged (meV)	Φ_B, initial (meV)	Φ_B, aged (meV)
	Sample 2		Sample 3	
$V_{oc}(T)$	276	388	240	320
AS	47	370	120	290

In order to "verify" this method, two different devices were chosen. Device A was deposited by a slightly different process compared to device B which could have affected the back contact (plausible but not proven). The difference in the processing of the CIGS absorber layer is not within the scope of this article. In Fig. 2.22 the temperature dependent IV-characteristics of device A and B are shown. Device A behaves like a solar cell down to 90 K whereas device B clearly shows this blocking of the IV-characteristics which is typical for this transistor like behavior at lower temperatures. Admittance data of these two devices are illustrated in Fig. 2.23. Device A exhibits almost over the entire temperature range no capacitance step whereas device B clearly shows a pronounced step at lower temperatures. The combination of these two measurements indicates that for device B a back contact barrier exists resulting in a capacitance step and in a transition to a transistor at low temperatures.

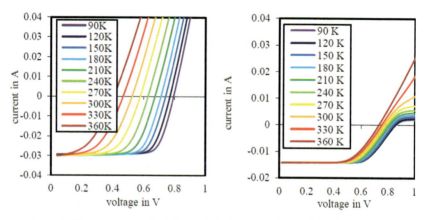

Figure 2.22 *IV(T)*-characteristics of device A (left) and device B (right) from (Ott, Walter, and Schäffler, On the Interpretation of Admittance and *IV(T)* Measurements of CIGS Thin Film Solar Cells, 2015).

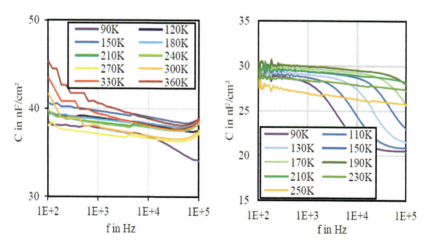

Figure 2.23 Corresponding capacitance spectra: device A (left), device B (right) from (Ott, Walter, and Schäffler, On the Interpretation of Admittance and *IV(T)* Measurements of CIGS Thin Film Solar Cells, 2015).

2.8 Discussion

Admittance spectroscopy is a fairly "easy" characterization technique for solar cells with respect to the required setup and infrastructure. However, the interpretation of admittance spectra might be ambiguous with respect to the underlying mechanisms. In this chapter, it was pointed out that a capacitance step can have different origins. A series resistance, deep traps, carrier freeze-out or a back contact diode might be responsible for this signature. Two approaches were suggested to reduce these interpretation ambiguities. The first approach addresses the low- and high-frequency capacitance values which differ for these models and which should be checked quantitatively. In case of the carrier freeze-out the high-frequency limit approaches the geometric capacitance which is an accessible parameter in a solar cell. The second approach requires the co-simulation of other device parameters based on the assumed model for the admittance spectra. This second approach was illustrated for a back contact diode which also gives rise to a temperature dependent capacitance step. Such a back diode should impact the temperature-dependent *IV*-characteristics severely. It was pointed out that there are algorithms to extract the height of the back contact barrier from *IV*(*T*) measurements (in the dark (Li, Crandall, Repins, Nardes, and Levi, 2011) and under illumination ($V_{oc}(T)$ (Ott, et al., 2015)). If—as was published in a few publications—the so-determined barrier height agrees well with the activation energy of the capacitance step, the plausibility of this underlying mechanism is rather high. The so-called N1 feature in CIGS is certainly a candidate for checking if these devices exhibit a transition to a phototransistor regime at low temperatures which is a pretty strong indication that a back contact diode is responsible for the observed sharp capacitance step. Co-simulation of other device parameters seems to be also a viable tool to confirm carrier freeze-out or series resistance as both mechanisms involve the resistivity of the absorber layer having definitely a severe impact on the *IV*-characteristics of the device. Furthermore, it could and should be checked if the temperature dependence of capacitance step is in agreement with the temperature dependence of the resistivity.

Not always steep capacitance steps but rather broad features are observed in AS. In (Walter, Herberholz, and Schock, Distribution of defects in polycrystalline chalcopyrite thin films, 1996) a energetic distribution of traps/defects was suggested as origin for this AS signature. For CIGS this signature was labeled as N2. This defect distribution exhibited a dependence on the quality of the device and almost disappeared completely for nowadays high-efficiency devices. Looking at alternative interpretations for such broad signatures, some proposals involving Fermi level pinning, potential fluctuations of the bands or in depth gradients of a discrete (but depth dependent) defect level were published. However, looking at today's publications on AS the interpretation of the N2 signature as a defect distribution seems to be accepted especially based on the fact that this signature correlates pretty much with the quality of the measured devices which certainly is no proof for this interpretation but increases the plausibility.

2.9   Conclusion

AS is certainly an interesting characterization technique for the quality of solar cells due to its technical "simplicity". However, the interpretation of admittance spectra can be ambiguous due to several underlying mechanisms which lead to similar admittance signatures. Fortunately, simulation tools such as SCAPS1D or PVMOS are nowadays available allowing for the simulation of admittance spectra and for the co-simulation of other device characteristics. Such a co-simulation of other accessible device parameters and a comparison with measurements are considered as a valuable method to verify (or exclude) assumed models for measured admittance signatures. Furthermore, such a correlation between AS and other solar cell characteristics can strengthen the importance of AS not only as a research method but also as a measurement technique for the quality control of solar cells.

References

Eisenbarth, T., Unold, T., Caballero, R., Kaufmann, C., and Schock, H.-W. (2010). Interpretation of admittance, capacitance-voltage, and current-voltage signatures in Cu(In, Ga)Se$_2$ thin film solar cells. *Journal of Applied Physics* 107, p. doi: 10.1063/1.3277043.

Heath, J., Cohen, J., and Shafarman, W. (2004). Bulk and metastable defects in CuIn$_{1-x}$Ga$_x$Se$_2$ thin films using drive-level capacitance profiling. *Journal of Applied Physics* 95, p. doi: 10.1063/1.1633982.

Herberholz, R., Igalson, M., and Schock, H.-W. (1998). Distinction between bulk and interface states in CuInSe$_2$/CdS/ZnO by space charge. *Journal of Applied Physics* 83, p. doi: 10.1063/1.366686.

Herberholz, R., Walter, T., Müller, C., Friedlmeier, T., Schock, H.-W., Saad, M., Alberts, V. (1996). Meyer-Neldel behavior of deep level parameters in heterojunctions to Cu(In, Ga)(S, Se)$_2$. *Applied Physics Letters* 69, p. doi: 10.1063/1.117352.

Lee, J. W., Cohen, J. D., and Shafarman, W. (2005). The determination of carrier mobilities in CIGS photovoltaic devices using high-frequency admittance measurements. *Thin Solid Films,* 480–481, pp. 336–340.

Kneisel, J., Siemer, K., Luck, I., and Bräunig, D. (2000). Admittance spectroscopy of efficient CuInS$_2$ thin film solar cells. *Journal of Applied Physics,* 88(9), pp. 5474–5481.

Leon, C., Martin, J., Santamaria, J., Skarp, J., Gonzalez-Diaz, G., and Sanchez-Quesada, F. (1996). Use of Kramers–Kronig transforms for the treatment of admittance spectroscopy data of p-n junctions containing traps. *Journal of Applied Physics* 79(10), pp. 7830–7836.

Li, J., Johnston, S., Li, X., Albin, D., Gessert, T., and Levi, D. (2010). Discussion of some "trap signatures" observed by admittance spectroscopy in CdTe thin-film solar cells. *Journal of Applied Physics* 108, p. doi: 10.1063/1.3475373.

Li, J., Crandall, R., Repins, I., Nardes, A., and Levi, D. (2011). Applications of admittance spectroscopy in photovoltaic devices beyond majority-carrier trapping defects. *37th IEEE PVSC.* Seattle, Washington.

Losee, J. (1975). Admittance spectroscopy of impurity levels in Schottky barriers. *Journal of Applied Physics* 46, p. http://dx.doi.org/10.1063/1.321865.

Mansfield, L., Kuciauskas, D., Dippo, P., Li, J., Bowers, K., To, B., Ramanathan, K. (2015). Optoelectronic Investigation of Sb-Doped

Cu(In,Ga)Se$_2$. *IEEE Journal of Photovoltaics,* 5(6), p. doi: 10.1109/ JPHOTOV.2015.2470082.

Niemegeers, A., and Burgelman, M. (1996). Numerical modelling of AC-characteristics of CdTe and CIS solar cells. *5th IEEE PVSC,* (p. doi: 10.1109/PVSC.1996.564274). Washington.

Ott, T. (2015). On the interpretation of admittance and *IV*(T) measurements of CIGS thin film solar cells. *Presented at the 30th EU PVSEC.* Hamburg.

Ott, T., Lavrenko, T., Walter, T., and Schäffler, R. (2014). On the Importance of the Back Contact For Cu(In, Ga)Se$_2$ Thin Film Solar Cells. *29th EU PVSEC.* Amsterdam.

Ott, T., Schönberger, F., Walter, T., Hariskos, D., Kiowski, O., Salomon, O., and Schäffler, R. (2015). Verification of phototransistor model for Cu(In, Ga)Se$_2$ thin film solar cells. *Thin Solid Films* 582, p. doi:10.1016/j.tsf.2014.09.025.

Ott, T., Walter, T., and Schäffler, R. (2015). On the Interpretation of admittance and *IV*(T) measurements of CIGS thin film solar cells. *30th EU PVSEC.* Hamburg.

Ott, T., Walter, T., and Unold, T. (2013). Phototransistor effects in Cu(In, Ga)Se$_2$ solar cells. *Thin Solid Films* 535, pp. 275–278.

Pieters, B. (2014). A free and open source finite-difference simulation tool for solar modules. *40 th IEEE PVSC,* (p. doi: 10.1109/ PVSC.2014.6925173). Denver.

Rau, U. (2007). Reciprocity relation between photovoltaic quantum efficiency and electroluminescent emission of solar cells. *Phys. Rev. B* 76, p. DOI:http://dx.doi.org/10.1103/PhysRevB.76.085303.

Seymour, F., Kaydanov, V., and Ohno, T. (2006). Simulated admittance spectroscopy measurements of high concentration deep level defects in CdTe thin-film solar cells. *Journal of Applied Physics* 100, p. doi: 10.1063/1.2220491.

Shockley, W., and Queisser, H. (1961). Detailed balance limit of efficiency of p-n junction solar cells. *Journal of Applied Physics* 32, pp. 510–519.

Walter, T., Herberholz, R., and Schock, H. (1996). Distribution of defects in polycrystalline chalcopyrite thin films. *Solid State Phenom.* 51–52, pp. 309–316.

Walter, T., Herberholz, R., Müller, C., and Schock, H.-W. (1996). Determination of defect distributions from admittance measurements and application to Cu(In, Ga)Se$_2$ based heterojunctions. *Journal of Applied Physics* 80, p. doi: 10.1063/1.363401.

Walter, T., Menner, R., Köble, C., and Schock, H.-W. (1994). Characterization and junction performance of highly efficient ZnO/CdS/CuInS$_2$ thin film solar cells *Proc. of the 12th EU PVSEC*, (pp. 1755–1758).

Yelon, A., Movaghar, B., and Branz, H. (1992). Origin and consequences of the compensation (Meyer-Neldel) law. *Phys. Rev. B* 46, p. 12244.

Chapter 3

Deep-Level Transient Spectroscopy

Johan Lauwaert and Samira Khelifi

Department of Electronics and Information Systems (ELIS),
Ghent University, Gent, Belgium

Johan.Lauwaert@UGent.be

3.1 Introduction

For a semiconductor that is in thermal equilibrium and neutral, the Fermi level is positioned between the shallow dopant and the intrinsic Fermi level. In conventional deep-level transient spectroscopy (DLTS) in principle any deep level with a position in between the Fermi level in the neutral semiconductor and the intrinsic Fermi level can be observed.

Figure 3.1 shows the band diagram of the neutral n-type semiconductor with a deep acceptor level. The four transitions for electrons involving this deep level are indicated by arrows.

(1) The first mechanism that can result in a change in electron occupation of the deep level is capture of an electron from the conduction band. The capture rate equals $c_n n$, which is proportional with the electron concentration (n) at that position and proportional to the capture rate constant c_n.

Capacitance Spectroscopy of Semiconductors
Edited by Jian V. Li and Giorgio Ferrari
Copyright © 2018 Pan Stanford Publishing Pte. Ltd.
ISBN 978-981-4774-54-3 (Hardcover), 978-1-315-15013-0 (eBook)
www.panstanford.com

By describing capturing of carriers by deep levels using standard collision theory, it can be shown that this capture rate constant c_n equals $\sigma_n v_{th}^n$. With σ_n the capture cross section for electrons in the neutral state, and v_{th}^n the thermal velocity for electrons.

(2) The emission rate for electrons to the conduction band is written as e_n. Because the fractional occupation of this deep level by electrons $f = \frac{c_n n}{(c_n n + e_n)}$ in the neutral situation should be equal to the Fermi-Dirac distribution the emission rate can be written as a function of the capture rate constant and a Boltzmann factor: $e_n = c_n N_C \exp - \frac{(E_C - E_T)}{k_B T}$. Here N_C is the density of states of the conduction band, k_B the Boltzmann constant, and T the temperature.

(3) The emission rate for holes to the valence band is written as e_p. As for the emission rate of electrons the emission rate for holes can also be written as $e_p = c_p N_V \exp - \frac{(E_T - E_V)}{k_B T}$. With N_V the density of states of the valence band.

(4) And finally the capture rate for holes $c_p p$ can also be described by the classic collision theory and can therefore be written as $\sigma_p v_{th}^p$. Here, σ_p is the hole capture cross section in the negatively charged state and v_{th}^p the thermal velocity for holes.

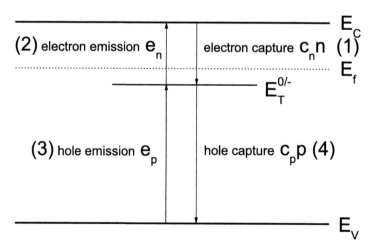

Figure 3.1 Energy diagram from a semiconductor with a deep level E_T. The possible transitions for electrons are shown by the different arrows.

Combining these four processes in one rate equation the fractional occupation f as a function of time is a solution of

$$\frac{df(t)}{dt} = (c_n n + e_p)(1 - f(t)) - (c_p p + e_n)f(t). \qquad (3.1)$$

Integrating Eq. 3.1 gives the fractional occupation $f(t)$ as a function of time:

$$f(t) = (f(0) - f(+\infty))\exp(-(c_n n + e_n + c_p p + e_p)t) + f(+\infty), \qquad (3.2)$$

where $f(0)$ and $f(+\infty) = \frac{c_n n + e_p}{c_n n + e_n + c_p p + e_p}$ are the initial and final fractional occupation, respectively.

To observe a transition that is described by Eq. 3.2 the charges within the depletion layer of a diode at reverse bias $V_r < 0$ are changed by a pulse V_p, with $V_p - V_r > 0$. Figure 3.2 shows the band diagram of a p⁺n diode. For an abrupt junction and in depletion approximation for a reverse bias $V_r < 0$, the depletion region is extended until W_r. In the region between the interface and W_r the semiconductor is not neutral. In this depleted region, there is no neutrality and the band is bended while the Fermi level is approximately flat since we have assumed that the current can be neglected. Consequently the deep acceptor level is neutral for $x < L_r = W_r - \lambda$ and negatively charged deeper in the bulk of the n-type semiconductor. For a uniform doping profile the distance λ can be approximated by

$$\lambda = \frac{\sqrt{2\varepsilon(E_f - E_T)}}{q^2 n}. \qquad (3.3)$$

When applying a pulse V_p with $0 \geq V_p > V_r$ the diode is still in reverse bias but the depletion layer is smaller $W_p < W_r$. By injecting majority carriers the region in between W_r and W_p becomes neutral. The free charges will respond on this pulse with the dielectric relaxation time constant $\tau_D = \varepsilon \rho$ (with ρ the resistivity of the semiconductor and ε the permittivity) which is usually much faster than the time constant describing the occupation changes of the deep level. From Eq. 3.2 we can calculate the time dependence of the fractional occupation of the

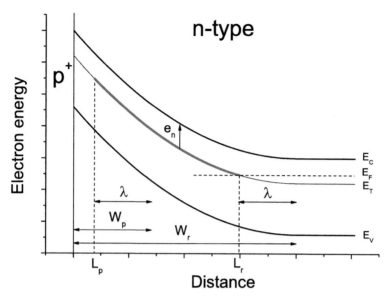

Figure 3.2 Band diagram of a p⁺n diode indicating the distances necessary to describe the region where DLTS is probing the deep levels.

deep levels in the region between $L_p = W_p - \lambda$ and L_r. At the start all the deep levels in this region are neutral $f(0) = 0$, while in equilibrium at pulse $f(+\infty)$ will become $\frac{c_n n}{c_n n + e_n}$. The fractional occupation in this region becomes equal to the one in the neutral region, and because in an n-type semiconductor for a deep level $c_n n \gg e_n + e_p$ and $c_p p \ll c_n n$, the time evolution during the pulse becomes

$$f(t) = (1 - \exp(-c_n nt)). \tag{3.4}$$

Starting from an equilibrium situation for V_p and going back to the reverse bias V_r, the region will be again depleted and the free carriers are going away. Due to the depletion of the carriers $e_p + e_n \gg c_n n$ and $e_p + e_n \gg c_p p$. The capture rates are indeed strongly reduced due to the absence of carriers. Consequently the evolution of the fractional occupation during reverse bias can be written as

$$f(t) = \exp(-(e_n + e_p) t). \tag{3.5}$$

In general, the emission of carriers happens with a rate constant $e_n + e_p$ independent of the type of semiconductor and independent of the majority or minority carrier concentration. However, in an n-type semiconductor we typically change occupations of deep levels in the upper half of the band gap and therefore $e_n > e_p$ and the thermal activation is mainly determined by the energy distance between the deep level E_T and the conduction band E_C.

In DLTS, the emission kinetic of the deep levels with a rate constant

$$e_n = \sigma_n v_{th} N_C \exp\left(-\frac{E_C - E_T}{k_B T}\right) \tag{3.6}$$

is observed by recording the exponential decay of the capacitance during reverse bias after applying a pulse. Because this rate constant is a characteristic of the deep level, and usually, is independent of the carrier concentration this can be used as a spectroscopic technique.

Although this signature can be used to identify certain defects, the capture cross section can also be thermally activated and the thermal emission energy is actually a Gibbs free energy. Therefore, estimation of the enthalpy change ΔH that correspond with the energy distance between the majority carrier band and the defect level can only be estimated by correction for the entropy ΔS (Schroder, 1998).

3.2 The Principle

In the previous introductory section, it is explained that in DLTS we change the occupation of deep traps by alternating between reverse bias and pulse. Now we will explain how the filling of and emission from these levels is reflected in the high frequency capacitance and how these capacitance transients are analyzed.

Solving Poisson's equation for the situation of a deep acceptor level (see Fig. 3.2) for V_r we can find

$$V_r = -\frac{q}{\varepsilon}\left(-N_T \frac{(W_r - \lambda)^2}{2} + n\frac{W_r^2}{2}\right) + U_b, \tag{3.7}$$

where U_b is the barrier height of the diode under study. In a similar way the total charge Q in the semiconductor with A the area of the diode can also be calculated:

$$Q = Aq(nW_r - N_T(W_r - \lambda)) \tag{3.8}$$

From Eq. 3.3, it can be seen that the distance λ is independent of bias for a uniform doping profile, therefore the high frequency capacitance becomes

$$C_r = \frac{\dfrac{dQ}{dW_r}}{\dfrac{dV_r}{dW_r}} = \frac{A\varepsilon(n - N_T)}{(n - N_T)W_r + N_T\lambda}. \tag{3.9}$$

This equation for a low deep level concentration $N_T \ll n$ becomes the well-known formula:

$$C_r = \frac{A\varepsilon}{W_r} \tag{3.10}$$

A transient in the high frequency capacitance will therefore be the effect of a change in depletion width. This is what actually happens during the emission carriers while the reverse bias is kept constant. Shortly after the voltage pulse, we assume the free carriers respond instantaneously while the deep levels are still occupied with electrons. This means for example that the single acceptors are negatively charged while the single donors are neutral. By integrating the Poisson equation gives a voltage drop V_r:

$$V_r = -\frac{q}{\varepsilon}\left(n\frac{(W_r + \Delta W_r)^2}{2} - N_T\left(\frac{W_r^2}{2} - \frac{(W_p - \lambda)^2}{2}\right)\right) + U_b, \tag{3.11}$$

Herein is ΔW_r the additional distance that is necessary to keep the potential drop over the semiconductor constant V_r. With the total charge in the semiconductor:

$$Q = Aq\left(n\left(W_r + \Delta W_r\right) - N_T\left(W_r - W_p + \lambda\right)\right) \tag{3.12}$$

The capacitance directly after the pulse ($t = 0$) can therefore be written as

$$C(0) = \frac{-\dfrac{dQ}{d\Delta W_r}}{\dfrac{dV_r}{d\Delta W_r}}. \tag{3.13}$$

Combining Eqs. 3.11 and 3.7 shows that $\Delta W_r > 0$ and therefore the capacitance transient independent of the type of deep level (acceptor or donor) for emission of majority carriers is always increasing. From the capacitance transient amplitude $\Delta C = |C(0) - C_r|$ it is possible to estimate the concentration of deep levels comparing Eqs. 3.13, 3.11, 3.10, and 3.7 gives us

$$\left(\left(1 - \frac{\Delta C}{C_r}\right)^{-2} - 1\right)\frac{n}{N_T} = \frac{(W_r - \lambda)^2 - (W_p - \lambda)^2}{W_r^2}. \tag{3.14}$$

Herein the left side of the equation can be approximated as $2\frac{\Delta C}{C_r}\frac{n}{N_T}$ for $\Delta C \ll C_r$. The right side of Eq. 3.14 is called the pulse correction factor, which is correcting for the fact that by alternating between reverse bias and pulse voltage not all the deep levels within the depletion layer are responding. Because this transient with amplitude ΔC is induced by the emission of majority carriers for $\Delta C \ll C_r$, the time constant of this transient will be equal to one over the emission rate for the electrons:

$$C(t) = C_r - \Delta C \exp(-e_n t) \tag{3.15}$$

This emission rate is characteristic for the deep level and is given by Eq. 3.6. Because the thermal velocity v_{th} is proportional to $\sqrt{T}$ and the density of states of the conduction band is proportional to $T^{3/2}$ the emission rate is often written as

$$e_n = K_T T^2 \exp\left(-\frac{\Delta E_T}{k_B T}\right). \tag{3.16}$$

Here, ΔE_T is the apparent activation energy and K_T the pre-exponential factor that are both independent of temperature and form the electronic signature for the deep level. The temperature dependence of this rate is mainly determined by the Boltzmann factor, therefore it can be seen that the rate is larger for higher temperatures while it is slower for lower temperatures. An example of different transients for a deep level with signature ($K_T = 10^7$ K^{-2} s^{-1}, $\Delta E_T = 0.34$ eV) is shown in Fig. 3.3a. Different observation windows $[t_0, t_0 + T_w]$ are also indicated on the graph. For the shortest observation window $T_w = 2$ ms the transient is within the observation period for the highest temperature and therefore the DLTS peak is also observed for higher temperatures (see Fig. 3.3b). By using longer observation windows $T_w = 20$ ms the transient is within this period for lower temperatures and the peak shifts to lower temperatures. To calculate the DLTS signal $S(T)$ from the transients originally the difference between two fixed points in time (t_1, t_2) was used $S(T) = \Delta C(\exp(-e_n t_2) - \exp(-e_n t_1))$, which has a maximum for $e_n = \dfrac{\ln\left(\frac{t_2}{t_1}\right)}{t_2 - t_1}$

However, different methods to analyze the transients are very common currently and are discussed in the following subsections.

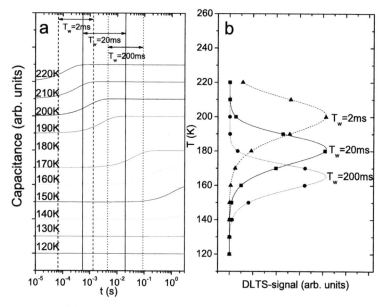

Figure 3.3 (a) Example of different transients calculated for a defect level $E_C - E_T = 0.34$ eV for different temperatures (transients are vertically shifted for clarity) (b) corresponding DLTS spectra calculated for three different observation windows (The transients have been analyzed using the b_1 coefficient, as explained later).

3.2.1 Fourier-Transform Analysis

A method to analyze the transient based on Fourier Transforms has been developed by Weiss and Kassing (Weiss, 1988). The main advantage over the original method is the improvement of the signal to noise ratio for the DLTS spectra. Secondly, this method is based on digitalization of the transient where, the observation window $[t_0, t_0 + T_w]$ is divided in N equidistant points with interval Δt which leads to a strong reduction in temperature scans. The measured data-points from the transient $C(t)$ can then be made periodic with a period T_w. Therefore, the transient can be analyzed using Fourier series with coefficients a_n and b_n:

$$a_n = \frac{2}{T_w} \int_0^{T_w} C(t)\cos(\omega n t)dt \quad n = 0, 1, 2, \dots \tag{3.17}$$

$$b_n = \frac{2}{T_w} \int_0^{T_w} C(t)\sin(\omega n t)dt \quad n = 0, 1, 2, \dots \tag{3.18}$$

with $\omega = \frac{2\pi}{T_w}$. As shown in Fig. 3.3 for the b_1 coefficient, these coefficients each form a different DLTS spectrum with a different maximum. Based on this Fourier transform coefficients two different methods exist to form an Arrhenius diagram or in other words to estimate the time constant for each temperature. Both methods have been combined to form the Arrhenius diagram shown in Fig. 3.4 that corresponds with the data from Fig. 3.3 ($K_T = 10^7$ K^{-2} s^{-1}, $\Delta E_T = 0.34$ eV).

A first method calculates the time constant via two Fourier components. For an exponential transient, the DLTS-signals Eqs. 3.17 and 3.18 with amplitude ΔC can be written as

$$a_n = \frac{2\Delta C}{T_w} \exp(-e_n t_0)[1 - \exp(-e_n T_w)]\frac{e_n}{e_n^2 + n^2\omega^2} \tag{3.19}$$

$$b_n = \frac{2\Delta C}{T_w} \exp(-e_n t_0)[1 - \exp(-e_n T_w)]\frac{n\omega}{e_n^2 + n^2\omega^2} \tag{3.20}$$

Both equations are independent of any offset in the transients. In general, the time constant can be calculated directly by combining these individual components:

$$\tau(a_n, a_k) = \frac{1}{\omega}\sqrt{\frac{a_n - a_k}{k^2 a_k - n^2 a_n}}$$

$$\tau(b_n, b_k) = \frac{1}{\omega}\sqrt{\frac{kb_n - nb_k}{k^2 nb_k - n^2 kb_n}}$$

$$\tau(b_n, a_n) = \frac{1}{n\omega}\frac{b_n}{a_n} \tag{3.21}$$

Therefore, if only a single transient is predominantly present, the time constant can be calculated via Eq. 3.21. Figure 3.4 includes also datapoints that have been calculated using this method based on a_1 and a_2.

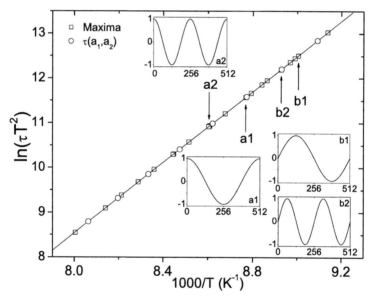

Figure 3.4 An example of an Arrhenius diagram made by different correlation functions using for each transient only one observation window width T_w.

The second method relies on maximum evaluation for different correlation functions. Although the basis for this methodology is Fourier analysis, a lot of other correlation functions are used. For a DLTS spectrum analyzed with such a digital filter function the maximum corresponds to a specific time constant. Consequently, combining different correlations functions gives us a number of $[\tau, T]$ couples that can make an Arrhenius diagram as shown in Fig. 3.4.

3.2.2 Laplace DLTS

Many methods to analyze the capacitance transients have been proposed, all of them are trying to separate the exponentials within the transient recorded. One of which is the high-resolution DLTS or Laplace DLTS methodology (Dobaczewski, 2006). It is obvious that the resolution is the major advantage of Laplace DLTS. In conventional DLTS single levels, especially those with the highest activation energy, can result in broader peaks. Therefore, using the inverse Laplace transformation could in principle result in a superior way of analyzing the recorded data. Nevertheless, numerical Laplace inversion is in ill-posed problem and suffer much more from the signal to noise ratio of the original data. Different numerical algorithm are available that showed their capability to be used in Laplace DLTS. One of which is the CONTIN package based on a Tikhonov regularization method (Provencher, 1982). Figure 3.5 shows that the same data as used in Fig. 3.3 can be interpreted using this Laplace DLTS methodology.

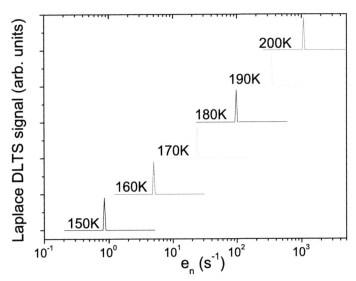

Figure 3.5 Calculated Laplace DLTS spectra corresponding with the spectra in Fig. 3.3, the different temperature are shifted vertically for clarity.

3.2.3 Coupling of Defect Levels

As discussed in the previous sections, DLTS observes the emission of carriers from deep levels. Therefore, it is often not straightforward to assign them to a certain defect structure or impurity. The assignment of a charge transition to the peak induced by emission often only relies on the dependence of the emission rate on the electric field. Poole and Frenkel showed how a uniform electric field can lower the activation energy for emission in a typical $1/r$ potential (Frenkel, 1938). A clear observation of the lowering of the apparent activation energy with higher electric field is therefore a good hint to assign the peak to an attractive center (acceptor levels in p-type material and donor levels in n-type). The absence of a peak shift is on the other hand not always sufficient to argue that the defect is neutral or repulsive. The charges onto deep defects are often strongly shielded that the interaction with the free carriers in to bands deviate strongly from a $1/r$ potential. A sequence of charge states for a defect level can tentatively indicate which levels are from the same defect structure. Of course, this can only be the case if the emission peaks have the same amplitude. Therefore, such an assignment although sometimes very convincing is never unambiguous. This can be circumvented by using optical stimulation. Optical stimulated emission allows one to observe the emission of two levels at the same temperature within the same observation window. Indeed by the modification of the light intensity for optical excitation of the carriers, the observed emission is enhanced. Therefore, this optical induced emission rate can be observed at lower temperature than the thermal emission (i.e. a temperature for which the thermal emission is too slow). This allows to demonstrate if two levels are coupled (i.e. originating from the same defect) or are uncoupled (i.e. originating from different chemical defects). An example of the acceptor levels of Chromium close to the valence band of Germanium where the method of photo-induced emission helped to assign them to the same defect structure is shown in Fig. 3.6.

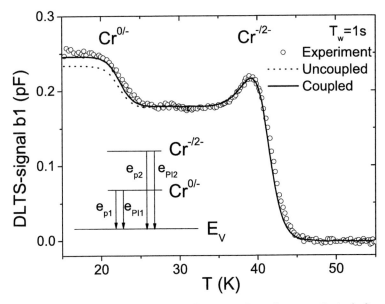

Figure 3.6 DLTS spectrum recorded for Cr implanted p-type Ge including photo-induced emission. Simulations for the two levels coupled and uncoupled are included in the figure. Data from Segers (2013).

3.3   Carrier Capture Cross Section

The capture rate is proportional with the capture cross section, thermal velocity, and carrier concentration. In principle the transient of the capture is induced during the filling pulse, however this transient is often too fast to be recorded with a conventional DLTS instrument. Indeed the capture rate is equal to the emission rate for the position where the Fermi-level is crossing the deep level, and during the filling pulse the Fermi level is coming closer to the majority carrier band giving a capture rate higher than the emission rate. Therefore, majority carrier capture cross sections are mostly measured by changing the filling pulse duration and measuring the DLTS amplitude as a function of this filling pulse duration. Although in this way DLTS instruments are often able to apply pulse lengths of only 10 ns, the detectability of the capture cross section mainly depends on the free carrier concentration. Therefore, the majority carrier capture cross section is often measured via variable pulse length experiments on dedicated specimens. Even for a small pulse length (10 ns), a capture cross-section of 10^{-14} cm^2 is expected to be only measurable in specimens with carrier concentrations lower than 10^{15} cm^{-3}. Nevertheless, this even for uniformly doped semiconductor substrates results in a non-uniform injection of majority carriers, because of the Debye tails. In these Debye tails the free carrier concentration is lower than in the neutral semiconductor, and gives therefore a smaller capture rate. This is theoretically discussed by Pons (1984). Figure 3.7a shows the fraction of defects that contribute to the DLTS signal for different pulse durations. It can be seen that for longer pulse durations, the deep levels are probed closer to the surface. These defects closer to the surface respond slower due to the fact that the free carrier profile during the pulse decreases towards the junction interface. This effect can also be seen from Fig. 3.7b were the DLTS signal is plotted as a function of filling pulse duration. Besides the initial rise for short pulse lengths that is often still exponential, also the slow capture regime can be seen for longer pulses. Consequently accurate determinations of the capture cross sections are based on models that include the capture rate in these Debye tails. A model that has proven its applicability on a wide range of specimens is based on a rectangular approximation for the profile of deep traps that contribute to the DLTS signal. This approximation can result in an analytical form that has been implemented in a fitting routine (Lauwaert, 2008). The result of this fitting routine is also shown in Fig. 3.7b. The method for fitting the evolution of the DLTS-signal as a function of filling pulse can take into account small variations in the deep level concentration. Within the region with normal capture rate $c_n n_0$ the defect profile will not induce variations in the DLTS amplitude with varying pulse length. Thus only a variation in amplitude is induced by non-uniform defect concentration in the region of slow capture (i.e. the Debye tails). Fortunately, this region is often small in comparison with the region with normal capture. Therefore, this variation can be included in the fitting routine by approximating this by a linear variation introducing only one extra fitting parameter. On the other hand, the accuracy of this fitting technique can be improved by fitting measurements with different reverse bias simultaneously as is demonstrated in Lauwaert (2016). The region of slow capture is the same for each reverse bias, indeed the pulse voltage stays the same. Correcting the data for C_r^3 makes it possible to have a series of measurements that are disturbed with exactly the same evolution of a slow capturing regime.

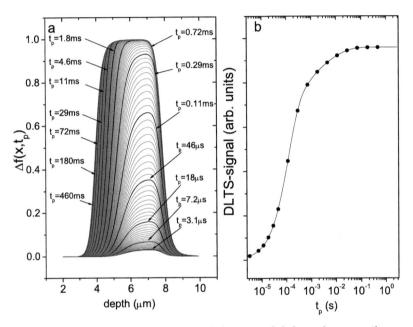

Figure 3.7 (a) Numerically integrated fraction of defects that contribute to the DLTS signal as a function of distance from the interface for different pulse durations. ($n = 2 \times 10^{13}$ cm^{-3}, $c_n n = 10^4$ s^{-1}, $e_n = 10^2$ s^{-1}, $V_r = -1$ V and $V_p = 0$ V) (b) Corresponding DLTS signal as a function of filling pulse duration.

This fitting method uses the data over a wide filling pulse time range, but often this variation is perturbed by an additional time-constant. This can be induced by an additional defect level responding within the same temperature and time range, but also from an extra transient induced by contacts or mobility. Therefore, it was suggested by Segers et al. (2014) to make an estimate of the capture cross section based on a narrow range of filling pulses. This method allows avoiding the time regions that are strongly perturbed by the extra time constant. To improve the accuracy of this rough estimate, measurements with different reverse biases have to be taken into account. In general the region of slow capture is negligible if DLTS is probing a large region, this means for small reverse capacitance. Consequently the most accurate value for the capture rate is obtained by extrapolating the capture data for $C_r \rightarrow 0$. A comparison of these methods is presented in Lauwaert (2016).

3.4 Minority Carrier Traps

In general, for a majority carrier trap conventional DLTS signals result in an increase in the capacitance. Actually independent of the charge state change of the deep trap, a majority carrier trap is emitting majority carriers resulting in a reduction of the depletion region and consequently an increase in the capacitance. Therefore, without injection of minority carriers it is not expected that in conventional DLTS one can observe minority carrier traps, and negative signals (decrease of capacitance) are thus uncommon. Thus conventional DLTS probes only half the band gap, information on the other half of the band gap will only be obtained by injecting minority carriers.

For many diodes, the injection of minority carriers can be obtained by a pulse in forward. These injected minority carriers are than captured by the minority carrier traps. An estimate of this capture cross section is not straightforward because this injection is not uniform and the resulting capture transient or the DLTS signal as a function of filling pulse is a sum of different rates. Nevertheless, the emission of minority carriers is independent on the concentration of minority carriers leading to a emission signature that is characteristic for the defect. This means that DLTS peaks in conventional DLTS in certain type of semiconductor (n-type or p-type) from majority carrier traps fall on the same position in the spectrum and have the same signature as when these defects are present in the other type of semiconductor where they behave as a minority carrier trap. This is why it is more convenient in DLTS studies to have both p- and n-type specimens. This allows to scan both the upper and lower half of the band gap using conventional DLTS. Nevertheless, often this is not possible and researchers want to study a certain specific electronic device and then probing using minority carrier injection becomes a very useful tool. Moreover, the results are easy to compare with previous studies on other specimens.

Besides injection via an electric pulse, filling the traps with minority carriers is often done by an optical pulse. Optical DLTS is a common DLTS variant to inject both majority and minority carriers. Except for solar cells, this technique often needs dedicated diodes or at least front contacts that are transparent for the wavelength over the band gap. This is different from deep-level optical spectroscopy, in which a wavelength with energy below the band gap is used to optically stimulate the emission of carriers as was also discussed in the section about coupling of defects.

When illuminating with light over the band gap, two methods are possible, illumination via the front contact and from the ohmic back contact. The generation of carriers that can be captured by the deep traps within the depletion layer goes together with a diffusion mechanism and are not necessary traps at the location where the photon is absorbed. During this process, not only purely the capture of minority carriers is possible, but also capturing majority carriers which often leads to a complicated spectrum of negative and positive peaks.

3.5   Extended Defects

Extended defects can induce deep lying states within the band gap of the semiconductor. These deep states can be band-like due to the atomic structure of the defect or can interact with the surrounding point defects and act as localized states. Consequently, the density of states determines the line shape of the DLTS signal. In contrast to point defects that induce a single level, these extended defects often induce odd signals, where the position of the band is dependent of the filling pulse duration t_p. In Schroter (1995), it is demonstrated that DLTS is a powerful tool to estimate the defect density of states based on the line shape as a function of filling pulse duration. As shown in Fig. 3.8a. the charge carrier kinetics were described in terms of different rates. R_i is the rate at which the levels reach internal equilibrium, while R_e and R_c are the emission and capture rate, respectively. For the electric repulsion case, an electric repulsion is included by a barrier in de capture rate δE_c, which is often proportional to the fractional occupation of the defect. This general description can explain the odd line shape and the logarithmic capture law and falls into two limiting cases of bandlike $(R_i \gg R_c, R_e)$ or localize $(R_i \ll R_c, R_e)$ states.

This model proposed by Schröter et al. (1995) can figure out how the density of states distribution for the extended defects look like. This is done by comparing the simulated DLTS signals as shown in Fig. 3.8b with the measurements.

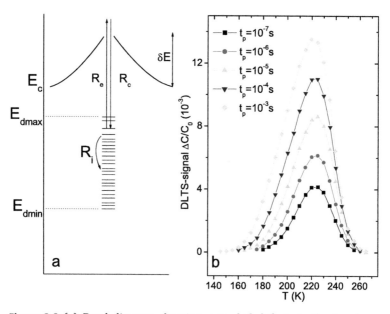

Figure 3.8 (a) Band diagram showing extended defects in the band gap. The extended defect is exchanging electrons with the conduction band with an emission rate R_e, a capture rate R_c and an internal equilibrium rate R_i. (b) Computer simulated DLTS lines for a point defect cloud formed in the elastic stress field of 60° dislocations in n-type silicon. Data from Schroter (1995).

3.6 Heavily Compensated Semiconductors and Non-Uniform Doping Profile

For defect concentrations, more than 10% of the majority carrier concentration the standard DLTS theory as discussed in one of the previous sections is not valid anymore. This is because the transient induced by the defects has a large impact on the depletion layer, and therefore the capacitance transient is not purely exponential.

To avoid this non-exponential transient constant capacitance DLTS can be used. Here the voltage transient is recorded while the capacitance is kept fixed, thus there is no depletion layer width change and no exponential transient. This is a valuable method when the capacitance does not change strongly with temperature, because this could lead to too high negative reverse biases for the highest temperature. Due to this non-exponential behavior of the transient, estimating the concentration of the defects is not so straightforward anymore. Indeed the amplitude ΔC is usually calculated by extrapolating the transient including an estimate of its time constant to $t = 0$ (see for example Eq. 3.21). However, not only the determination of the concentration will suffer from the non-exponential decay but also the determination of the emission rate will become dependent on the bias voltage and pulse height. A simulation of this effect in isothermal measurement in a semiconductor with free carrier concentration $n_0 = 1 \times 10^{16}$ cm^{-3} for a deep trap with activation energy for the emission $E_T = 580$ meV and concentration $N_T = 2 \times 10^{17}$ cm^{-3} is shown in Fig. 3.9. From this simulation, it can be seen that the emission rate becomes slower for larger pulse amplitude. This is opposite to what is expected for peak shifts induced by an electric field. Therefore, peak shifts induced by high deep level concentrations are often easily distinguished from those induced by electric field enhanced emission. The high frequency capacitance transient is a change in the depletion layer width induced by the conservation of bias voltage during the emission of carriers from the deep traps. As discussed above, this transient can become non-exponential if the concentration of deep traps is not negligible in comparison with the free carrier concentration. However, of course, this depletion width is strongly dependent on the free carrier concentration. Therefore, a steep free carrier profile can also deform the exponential transient.

This effect is discussed by Ito and Y. Tokuda (Ito, 2002). Figure 3.10 shows a simulated example of a DLTS peaks influenced by a steep carrier profile. It can be seen from the example and it can be easily understood that this steep profile can induce any kind of peak shift. Indeed, such a steep profile can both increase or decrease toward the junction interface.

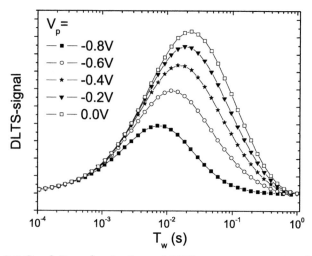

Figure 3.9 Simulation of an isothermal DLTS measurement using different pulses $V_r = -1$ V.

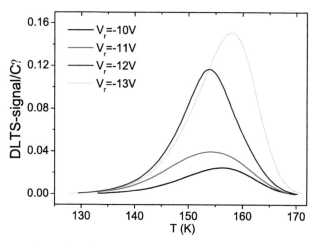

Figure 3.10 A simulated example to demonstrate the influence of a steep free carrier profile on the peak position of the DLTS signal. Data taken from Ito (2002).

3.7   Anomalous Signals

In general, DLTS is invented to probe deep states in semiconductors at low concentrations. Nevertheless, due to the evolution of using more complicated devices in electronics or solar cell technology, this methodology often encounters anomalous DLTS signals. This is not unexpected because any process with a certain time constant that vary with temperature can in principle induce such a transient. Because the sensitivity of DLTS instruments these transients are often detected.

Anomalous signals are induced by the fact that for the diode under study the voltage steps switching between reverse bias and pulse are not fast enough. Therefore, this varying of the potential drop over the diode can generate an unwanted transient in the capacitance that is recorded within the time window of the measurement (Lauwaert, 2011). The model that has been used to describe this process and to allow distinction between emission and capture of carriers is based on adding an additional RC (Resistor and capacitor in parallel) to the circuit.

Since this anomalous transient is induced by a voltage drop variation over the junction, the amplitude of the DLTS signal can be calculated from the variation of the capacitance as a function of bias over the main junction. This difference in origin makes it possible to estimate the DLTS-signal amplitude as a function of the reverse capacitance. In contrast to the conventional signals the sign of these anomalous signals can be both positive (increase in the capacitance) and negative (decrease of the capacitance). Because the main junction is often a good junction that is biased in reverse the leakage current is small so that the time constant of the RC circuit describing the main junction is much larger than the RC time constant of the perturbing back contact. These anomalous signals have mostly the opposite sign as the conventional signals. Nevertheless, both positive and negative signals are possible. Based on this model of different RC-like contacts three different properties apart from the sign have been proposed that make it possible to discriminate between a conventional defect signal and an anomalous signal induced by an RC mechanism. (1) For small pulse amplitudes the time constant of the signals with $\Delta V > 0$ and $\Delta V < 0$ converge to one another. Indeed for a conventional signal we have capture or emission depending on the sign of the pulse, while for an RC-like contact we have the same mechanism for both signs of pulses. (2) It is also typical that for these RC-signals the largest amplitude is observed for $\Delta V > 0$. And finally (3) the amplitude of such a signal scales with the reverse capacitance with the factor four $\Delta C \propto \| \Delta V \| C_r^4$. Of course this last property is only valid for a uniform doping and the additional contact less resistive in comparison to the main junction.

The properties described above are for an RC-like perturbation of the main junction that induces a DLTS signal. Although the physical origin is often a back contact barrier the assignment is not straightforward to make. Nevertheless, besides these properties that can be verified to be sure that this signal is not a pure defect, the time constant and amplitude as a function of temperature contains information on the current and capacitance change as a function of bias voltage and pulse. Indeed numerical calculations including $I(V)$ and $C(V)$ for the back contact have shown the electronic parameters for this back contact (Lauwaert, 2014). Figure 3.11 shows an example where numerical regression demonstrated that this anomalous signal observed in CIGS solar cells can be assigned to a barrier most probably at the back contact.

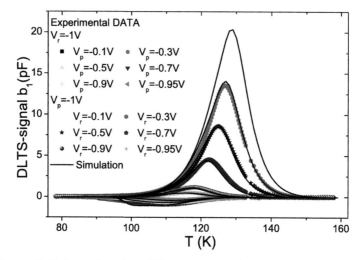

Figure 3.11 Example of a DLTS spectrum for different pulses and reverse bias of a RC-like contact, recorded on a thin film solar cell with a Cu(In, Ga)Se$_2$ absorber.

3.8 Summary

In this chapter, deep-level transient spectroscopy has been discussed. After describing the DLTS theory, we discuss different methods to analyze the recorded transients and how background illumination can help to assign levels to the same defect structure. The following sections cover the determination of the capture cross sections and the observation of minority carrier traps. Finally, this chapter discusses different signals that deviate from a single carrier trap. These are the extended defects, signals deformed by high defect concentrations or a non-uniform doping profile and the real anomalous signals induced by parasitic effect in the electronic device.

References

P. Blood and J. W. Orton, *The Electrical Characterization of Semiconductors: Majority Carriers and Electron States*, Academic Press (1992).

M. Burgelman, P. Nollet, and S. Degrave, *Thin Solid Films,* 361–362, 527–532 (2000).

L. Dobaczewski, A. R. Peaker, and K. Bonde Nielsen, *J. Appl. Phys.*, **94**, 4689 (2004).

J. Frenkel, *Phys. Rev.*, **54**, 647 (1938).

A. Ito and Y. Tokudo, *J. Crystal Growth*, **210**, 384 (2000).

A. Ito and Y. Tokuda, *Solid-State Electronics*, **46**, 1307 (2002).

D. V. Lang, *J. Appl. Phys.*, **45**, 3023 (1974).

J. Lauwaert, L. Callens, S. Khelifi, K. Decock, M. Burgelman, A. Chirila, F. Pianezzi, S. Buecheler, A. N. Tiwari, and H. Vrielinck, *Prog. Photovoltaics*, **20**(5) 588594 (2012).

J. Lauwaert, S. Khelifi, K. Decock, M. Burgelman, and H. Vrielinck, *J. Appl. Phys.*, **109**, 063721 (2011).

J. Lauwaert, S. Khelifi, and H. Vrielinck, *ECS J. Solid State Sci. Technol.*, **5(4)**, N1–N7 (2016).

J. Lauwaert, J. Lauwaert, L. Van Puyvelde, J. W. Thybaut, and H. Vrielinck, *Appl. Phys. Lett.*, **104**, 053502 (2014).

J. Lauwaert, J. Van Gheluwe, and P. Clauws, *Rev. Sci. Instrum.*, **79**, 093902 (2008).

D. Pons, *J. Appl. Phys.*, **55**(10), 3644 (1984).

S. W. Provencher, *Comput. Phys. Commun.*, **27**, 229 (1982).

D. K. Schroder, *Semiconductor Materials and Device Characterization*, John Wiley & Sons Inc, 2nd edition (1998).

W. Schröter, J. Kronewitz, U. Gnauert, F. Riedel, and M. Seibt, *Phys. Rev. B*, **52**(19), 13726 (1995).

S. H. Segers, J. Lauwaert, P. Clauws, E. Simoen, J. Vanhellemont, F. Callens, and H. Vrielinck, *J. Phys. D: Appl. Phys.*, **46**, 425101 (2013).

S. H. Segers, J. Lauwaert, P. Clauws, E. Simoen, J. Vanhellemont, F. Callens, and H. Vrielinck, *Semicond. Sci. Technol.*, **29**, 125007 (2014).

D. Stievenard, M. Lannoo, and J. C. Bourgoin, *Solid-State Electronics*, **28**(5), 485–492 (1985).

S. Weiss and R. Kassing, *Solid State Electronics*, **31**(12) 1733 (1988).

Chapter 4

Capacitance-Voltage and Drive-Level–Capacitance Profiling

Jennifer T. Heath

Department of Physics, Linfield College, McMinnville, Oregon 97128, USA

jheath@linfield.edu

The capacitance–voltage (CV) and drive-level–capacitance-profiling (DLCP) techniques yield information about the density of states in, or near, the depletion region of a diode. The CV technique provides a depth profile, while the DLCP technique adds energy resolution. These techniques are typically applied to one-sided diodes for simplicity of analysis, and the reader should assume an n^+-p junction in the following discussion, unless otherwise specified.

The CV technique is an analysis of capacitance as a function of DC bias, V_{dc}, to yield a density of states, N_{CV}, which is normally assumed to be the shallow doping density at the depletion edge [1]. This approach relies on the depletion approximation, and is very good as long as the change in capacitance with V_{dc} is primarily due to change in the depletion width; thus, it works

Capacitance Spectroscopy of Semiconductors
Edited by Jian V. Li and Giorgio Ferrari
Copyright © 2018 Pan Stanford Publishing Pte. Ltd.
ISBN 978-981-4774-54-3 (Hardcover), 978-1-315-15013-0 (eBook)
www.panstanford.com

well for semiconductors with fully ionized dopant states and no deeper states within the bandgap, measured within appropriate bias, frequency, and temperature ranges. For less ideal semiconductor materials, which may have a significant density of deeper states, charge trapped inside the depletion region can contribute significantly to the capacitance, and the CV density may be misleading [2]. The admittance spectroscopy (AS) technique discussed in Chapters 1 and 2 can be helpful in identifying an appropriate parameter space.

In less ideal semiconductor materials, the DLCP density N_{DL} becomes particularly useful, as it is a purely dynamic measurement depending only on the ac response of the device. Instead of varying the dc bias to measure carrier density, as in CV, in DLCP the amplitude of the ac bias is varied. The energetic distribution of the density of states can be probed by observing the frequency and temperature dependence of N_{DL}. In both DLCP and CV, varying V_{dc} adjusts the spatial location being measured.

The DLCP technique was developed by J. D. Cohen and C. E. Michelson to study trap states in a-Si films for photovoltaics [3]. The technique was inspired by its predecessors, including AS, CV, and DLTS (discussed in Chapter 3). Cohen wished to find a quicker way to obtain densities of states similar to those found from DLTS, by moving the measurement from the time domain (transients) to the frequency domain. Compared to DLTS, DLCP is a quasi-equilibrium measurement, not requiring "filling pulses" to change the state of the system. However, DLCP does not yield information about the majority/minority carrier nature of the trap states. The DLCP technique can also be considered a more direct way to obtain densities from AS-like data without needing prior knowledge or estimates of material parameters (other than the dielectric constant).

The DLCP technique thus retains and builds on some advantages of its predecessors—it retains the spatial profiling capability of CV, the energy profiling capability of AS, and the ability to identify densities of carriers responding dynamically as in DLTS. However, it also retains some of the shortcomings of these techniques. Any shift in quasi-Fermi energy with temperature will affect the position and/or energy of the measured response. The energy scale is also likely to be shifted and distorted by the Meyer–Neldel rule [4, 5]. And, at best, measured positions only indicate a spatial average of the response locations, indicating the first moment of charge response $<x>$ averaged laterally over the area of the device.

While the CV technique is very broadly applied, use of the DLCP technique remains primarily within the community studying semiconductor thin films for photovoltaics, especially a-Si, the Cu(In,Ga)(Se,S)$_2$ alloys, and CdTe, where it has been useful for optimizing these materials [6–8].

In this chapter, we detail the theoretical derivation, assumptions, strengths and limitations, application and analysis of these two related techniques, CV and DLCP.

4.1 Depletion Capacitance and CV Profiling

4.1.1 The Ideal One-Sided Diode with an Abrupt Junction

The variation of capacitance with bias in the depletion regime is fairly straightforward, assuming (i) the change of charge with voltage occurs at the depletion edge, i.e., only shallow, fully depleted states change their charge with dc bias, and (ii) the depletion region is precisely defined, ends abruptly, and is fully depleted of free carriers. The second assumption is known as the depletion approximation.

Then, for a one-sided junction, the depletion width is

$$W = \sqrt{\frac{2\varepsilon\varepsilon_0(V_{bi} - V_{dc} - 2kT/e)}{eN_B}}, \tag{4.1}$$

where N_B is the doping on the lightly doped side of the junction, e is the elementary charge, and V_{dc} is the applied dc bias, and is positive for forward bias. For a two-sided abrupt junction, i.e., with similar doping densities on each side such that neither can be neglected, $N_B = N_A N_D/(N_A + N_D)$. The term $2kT/e$ comes from a majority-carrier contribution to the charge distribution, and is often neglected [1]. Equation 4.1, of course, only applies when $V_{dc} < V_{bi}$; more accurately, it applies for values of V_{dc} where $W > L_D$, the Debye length, discussed further below (Eq. 4.6).

Such data are commonly analyzed by graphing C^{-2} vs. V_{bi}, known as a Mott–Schottky plot, where

$$C^{-2} = \frac{2(V_{bi} - V_{dc} - 2kT/e)}{e\varepsilon\varepsilon_0 A^2 N_B}. \tag{4.2}$$

For a sample with constant doping density, these data should fit a straight line whose slope and intercept yield the values of N_B and V_{bi}.

The density N_B need not be uniform throughout the film. The small signal capacitance

$$C = \frac{\delta Q}{\delta V} \tag{4.3}$$

originates at the depletion edge, due to the change of depletion width, and so the instantaneous slope of C^{-2} will approximately give N_B at the location W. Since W is a function of V_{dc}, a changing slope with bias typically indicates spatial variations in N_B.

$$N_{CV} = -\frac{2}{e\varepsilon\varepsilon_0 A^2}\left[\frac{d(C^{-2})}{dV_{dc}}\right] \approx N_B \tag{4.4}$$

While the slope of the Mott–Schottky plot is well defined, the intercept is strongly affected by any non-uniformity in N_B, as well as by any other source of static charge. This means that V_{bi} cannot usually be directly extracted from the intercept without further analysis. This sensitivity opens up the possibility of using the intercept to investigate other sources of static charge, for example, at interface impurities and traps.

The approach to such analysis has continued largely unchanged from the work of early pioneers in the field. In addition to standard texts such as [1], it can be interesting to read early perspectives, such as by Schottky, summarized in [9], and by Shockley [10]. Already, these authors anticipate and attempt to analyze factors which complicate and deepen the simple depletion analysis described above, including the diffusion of minority carriers, the role of resistive regions, frequency dependence, spatial non-uniformities, and additional barriers at the contacts.

The CV profiling technique is typically defined by Eq. 4.4, and the approximate location of the response is more accurately described as the first moment of charge response

$$\langle x \rangle = \frac{\varepsilon \varepsilon_0 A}{C} = \frac{\int_0^\infty x \delta \rho(x) dx}{\int_0^\infty \delta \rho(x) dx} \approx W \qquad (4.5)$$

The accuracy of the approximations $N_{CV} \approx N_B$ and $<x> \approx W$ depends largely on the depletion and shallow doping assumptions discussed above, and such assumptions are further investigated below. Even in ideal crystalline semiconductors, the depletion edge is never precisely abrupt, but is smeared on the scale of the Debye length [11]

$$L_D = \sqrt{\frac{\varepsilon \varepsilon_0 kT}{q^2 N_B}} \qquad (4.6)$$

A sample with $\varepsilon \approx 10$ would typically have L_D in the range of tens to hundreds of nm; for example, it is 400 nm at $(T, N_B) = (300\ K, 10^{14}\ cm^{-3})$ and 20 nm at $(100\ K, 10^{16}\ cm^{-3})$.

The simulated data in Fig. 4.1 demonstrates the way in which CV data is normally displayed. These CV results are from a simple model of a silicon diode at 300 K, using the modeling program SCAPS [12]. This model was designed to demonstrate the impact of spatial variation on the CV profile. In this model, the first 0.6 μm of the Si film contains $10^{16}\ cm^{-3}$ acceptors giving $L_D = 0.04$ μm, while the rest of the film contains $10^{15}\ cm^{-3}$ acceptors with $L_D = 0.12$ μm. Limitations to the spatial resolution are evident, but far enough from the transition region the approximation $N_{CV} \approx N_B$ is good.

Metal-semiconductor diodes behave as described above, where the metal can be seen as an extreme example of a highly doped layer.

4.1.2 Influence of Series Resistance

The first, and simplest, refinement to the capacitance model is to include the effect of series resistance R_s as discussed in Chapter 1 and illustrated in Fig. 1.3c. The series resistance could originate from sample specific mechanisms, such as interfaces and contacts, but all diodes will have some series resistance due to conduction through the undepleted part of the diode. The influence of the series resistance is most easily checked using impedance spectroscopy as described in Chapter 1. Then the voltage across the capacitor is reduced from the applied voltage by IR_S, where I is the total current into the device. Normally the effect is small, and can easily be calculated by finding the total admittance Y, since the apparent measured capacitance is $C_m = \mathrm{Im}(Y)/\omega$. In the series/parallel model of Fig. 1.3c,

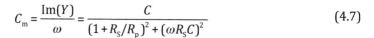

$$C_m = \frac{\mathrm{Im}(Y)}{\omega} = \frac{C}{(1+R_S/R_p)^2 + (\omega R_S C)^2} \tag{4.7}$$

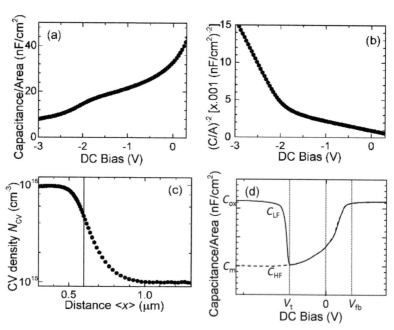

Figure 4.1 Characteristic diode capacitance response. (a) Simulated data for a Si diode with two distinct regions: 0.6 μm with $N_A = 10^{16}$ cm^{-3}, and 3 μm with $N_A = 10^{15}$ cm^{-3}. (b) Mott–Schottky plot of the simulated data. (c) Calculated N_{CV} of the simulated data. The two regions are clearly distinct, as are the limits to the spatial resolution. (d) Schematic response for a MOS device showing low frequency (solid line) and high frequency (dashed line) response as discussed in the text.

Writing $R_x^{-1} = R_s^{-1} + R_p^{-1}$, then for low frequencies $\omega \ll 1/R_x C$, a voltage-divider effect reduces the measured value to $C_m \approx C/(1 + R_s/R_p)^2$, and for high frequencies $\omega \gg 1/R_x C$, a low-pass filter effect attenuates the capacitance to $C_m \approx C/(\omega R_s C)^2$. Typically, $R_s \ll R_p$ and this effect can be neglected. However, a red flag is raised if C_m falls off as ω^{-2} with increasing frequency [13]. Once identified, the effect of series resistance can be corrected for in the data [14]. These issues were also discussed in Chapter 2.

4.1.3 Influence of Interface States

Interface states similarly provide a layer of the sample which can accumulate charge and affect the measured capacitance, and their influence has long been considered in CV analysis [15]. A significant number of interface states can even pin the quasi-Fermi energy at the interface, such that the capacitance changes little with V_{dc} and the doping density appears to be artificially large. Another common effect is an apparent spike in N_{CV} at a value of V_{dc} where the charge trapped at the interface changes abruptly [16], for example, as E_F moves with V_{dc} and populates or empties an interface trap. More minor effects include the presence of a constant sheet of charge or dipole layer at the interface, which changes the voltage intercept and the apparent value of V_{bi}, but does not significantly affect N_{CV}.

Although the charge trapped at interface states can strongly influence CV results, details vary significantly between systems. An in-depth understanding normally requires application of a several complementary techniques which can together support the analysis. Typically, researchers have other reasons to suspect that interface states may significantly influence the capacitance or may even dominate the results, and interpret the CV data accordingly. Or, the CV data look unusual in some way, prompting a search for explanations. Possible factors include: Are the interface traps distributed broadly in energy, or are they primarily at discrete energies? Is the Fermi energy pinned with voltage? Do the interface traps respond dynamically with ac bias? Do they trap minority or majority carriers? Does the charge density or the trap characteristics change in a metastable way? How do these properties depend on sample preparation? A few examples include Refs. 17–19.

4.1.4   Capacitance in Forward Bias

During the CV measurement, it is tempting to scan further and further in forward bias, to collapse the depletion width and measure material properties closer to the interface. However, when $W \sim L_D$, and/or significant currents flow, the depletion approximation ceases to apply and the standard CV analysis becomes invalid. Understanding the ac response in forward bias is not intuitive, as many mechanisms can influence the magnitude and phase of the current.

In forward bias, the best known capacitive mechanism is that of diffusion capacitance [10, 20]. Since diffusion current increases exponentially with bias, in forward bias the diffused carriers represent a significant separation of charge which is very sensitive to the applied bias. This separation of charge results in the diffusion capacitance, which is proportional to current and so also increases exponentially with bias. The diffusion capacitance is most significant at low frequencies.

However, additional effects also occur in forward bias that can influence the measured capacitance. Consider, for example, the voltage drop V_s across any series resistance R_s. In forward bias, the increased current can significantly affect the value of the series resistance itself, due to carrier injection into the undepleted part of the film. At high currents the series resistance can thus become strongly voltage-dependent, decreasing exponentially with increasing bias. When R_s is measurably changed by the ac component of current, the effect known as modulated conductivity becomes apparent. Then, the ac variation *in R_s itself* contributes an unexpected negative ac component to the series voltage drop.

In a typical capacitance measurement, the source drives both a dc current and an ac current at frequency ω. If we explicitly separate out the dc ($\overline{V}_s$) and the first order ac ($\tilde{V}_s$) parts of the series voltage drop V_s, we have

$$V_s = \overline{V}_s + \tilde{V}_s \exp(j\omega t) + \text{higher order terms} \tag{4.8}$$

and similarly the total current is

$$I = \overline{I} + \tilde{I}\ \exp(j\omega t) + \text{higher order terms} \qquad (4.9)$$

In a typical measurement with a linear series resistance, we would expect that $\tilde{V}_s = \tilde{I}R_S$, where R_S has a constant value. However, because R_S also oscillates at frequency ω, out of phase with the current, due to the carrier injection described above,

$$R = \overline{R}_s - \tilde{R}_s \exp(j\omega t) + \text{higher order terms} \qquad (4.10)$$

and $\tilde{V}_s$ has an additional component [21]:

$$\tilde{V}_s = \overline{I}\tilde{R}_s - \tilde{I}\overline{R}_s \qquad (4.11)$$

Note there are also higher order terms; the most obvious is a contribution to V_S at frequency 2ω, due to the product $\tilde{I}\tilde{R}_S$.

Carrier injection effects, including (but not limited to) modulated conductivity, can result in a measured capacitance that decreases with increasing V_{dc} and even becomes negative, in stark contrast to the expected depletion or diffusion capacitance behavior. This is a longstanding observation, and has been seen in a wide range of material systems. A much more careful analysis of carrier motion, starting with a general treatment of the ac current, is necessary to fully understand the negative capacitance effect, which is not necessarily easily categorized to a single mechanism or lumped into an equivalent circuit element. Early work on negative capacitance focused on the mechanism of modulated conductivity discussed above [21], while more recent work by Laux and Hess brings out the range of mechanisms that could contribute to a decreasing or even negative capacitance [22]. Note that mechanisms decreasing C would also, in general, cause a decreased or negative Z''.

The back contact of the diode is typically assumed to be an ideal Ohmic contact. However, the presence of a blocking contact would also cause the capacitance to decrease with increasing forward bias, as the main junction becomes conducting and the applied bias is dropped across a depletion region at the back. Such an effect would also likely be visible as phototransistor behavior in the current–voltage curve [23]. In this case, the capacitance would decrease with increasing dc bias, approaching a minimum value of $C = \varepsilon A/t$; it would not become negative. This was discussed more extensively in Chapter 2.

4.1.5 Metal–Insulator–Semiconductor Devices

In metal–insulator–semiconductor (MIS or MOS) devices, the discussion of depletion capacitance in Section 4.1.1 also holds in the depletion regime, as the insulator capacitance is in series with the depletion capacitance, and is constant with V_{dc}. Thus, the insulator capacitance does not appreciably affect the value of N_{CV} found using Eq. 4.4, though it does affect $<x>$ [1, 24].

In MIS devices the current cannot flow freely, but is blocked by the insulator layer. Thus, the sample is in quasi-equilibrium with constant Fermi energy throughout the film, even when bias is applied, and carriers can accumulate at the insulator surface. This affects the CV behavior for larger values of reverse and forward bias. The larger reverse and forward bias regimes, known as the inversion and accumulation regimes, are interesting in their own right for the study of fixed charges on the insulator and minority carrier dynamics [25].

As the reverse bias increases, it reaches the threshold voltage V_t where the intrinsic energy drops below the Fermi energy (for a p-type semiconductor) at the surface, and a large number of minority carriers (electrons) invert the surface to n-type. This is the inversion regime. The depletion width reaches a maximum and does not vary further with dc bias, as any additional bias is dropped across the insulator layer. In the inversion regime, the capacitance is frequency dependent, as the generation of minority carriers is relatively slow. At low frequencies, minority carriers have time to respond and so the additional bias is dropped entirely across the insulator layer, giving $C = C_{ox}$. At high frequencies, the capacitance reflects the maximum depletion layer width in series with the insulator layer, and reaches a constant minimum value $C = C_{min}$.

In forward bias, as the voltage increases, the bands flatten. The flat-band voltage is the bias at which the depletion region entirely disappears, and at higher voltages the device is in the accumulation regime. In this regime, majority carriers accumulate at the surface of the insulator, and $C = C_{ox}$. The MIS capacitance is schematically illustrated in Fig. 4.1d.

MIS devices are further discussed in Chapter 10.

4.1.6 Experimental Details of CV

The CV experiment requires the ability to measure capacitance as a function of dc bias. Temperature and frequency are also useful parameters to control. This functionality is available with most LCR meters, or by using a lock-in based system as described in Ref. 26. Experimental techniques are treated in Chapters 5–7.

In general, AS and/or IS data should be collected together with CV to ensure that the sample is in a productive measurement regime. Carriers should be thermally activated and the sample should not be fully depleted. In some cases, CV is applied to materials which have hopping conduction, poor contacts, known interface states, deep trap states in the bandgap, or other limitations to their response. It can still be a useful measurement, but the interpretation must reflect these characteristics of the device [27, 28].

4.2 Drive-Level Capacitance Profiling

The DLCP technique moves away from the small signal limit used in all other admittance techniques. Instead, in DLCP, the dependence on C of the ac amplitude dV provides necessary information to calculate the density of dynamically responding states:

$$C = C_0 + C_1 dV + C_2 dV_2 + ...,$$

where the usual small signal capacitance is C_0. Note that each prefactor does not have the same units, i.e., C_1 has units of F/ V.

As a side note, typical capacitance measurements assume that for a small enough ac amplitude, perhaps 20 mV, $C \approx C_0$. DLCP gives the actual limiting small-signal value of capacitance, yielding a more accurate value. For the one-sided junction analyzed here, C_1 is negative, so C_0 is always slightly larger than the typically reported value of C.

4.2.1   Theoretical Development of DLCP

The calculation of the density, N_{DL}, from the capacitance response is derived here based on the depletion approximation in a one-sided diode, following the approach of Ref. [29, 30]. In devices which are not one-sided, the resulting response will be influenced by the carrier and trap densities on each side with appropriate weighting. In devices with additional series contributions to capacitance, which could include the i-layer in a p-i-n diode, or the depletion region at a blocking back contact, the additional series capacitance will shift the total capacitance, and hence the apparent location of the response calculated from Eq. 4.4, but this shift will not normally affect the density N_{DL}. Only aspects of the device that respond dynamically, i.e., those that capture and emit charge during the ac cycle, will affect N_{DL}.

The experimental frequency and temperature define a limiting energy E_e, below which states can respond dynamically, and this in turn defines a limiting location x_e where traps at the energy E_e would cross E_F; x_e is the point where $E_F - E_V = E_e$. These values are schematically illustrated in Fig. 4.2. These are the states measured in AS as well as in DLCP. Using the low-temperature step function approximation for the Fermi function, the total charge density of states detected by DLCP is then approximately

$$N_{DL} = \frac{\rho_e}{q} = p + \int_{E_F^0}^{E_e} g(E)dE \qquad (4.12)$$

for an n$^+$-p device with density of sub-bandgap states $g(E)$, ionized shallow acceptors p, and Fermi energy in the bulk, E_F^0, measured in relation to the valence band edge. A similar expression can be written for a p$^+$-n diode. The DLCP density is thus the total charge density that can respond dynamically at the experimental frequency and temperature.

In general, the characteristic energy of the response is assumed to reflect the carrier emission rate of a trap state, however, in this quasi-equilibrium measurement, capture and emission times must be in balance, and in some cases the capture rate can be limited by the position of the quasi-Fermi energy or by carrier transport mechanisms, just as in AS [31].

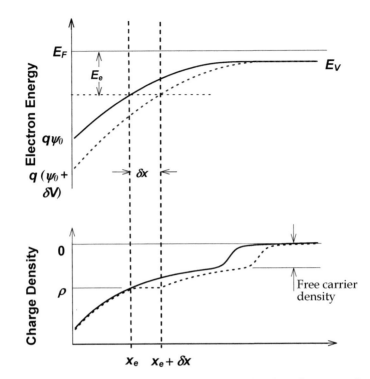

Figure 4.2 Upper curves indicate schematically the change in band bending when a small bias, δV, is applied: the solid line and dashed lines indicate the band bending before and after the application of δV, respectively. Lower curves show the corresponding charge densities also as solid and dashed lines. This assumes a very broad density of trap states across the gap. The time scale of the measurement determines E_e and x_e, such that the trapped charge has time to equilibrate only at positions where $E_F - E_V < E_e$. The value of ρ would be measured by DLCP for these parameters. Adapted with permission from Ref. 30, Fig. 2. Copyrighted by the American Institute of Physics.

The DLCP density is normally derived by first considering the band bending within the depletion region. Assuming a one-dimensional n⁺-p diode, from Poisson's equation one can write the equilibrium potential at the interface, ψ_0, to within a constant factor as

$$\psi_0 = \int_0^\infty x \frac{\rho_0(x)}{\varepsilon \varepsilon_0} dx, \qquad (4.13)$$

where the interface is located at $x = 0$ and $\rho_0(x)$ is the equilibrium charge density through the depletion region.

When an additional ac bias of amplitude δV is added, an important change occurs around the location x_e. For $x < x_e$, the charge density is static and is not affected by δV on the time scale of the frequency. For $x > x_e + \delta x$, the charge density has time to equilibrate on the time scale of the measurement. But in the region $x_e < x < x_e + \delta x$, traps right at the demarcation energy E_e cross the Fermi energy and their occupation is emission limited. The charge density in this region is approximately constant and equal to the density, ρ_e, at x_e before δV was applied.

Thus, we write the potential at the interface as

$$\psi(0) = \psi_0 + \delta V$$

$$\approx \int_0^{x_e} x \frac{\rho_0}{\varepsilon \varepsilon_0} dx + \int_{x_e}^{x_e + \delta x} x \frac{\rho_e}{\varepsilon \varepsilon_0} dx + \int_{x_e + \delta x}^{\infty} x \frac{\rho(x)}{\varepsilon \varepsilon_0} dx. \qquad (4.14)$$

Assuming that the distance δx is small compared to the length scale of any non-uniformities in the film, then the region $x > x_e + \delta x$ is in quasi-equilibrium and has the same boundary conditions as the region $x > x_e$ had before δV was applied. Thus, for $x > x_e + \delta x$, $\rho(x) \approx \rho_0(x - \delta x)$. With a change of variables, we can then rewrite Eq. 4.14 as

$$\psi(0) \approx \int_0^{x_e} x \frac{\rho_0}{\varepsilon \varepsilon_0} dx + \int_{x_e}^{x_e + \delta x} x \frac{\rho_e}{\varepsilon \varepsilon_0} dx + \int_{x_e}^{\infty} (x + \delta x) \frac{\rho_0(x)}{\varepsilon \varepsilon_0} dx$$

$$(4.15)$$

and

$$\psi(0) \approx \int_0^{\infty} x \frac{\rho_0}{\varepsilon \varepsilon_0} dx + \frac{\rho_e[(x_e + \delta x)^2 - x_e^2]}{2 \varepsilon \varepsilon_0} + \int_{x_e}^{\infty} \delta x \frac{\rho_0(x)}{\varepsilon \varepsilon_0} dx \quad (4.16)$$

This allows δV to be related to δx, and ultimately determines the capacitance response. First, solving for δV,

$$\delta V = \psi(0) - \psi_0 \approx \frac{\rho_e(2 x_e \delta x + \delta x)^2}{2 \varepsilon \varepsilon_0} + \delta x F_e, \qquad (4.17)$$

where F_e is the electric field magnitude at location x_e:

$$F_e = \int_{x_e}^{\infty} \frac{\rho_0(x)}{\varepsilon\varepsilon_0} dx \tag{4.18}$$

Then, using the quadratic equation to relate δV and δx yields

$$\delta x \approx \left(\frac{\varepsilon\varepsilon_0}{\rho_e} F_e - x_e \right) \left[1 - \left(1 + \frac{2\varepsilon\varepsilon_0\rho_e\delta V}{(\varepsilon\varepsilon_0 F_e - \rho_e x_e)^2} \right)^{\frac{1}{2}} \right] \tag{4.19}$$

Expanding this expression with the binomial approximation,

$$\delta x \approx \frac{\varepsilon\varepsilon_0}{\varepsilon\varepsilon_0 F_e - \rho_e x_e} \delta V - \frac{\rho_e(\varepsilon\varepsilon_0)^2}{2(\varepsilon\varepsilon_0 F_e - \rho_e x_e)^3} \delta V^2 + \cdots \tag{4.20}$$

$$C = \frac{\delta Q}{\delta V} \approx A\rho_e \left[\frac{\varepsilon\varepsilon_0}{\varepsilon\varepsilon_0 F_e - \rho_e x_e} - \frac{\rho_e(\varepsilon\varepsilon_0)^2}{2(\varepsilon\varepsilon_0 F_e - \rho_e x_e)^3} \delta V + \cdots \right] \tag{4.21}$$

And, using Eq. 4.12, which defines N_{DL}, we have

$$N_{DL} \approx \frac{-C_0^3}{2q\varepsilon\varepsilon_0 A^2 C_1} \tag{4.22}$$

4.2.2 Experimental Details of DLCP

The DLCP experiment requires the ability to measure capacitance as a function of temperature, frequency, dc bias and ac bias amplitude. This functionality is available with some LCR meters, or by using a lock-in based system as described in Ref. 32. Capacitance measurement is also discussed in Chapters 5–7. Data collection software is not currently commercially available and must be customized.

In setting up the experiment, it is important to note that x_e is a limiting value corresponding to $\delta V = 0$, where δV sends the device further into reverse bias as shown in Fig. 4.2. Hence, δV corresponds to the *peak-to-peak* amplitude of the ac bias, with x_e being measured at the most positive extent of the ac signal.

For this reason, as the amplitude x_e is varied, to remain consistent with this derivation its dc offset must also be slightly varied so that the peak of the wave, rather than its center, is aligned. This is illustrated in Fig. 4.3.

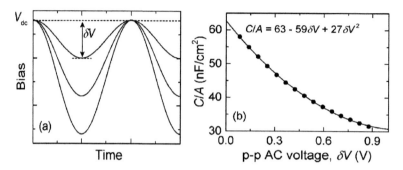

Figure 4.3 (a) Schematic of three applied ac signals for the DLCP measurement, as described in the text. The applied dc bias is adjusted to align the peaks of the ac voltage, and the peak-to-peak amplitude δV is used to calculate N_{DL}. (b) Typical data shows a good fit to a quadratic function, yielding the constants C_0, C_1, and C_2 described in the text.

Typical experiments have historically used ac bias with rms values from about 20–300 mV, with 10 or so data points. This gives a rather lengthy data collection time, especially at lower frequencies, which may be problematic for some devices that exhibit metastable changes under bias stress. Truncation of the expansion of Eq. 4.21 is justified when $\delta V \ll C_0/C_1$, and this criterion can help determine the number of data points, and order of the polynomial, needed for a good curve fit. Normally, a second-order polynomial curve fit is sufficient, and even though the coefficient C_2 is unnecessary for finding N_{DL} it is often required for an accurate fit. In some cases, only a few data points and a first-order polynomial are sufficient for a reasonable extraction of C_0 and C_1.

Some typical raw data are illustrated in Fig. 4.2b. A quick "reasonableness" check should find $C_0 > 0$ and $C_1 < 0$. As in all capacitance measurements, it is also wise to check that the capacitance can be well separated from conductivity. Usually this means $G/\omega \lesssim 10C$, with the exact confidence limit depending on the instrumentation.

Initial AS experiments can help hone in on a productive (ω, T) regime, to avoid carrier freeze-out or transport issues, and select a parameter space that focuses on interesting trap transitions. In general, because changing T can cause a shift in E_F, it is best to find an optimal temperature that allows $g(E)$ to be profiled through a range of energies while only varying ω. However, adjusting T more readily changes E_e over a larger range, and many data reported in the literature assume that any resulting change of E_F is inconsequential. A quick CV scan can help ensure that the chosen dc bias and measurement protocol are optimized, to avoid excessive dc current, identify any metastable drifts, and ensure that diffusion capacitance is not affecting the results. These preliminaries set the stage for a successful DLCP measurement.

4.2.3 Analyzing DLCP Data

Once the C vs. δV data is collected as a function of (ω, T, V_{dc}), and N_{DL} is calculated using Eq. 4.22, the task of understanding the results begins. Figure 4.4 demonstrates several approaches to graphical analysis of the DLCP results.

Typically N_{DL} is plotted versus the first moment of charge response $\langle x \rangle$, also called the center of gravity of charge response. The value of $\langle x \rangle$ is defined as in Eq. 4.5, but unlike in a CV profile, we no longer assume $\langle x \rangle \approx W$; in fact, we assume that $\langle x \rangle$ and W could have very different values.

Although $\langle x \rangle$ is not an exact indication of the origin of charge response, plotting N_{DL} versus $\langle x \rangle$ helps overlay comparable locations, as V_{dc} and E_e both affect the location of x_e. Note that for increasing E_e (increasing T or decreasing ω), or for increasing V_{dc}, x_e moves closer to the interface. For higher values of E_e, not only does x_e decrease, but deeper energy states can respond and will then more strongly influence $\langle x \rangle$, causing $\langle x \rangle$ to move toward the interface, closer to the location x_e. On the other hand, for lower values of E_e, the trap response becomes frozen out and $\langle x \rangle$ approaches the depletion width W. In Fig. 4.4a it is clear that for increasing temperature, the value of $\langle x \rangle$ decreases significantly, as expected.

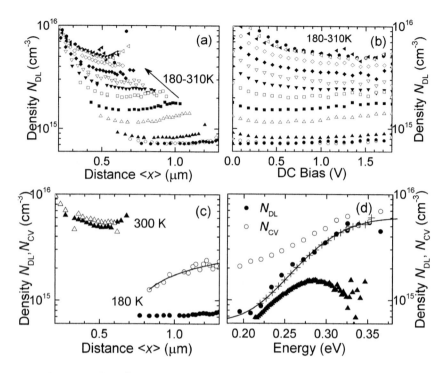

Figure 4.4 Visualizing DLCP results of a CuInSe$_2$ n$^+$-p device prepared as described in Ref. 30. Data was collected at f = 11 kHz and V_{dc} ranging from 0 to –1.8 V. (a) Since the measurement location varies with both E_e and V_{dc}, data can often be best compared when graphed with respect to $\langle x \rangle$. (b) The N_{DL} data with respect to reverse dc bias. (c) Comparison of N_{DL} (closed symbols) and N_{CV} (open symbols) at the two temperature extremes. Solid line is a fit to Eq. 4.23 yielding $\lambda \approx 0.6$ µm. (d) Values of N_{DL} (solid circles) and N_{CV} (open circles) averaged from V_{DC} = –1.0 V to –1.8 V, with (f, T) converted to an energy scale as described in the text. The AS (arbitrary scale, crosses) results are shown for comparison. Data agree well with an error function fit (solid line). The AS derivative data –ωdC/dω (arbitrary scale, triangles) more clearly shows the trap state energy and width.

The sample shown in Fig. 4.4 also appears to have a higher density of states near the interface, which is fairly common. Such a response can be real or an artifact of the contacts; these may be difficult to distinguish from each other. This is because, just as for *CV* profiles, the spatial dependence of N_{DL} is smeared on roughly the Debye length scale L_D (Eq. 4.3).

Measurements which have $\langle x \rangle$ within a few Debye lengths from the contact will pick up some of the unchanging capacitive contribution of the contact, resulting in an apparently very high density contribution to the signal. This signal can be difficult to distinguish from real spatial variation in the material, which is often most significant near contacts and interfaces. In forward bias, the capacitance can also be reduced for a variety of reasons summarized in Section 4.1.4. Considering these effects, which influence both CV and DLCP data, it is not surprising that a characteristic "U" shaped profile is often measured in lower doped samples. Studying thicker samples can help to distinguish the actual bulk properties of the film. Note that some results contain so much spatial variation it can be difficult to distinguish an increase in N_{DL} from a lateral shift in $\langle x \rangle$. Such data must be interpreted cautiously, as these would imply very different things about the device. A lateral shift in $\langle x \rangle$ could come from any static capacitive layer beginning to appear, perhaps from a contact barrier or another layer of the sample, and does not necessarily contain information about trap densities. One way to check the data is to ensure that changes in N_{DL} are closely related to changes in $\langle x \rangle$, approximately according to Eqs. 4.1 and 4.5.

4.2.4 History of DLCP

The DLCP technique was developed in the Cohen laboratory at University of Oregon, and first reported by Michelson, Gelatos, and Cohen in 1985 [3]. It was developed for the purpose of studying trap states in the bandgap of hydrogenated amorphous Si (a-Si:H), which is weakly n-type, and has a large, broad density of states roughly centered about mid-gap [33]. The a-Si:H films also have interesting—and, for photovoltaic applications, problematic—metastable change with optical exposure, known as the Staebler–Wronski effect [34].

The technique was then applied to tetrahedral amorphous carbon films [35], and subsequently to Cu(In,Ga)Se$_2$ alloys [30]. The latter material, also being developed for photovoltaic applications, is polycrystalline, *p*-type, and has significant but somewhat narrower and shallower defect responses. Other groups have used DLCP to study CdTe, as well as other semiconducting compounds. The DLCP technique has mainly been applied by the thin-film photovoltaics community, reflecting DLCP's utility in measuring amorphous and polycrystalline semiconductors with significant densities of sub-bandgap states, where CV techniques would be highly inaccurate.

The deepest measurable energies in DLCP are limited by E_F, which cannot shift deeper than roughly mid-gap, and in a-Si:H the density of states extends across mid-gap. These mid-gap states could not be easily accessed, and, over about 380 K, metastable states in the device begin to anneal on the time scale of the measurement. Because the maximum change of N_{DL} with temperature (or frequency) was found to occur at about 11 kHz and 360 K, this data point approximates the peak of the broad distribution in a-Si:H. The N_{DL} measured using these parameters is then doubled to estimate the total density of the defect band in a-Si:H [36].

4.3 Comparing Results from Multiple Techniques

4.3.1 Comparisons between DLCP and CV

At low values of E_e, the DLCP data give a good approximation to the carrier density of the device, while if there are deeper trap states, the N_{CV} value will be skewed to larger values. A quick estimate of the influence of a single discrete trap state is given by [2]

$$N_{CV} = N_T(x_T)[1 - \lambda/W] + N_A(W),\qquad(4.23)$$

where $\lambda = W - x_T$, and x_T gives the position at which the trap state crosses the quasi-Fermi energy, i.e., the trap adds to the depletion charge density for $x < x_T$ and is neutral for $x > x_T$. Fig. 4.4c includes a fit to Eq. 4.23 using $N_A = 6.5 \times 10^{14}$ cm^{-3} derived from the DLCP data. The value of $N_T = 3.4 \times 10^{15}$ cm^{-3} derived from the fit agrees well with the DLCP results, especially considering the differing spatial locations and temperatures of the two measurements.

For large enough E_e, the values of N_{CV} and N_{DL} should typically agree. This can be seen in the 300 K data of Fig. 4.4c. In some cases, a single DLCP measurement at a *low* value of E_e may be sufficient to estimate both the free carrier density and depletion width (from N_{DL}), as well as the total density of deeper states (from the difference $N_{CV} - N_{DL}$, where N_{CV} can be derived from the C_0 values collected during the DLCP measurement). For this to work, V_{dc} must be sufficiently large and changed slowly enough so that the effective value of λ/W is $\ll 1$ and the deep states have time to fully respond to changes in bias. Once the necessary experimental parameters are verified, this can be a quicker way to collect data for a fairly well understood sample, which may assist in studying (or reducing) complications due to metastability or other time-sensitive effects [37].

Interface states will only contribute to N_{CV} as they cross the Fermi or quasi-Fermi energy at the interface, which depends on V_{dc}. Such contributions will not appear in N_{DL}, except at the specific ω and T when they also coincide with E_e and therefore respond dynamically [38].

A contribution to capacitance that changes with the dc bias, but does not respond dynamically to the ac bias, can cause an apparent spike in N_{CV}, and will cause $<x>$ to shift laterally while N_{DL} remains constant. Such effects can be due to interface states or other capacitive layers in the sample, such as a depletion region at a blocking back contact [39].

4.3.2 Comparisons with AS

A clearer comparison between the AS, CV, and DLCP techniques can be made by converting the (ω, T) data to an energy scale using the prefactor ν_0 derived from AS measurements (as discussed in Chapter 2). In Fig. 4.4d, N_{DL} is plotted (solid circles), using average values for reverse bias between 1.0 and 1.8 V, i.e., in the region where N_{DL} does not have strong spatial dependence. These comparison to N_{CV} has been discussed above, with good agreement at higher E_e, where both measurements should give the total free carrier and trap response. These measurements, however, are very different from each other at low E_e where the broad 0.28 eV trap state will influence N_{CV} but will not impact

N_{DL}. This graph also shows the AS result, scaled to overlay with N_{DL}. Clearly the AS and DLCP measurements show response from the same trap state, as they should; however, the DLCP data directly yield a value for density without relying on knowledge of other sample properties beyond area and ε.

The energy and breadth of the trap state can be determined from an Error function fit to the DLCP or AS data, shown in the graph as a solid line, or from the derivative of the AS data, also shown. Either approach, for this sample, yields a broad energy density centered at 0.28 eV with density 4.5×10^{15} cm^{-3} and background doping 6.5×10^{14} cm^{-3}.

4.3.3 Other Quantitative Tests of DLCP

In its initial development, several efforts were made to assess the accuracy of DLCP by comparing results to other techniques. Carbon impurities in a-Si:H were particularly effective for this purpose, as they seemed to have a strong impact on the metastable defect response seen in DLCP, and were also detectable by other means. In 1991, Unold et. al. intentionally varied the carbon content of films, and compared the secondary-ion mass spectroscopy (SIMS) profiles of the carbon impurity density with DLCP derived values of the defect density as shown in Fig. 4.5 [40].

In Fig. 4.5, the N_{DL} values clearly oscillate over the same characteristic length as the variation in carbon content. The higher N_{DL} (higher carbon) regions also have much more significant metastable state creation with light exposure. The light-exposed sample is labeled as state B in Fig. 4.4, while the annealed sample is labeled as state A.

The shape of the SIMS and DLCP profiles are noticeably different from each other, with the SIMS profile being more abrupt, while DLCP yields something closer to a triangle shape. This is not surprising, and is an important illustration of the spatial accuracy of the DLCP method, which averages over a distance roughly equal to the Debye length. In this sample, the Debye length is about 0.1 μm. The SIMS profiles were input to a numerical model which predicted the expected DLCP profiles, and the predicted shape and periodicity is in close agreement withthe actual results, as is illustrated in Fig. 4.5b.

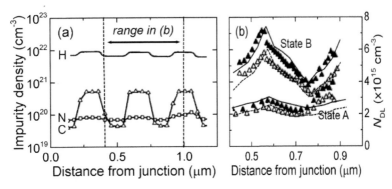

Figure 4.5 Spatial profiles of a-Si:H diodes with intentionally introduced carbon impurities. (a) Secondary ion mass spectroscopy depth profile of the film composition. (b) Measured N_{DL} at two temperatures: 375 K (open triangles) and 385 K (closed triangles), for the annealed state (state A) and the metastable light-soaked state (state B). A numerical model, using the SIMS data to calculate the spatial dependence of N_{DL}, predicts a similar result (solid and dashed lines). Adapted with permission from Ref. 40, Fig. 1. Copyrighted by the American Physical Society.

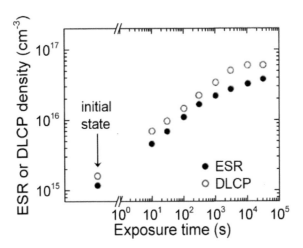

Figure 4.6 Electron Spin Resonance (ESR) and DLCP results are in close agreement for a-Si:H diodes with intentionally introduced carbon impurities. The metastable trap density increases as the sample is light soaked. Adapted with permission from Ref. 36, Fig. 3a. Copyrighted by the American Institute of Physics.

A second study looked more closely at the relationship between DLCP densities and the electron spin resonance (ESR) signal produced by carbon impurities, throughout the metastable defect creation and annealing process, shown in Fig. 4.6. They agree to within the uncertainty of the measurement, providing further confirmation, not only of the impact of carbon impurities in a-Si:H, but of the density values obtained from the DLCP experiment.

4.4 Cautions and Opportunities for CV and DLCP

The differential capacitance is sensitive to very small charge variations with bias, and this sensitivity is both a strength and a liability, as samples—especially the not-very-ideal samples of interest—can and do accumulate charge in various ways. If nothing else, hopefully the reader is clear that the values of C_0 and N_{CV} only give the depletion width and doping density in special situations, and in fact, these data can be quite misleading, as discussed above.

While early papers on CV, some cited here, are quite cautious about interpreting possible contributions to the sample capacitance, now that CV and DLCP are established techniques, these early cautions are sometimes left aside. It is important to remember that these techniques are both derived for the very simple case of a one-sided diode in the depletion regime, and, even in that case, understanding the multiple contributors to the capacitance is not necessarily a straightforward task. The DLCP technique is advantageous in that it is a dynamic measurement, and so adds an additional experimental parameter to the CV result. However, like all experimental data, these must be interpreted carefully.

With this in mind, understanding the results of both CV and DLCP requires checking repeatability, varying as many experimental parameters as possible, and preferably checking the interpretation using other experimental tests. This is particularly true for samples that appear to have strong spatial or voltage-dependent variations, or that otherwise seem to behave unusually. At minimum, AS and/or IS data are necessary to ensure that the sample is measured within an appropriate frequency, temperature, and bias range, where the depletion approximation is appropriate, and conductivity or series resistance effects are not a factor. The CV and DLCP measurements are also useful in other measurement regimes, but, of course, the analysis must take the departure from the depletion approximation into account.

With these cautions in mind, CV and DLCP can be powerful tools for understanding and comparing materials properties. There is also potential to further development of these techniques.

After the initial development of DLCP over 30 years ago, careful modeling and experimentation were applied to understand and verify the results, as discussed above. These efforts remained almost entirely constrained to the simple ideal one-sided junction in the depletion approximation, due to the simplicity of its analysis. While there are legitimate concerns about more complex models, which introduce additional variables and thus can have more than one unique solution, modern numerical modeling could potentially allow DLCP results to be understood more deeply. There is no longer a need to only use the depletion approximation; instead, possible contributions to I or to δQ could be analyzed more generally, as we have begun to do with CV.

Even within the depletion approximation, only simple estimates have been made for the contributions of trap states and interface states, and these are the main focus of the theoretical work that has been undertaken so far. The CV technique has been more widely modeled, and is included as an option in numerical packages like SCAPS. In their current state, both DLCP and CV profiling techniques are very useful, however, they would benefit from more sophisticated analytical and numerical approaches. This is especially true given the propensity of researchers to measure N_{CV} and N_{DL} in forward bias where the depletion assumption starts to fail. In addition, the DLCP technique in particular would benefit from further careful testing with well-understood calibrated samples, to test numerical models and gain a better qualitative and quantitative understanding of the technique's output, spatial and energetic resolution, uncertainties, and applicability to a wider range of samples.

For the last 30 years, the CV and DLCP techniques have together contributed greatly to our understanding of the electronic properties of semiconductors. Given modern analytical, numerical, and experimental capability, it will be exciting to see how these techniques develop further, and are applied to interesting new problems, in the future.

Acknowledgments

Much appreciation goes to Dr. J. D. Cohen, who developed the DLCP technique and was behind much of the work described here. I would also like to thank Dr. Jian Li and Mandip Sibakoti for useful feedback.

References

1. Sze, S. M. (2002) *Semiconductor Devices Physics and Technology*, 2nd ed. (Wiley, New York).

2. Kimerling, L. C. (1974) Influence of deep traps on the measurement of free-carrier distributions in semiconductors by junction capacitance techniques. *J. Appl. Phys.*, **45** pp. 1839–1845.

3. Michelson, C. E., Gelatos, A. V., and Cohen, J. D. (1985) Drive-level capacitance profiling: Its application to determining gap state densities in hydrogenated amorphous silicon. *J. Appl. Phys.*, **47** pp. 412–414.

4. Meyer, W., and Neldel, H. (1937) Über die beziehungen zwischen der energiekonstanten und der mengenkonstanten a in der leitwertstemperaturformel bei oxydischen halbleitern. *Zeitschrift für Technische Physik.*, **12**, pp. 588–593.

5. Yelon, A., Movaghar, B., and Crandall, R. S. (2006) Multi-excitation entropy: Its role in thermodynamics and kinetics. *Rep. Prog. Phys.*, **69** pp. 1145–1194.

6. Hugger, P., Cohen, J. D., Yan, B., Yang, J., and Guha, S. (2009) Insights and challenges toward understanding the electronic properties of hydrogenated nanocrystalline silicon. *Phil. Mag.*, **89** pp. 2541–2555.

7. Repins, I. L., Stanbery, B. J., Young, D. L., Li, S. S., Metzger, W. K., Perkins, C. L., Shafarman, W. N., Beck, M. E., Chen, L., Kapuer, V. K., Tarrant, D., Gonzalez, M. D., Jensen, D. G., Anderson, T. J., Wang, X., Kerr, L. L., Keyes, B., Asher, S., Delahoy, A., and Von Roedern, B. (2006) Comparison of device performance and measured transport parameters in widely-varying Cu(In,Ga)(Se,S) solar cells. *Prog. Photovolt: Res. Appl.*, **14** pp. 25–43.

8. Li, J. V., Duenow, J. N., Kuciauskas, D., Kanevce, A., Dhere, R. G., Young, M. R., and Levi, D. H. (2013) Electrical characterization of Cu composition effects in CdS/CdTe thin-film solar cells with a ZnTe:Cu back contact. *IEEE J. Photov.*, **3** pp. 1095–1099.

9. J. Joffe (1945) Schottky's theories of dry solid rectifiers. *Electrical Comm.*, **22** pp. 217–225.

10. Shockley, W. (1949) The theory of p-n junctions in semiconductors and p-n junction transistors. *Bell Syst. Tech. J.,* **28** pp. 435–489. Also reprinted in (1991) *Semiconductor Devices: Pioneering Papers*, ed. Sze, S. M. (World Scientific, New Jersey) pp. 7–61.

11. Johnson, W. C., and Panousis, P. T. (1971) The influence of Debye length on the C–V measurement of doping profiles. *IEEE Trans. Electron Devices*, **ED-8**, 965–973.

12. Burgelman, M., Nollet, P., and Degrave, S. (2000) Modeling polycrystalline semiconductor solar cells. *Thin Solid Films,* **361–362**, pp. 527–532.

13. Weiss, T. P., Redinger, A., Luckas, J., Mousel, M., and Siebentritt, S. (2013) Admittance spectroscopy in kesterite solar cells: Defect signal or circuit response. *Appl. Phys. Lett.,* **102** p. 202105.

14. Lauwaert, J., Decock, K., Khelifi, S., and Burgelman, M. (2010) A simple correction method for series resistance and inductance on solar cell admittance spectroscopy. *Sol. En. Mats. Sol. Cells,* **94** pp. 966–970.

15. Donnelly, J. P., and Milnes, A. G. (1967) The capacitance of p-n heterojunctions including the effect of interface states. *IEEE Trans. Electron Devices,* **ED-14** pp. 63–68.

16. Johnson, P. K., Heath, J. T., Cohen, J. D., Ramanathan, K., and Sites, J. R. (2005) A comparative study of defect states in selenized and evaporated CIGS(S) solar cells. *Prog. Photovoltaics,* **13**, pp. 579–586.

17. Li, J. V., Crandall, R. S., Young, D. L., Page, M. R., Iwaniczko, E., and Wang, Q. (2011) Capacitance study of inversion at the amorphous-crystalline interface of n-type silicon heterojunction solar cells. *J. Appl. Phys.,* **110** p. 114502.

18. Kita, K., Ihara, M., Sakaki, K., and Yamada, K. (1997) Estimation of the interface states of a Si/C$_{60}$ heterojunction by frequency-dependent capacitance-voltage characteristics. *J. Appl. Phys.,* **81** pp. 6246–6252.

19. Lim, T. H., Miller, T. J., Williamson, F., and Nathan, M. I. (1996) Characterization of interface charge at Ga$_{0.52}$In$_{0.48}$P/GaAs junctions

using current-voltage and capacitance-voltage measurements. *Appl. Phys. Lett.,* **69** pp. 1599–1601.

20. Yuan, J. S., and Liou, J. J. (1998) *Semiconductor Device Physics and Simulation* (Springer, New York).

21. Barna, A. A., and Horelick, D. (1971) A simple diode model including conductivity modulation. *IEEE Trans. Circuit Theory,* **CT-18** pp. 233–240.

22. Laux, S. E., and Hess, K. (1999) Revisiting the analytic theory of p-n junction impedance: Improvements guided by computer simulation leading to a new equivalent circuit. *IEEE Trans. Electron Devices,* **46** pp. 396–412.

23. Rockett, A., van Duren, J. K. J., Pudov, A., and Shafarman, W. N. (2013) First quadrant phototransistor behavior in CuInSe$_2$ photovoltaics, *Sol. En. Mats. Sol. Cells,* **118** pp. 141–148.

24. Grove, A. S., Snow, E. H., Deal, B. E., and Sah, C. T. (1964) Simple physical model for the space-charge capacitance of metal-oxide-semiconductor structures. *J. Appl. Phys.,* **35** pp. 2458–2460.

25. Nicollian, E. H., and Goetzberger, A. (1967) The Si-SiO$_2$ interface— Electrical properties as determined by the Metal-Insulator-Silicon conductance technique. *Bell Syst. Tech. J.,* **46** pp. 1055–1133.

26. Heath, J., and Zabierowski, P. (2011) *Advanced Characterization Techniques for Thin Film Solar Cells,* ed., Abou-Ras, D., Kirchartz, T., and Rau, U., capacitance spectroscopy of thin-film solar cells (Wiley-VCH, Weinheim, Germany), Chapter 4, pp. 81–105.

27. Li, J. V., Halverson, A. F., Sulima, O. V., Bansal, S., Burst, J. M., Barnes, T. M., Gessert, T. A., and Levi, D. H. (2012) Theoretical analysis of effects of deep level, back contact, and absorber thickness on capacitance-voltage profiling of CdTe thin-film solar cells. *Sol. En. Mater. Sol. Cells,* **100** pp. 126–131.

28. Kirchartz, T., Gong, W., Hawks, S. A., Agostinelli, T., MacKenzie, R. C. I., Yang, Y., and Nelson, J. (2012) Sensitivity of the Mott-Schottky analysis in organic solar cells. *J. Phys. Chem. C,* **116** pp. 7672–7680.

29. Michelson, C. E., Gelatos, A. V., and Cohen, J. D. (1985) Drive-level capacitance profiling: Its application to determining gap state densities in hydrogenated amorphous silicon films. *Appl. Phys. Lett.,* **47** pp. 412–414.

30. Heath, J. T., Cohen, J. D., and Shafarman, W. N. (2004) Bulk and metastable defects in CuIn$_{1-x}$Ga$_x$Se$_2$ thin films using drive-level capacitance profiling. *J. Appl. Phys.,* **95**, 1000–1010.

31. Cohen, J. D. (1984) *Semiconductors and Semimetals* **21C**, ed., Pankove, J., Density of states from junction measurements in hydrogenated amorphous silicon. (Academic Press, Orlando), Chapter 2, pp. 9–99.

32. Heath, J., and Zabierowski, P. (2011) *Advanced Characterization Techniques for Thin Film Solar Cells*, ed., Abou-Ras, D., Kirchartz, T., and Rau, U., Capacitance spectroscopy of thin-film solar cells (Wiley-VCH, Weinheim, Germany), Chapter 4, pp. 81–105.

33. Street, R. A. (1991) *Hydrogenated Amorphous Silicon* (Cambridge University Press, Cambridge).

34. Staebler, D. L., and Wronski, C. R. (1977) Reversible conductivity changes in discharge-produced amorphous Si, *Appl. Phys. Lett.*, **31** pp. 292–294.

35. Palinginis, K., Lubianiker, Y., Cohen, J., Ilie, A., Kleinsorge, B., and Milne, W. (1999) Defect densities in tetrahedrally bonded amorphous carbon deduced by junction capacitance techniques. *Appl. Phys. Lett.*, **74** pp. 371–373.

36. Unold, T., Hautala, J., and Cohen, J. D. (1994) Effect of carbon impurities on the density of states and the stability of hydrogenated amorphous silicon, *Phys. Rev. B*, **50** pp. 16985–16994.

37. Lee, J., Heath, J. T., Cohen, J. D., and Shafarman, W. N. (2005) Detailed study of metastable effects in the Cu(InGa)Se$_2$ alloys: Test of defect creation models. *MRS Proceedings* **865**, ed. Gessert, T., Niki, S., Shafarman, W., and Siebentritt, S. (Materials Research Society, USA) pp. F12.4.

38. Johnson, P. K., Heath, J. T., Cohen, J. D., Ramanathan, K., and Sites, J. R. (2005) A comparative study of defect states in selenized and evaporated CIGS(S) solar cells. *Prog. Photovoltaics,* **13**, pp. 579–586.

39. Eisenbarth, T., Unold, T., Caballero, R., Kaufmann, C. A., and Schock, H.-W. (2010) Interpretation of admittance, capacitance-voltage, and current-voltage signatures in Cu(In, Ga)Se$_2$ thin film solar cells. *J. Appl. Phys.*, **107** p. 034509.

40. Unold, T., and Cohen, J. D. (1991) Enhancement of light-induced degradation in hydrogenated amorphous silicon due to carbon impurities. *Appl. Phys. Lett.*, **58** pp. 723–725.

Section II: Instrumentation

Chapter 5

Basic Techniques for Capacitance and Impedance Measurements

Marco Carminati and Giorgio Ferrari

Dipartimento di elettronica, informazione e bioingegneria (DEIB),
Politecnico di Milano, Piazza Leonardo da Vinci 32, Milano, 20133, Italy

giorgio.ferrari@polimi.it

Impedance is a ubiquitous quantity: being the ratio between two fundamental electrical quantities, voltage and current, it can be leveraged in a wide range of applications from materials to devices, transducing measurable quantities, and in particular their variations in time, from the physical domain to the electrical domain. Impedance can be measured in several ways: in this chapter, we review the most common measurements approaches, with special focus on the detection circuitry and on the minimization of noise required to achieve high resolution, pivotal in modern micro- and nano-scale applications.

5.1 Definitions

The electrical resistance of a two-terminal element (Fig. 5.1a) is defined as the ratio between the voltage applied across the

Capacitance Spectroscopy of Semiconductors
Edited by Jian V. Li and Giorgio Ferrari
Copyright © 2018 Pan Stanford Publishing Pte. Ltd.
ISBN 978-981-4774-54-3 (Hardcover), 978-1-315-15013-0 (eBook)
www.panstanford.com

dipole and the current correspondingly flowing through it. In simple terms, resistance expresses the ease (more precisely the impediment) for the current to flow through a dipole given an applied potential difference: the higher is the resistance, the smaller is current. This definition makes sense only for linear dipoles, i.e., elements for which the relation between voltage and current is linear. Typically, if the system is non-linear for large excursions of the electrical parameters, a small range ΔV around a bias point is considered and a linearization is operated (Fig. 5.1b), so that resistance is defined under the "small signal assumption" (i.e., with a stimulation amplitude small with respect to ΔV) around a specific bias point (V_0, I_0).

Impedance [1] is the extension of the concept of resistance to the case of a sinusoidal stimulus. If we apply a sinusoidal voltage across a linear dipole, the current forced to flow will be sinusoidal as well, at the same frequency f. At that given frequency, the relation between the amplitude of the current sinusoid and its phase φ and the amplitude and phase of the externally applied voltage sinusoid is called the dipole impedance. Consequently, impedance, commonly indicated as Z, is a complex quantity, the ratio between the voltage and the current phasors, varying as a function of frequency. The real part of impedance is called *resistance* (R), while the imaginary part is called *reactance* (X).

$$Z \equiv \frac{V}{I} = Z(f) = R(f) + jX(f) \tag{5.1}$$

The reciprocal of impedance is *admittance* (Y), whose real part is *conductance* (G) and imaginary part is *susceptance* (B). The SI unit of impedance is Ohm $[\Omega]$, while admittance is measured in Siemens [S].

$$Y \equiv \frac{1}{Z} = G(f) + jB(f) \tag{5.2}$$

Capacitance is a particular type of impedance characterized by a purely reactive term. Its prominence among other types of reactive impedances is due to the key role of capacitors in electronic circuits and to the ubiquitous presence of capacitive coupling between two conductors separated by a dielectric material or between two layers of charge. Given a capacitance C, the purely imaginary impedance decreases linearly with frequency f:

$$Z = -j\frac{1}{2\pi f C} \tag{5.3}$$

Interestingly, the real part of impedance is always associated with energy dissipation and, consequently, with thermal noise due to the random motion of charge carriers (such as electrons in metals and semiconductors, and ions in electrolytic solutions). The power spectral density of the resulting voltage noise across the impedance Z is given by the fluctuation-dissipation theorem:

$$\overline{v_n^2} = 4kTRe\{Z\}, \tag{5.4}$$

where k is the Boltzmann constant and T the absolute temperature. On the contrary, the imaginary part is associated with energy storage (such as in capacitors and inductors) and is noise-free.

In order to fully characterize the impedance of a system, its magnitude and phase should be known for all the frequencies of interest. This set of data represents the impedance *spectrum*. Impedance spectroscopy is a powerful and widespread technique, used to characterize materials and devices by measuring their impedance in a given frequency span. Typically, measured spectra are then fitted with equivalent models composed of lumped basic electrical components (resistors, capacitors, inductors) as well as more sophisticated analytical blocks (constant-phase elements, Warburg terms, etc...) which should provide some insight in the physical mechanisms explaining the electrical response of the system.

Under the linearity assumption, operations between impedances (such as parallel and series composition) can be conveniently carried out in the Laplace domain. Being a complex quantity, impedance can be represented as a vector in a complex (Gauss) plane (Fig. 5.1c).

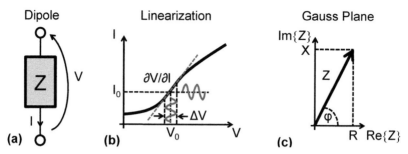

Figure 5.1 Impedance Z of a generic two-terminal element (a) whose non-linear I/V characteristic can be linearized (b) and treated as (c) a vector in the complex Gauss plane.

Impedance spectra can be displayed in two ways: in the Bode plot and in the Cole-Cole plot. As illustrated in Fig. 5.2, the Bode plot is composed of two separate plots: a log-log plot of the impedance magnitude as function of frequency (Fig. 5.2a) and a semi-log plot of the phase (Fig. 5.2b). The Cole-Cole is, instead, a single graph in the complex plane (where the sign of the imaginary axis is reversed, Fig. 5.2c). Each point corresponds to the apex of the impedance vector at a given frequency. As an example, in Fig. 5.2 the spectrum of a simple $R||C$ impedance is shown. The magnitude is flat and equal to R up to the pole frequency ($f_p = 1/(2\pi RC)$), above which it rolls off with a constant slope –20 dB/decade). Correspondingly, the phase starts from 0 (real resistance R at low frequency) and, after the pole, decreases reaching asymptotically –90° (the capacitor phase). The same spectrum has a semi-circular shape in the Cole-Cole complex plane. In DC ($f = 0$) the starting point is on the real axis at coordinate R (pure resistance). For increasing frequency the reactive component increases up to the maximum at the pole frequency. The use of the Bode plot is more common in the engineering community, while the Cole-Cole plot is mostly used in the electrochemistry and bio-sensing communities. From the point of view of information content, they are perfectly equivalent.

Note that the real and imaginary part of the impedance of a physical system are related by tight mathematical relations, called Kramers–Kronig transformations [2], that allow recovering one component from the other.

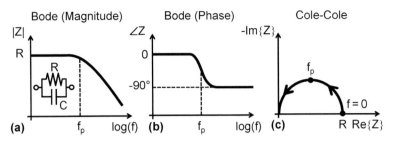

Figure 5.2 Graphical representations of impedance spectra: Bode magnitude (a) and phase (b) plot and Cole–Cole plot (c) in the complex plane.

5.2 Classification of Measurement Techniques

In order to measure the unknown electrical impedance Z_X of a system, a stimulation is applied to it and its response is recorded [3]. Thus, all the instruments comprise a section generating the excitation signal (Fig. 5.3a) and another section to readout the response signal.

Impedance measuring schemes can be classified according to different criteria; here we summarize the most important ones:

- Type of forcing signal and output signal: the excitation signal can be either a current (Fig. 5.3b) or a voltage (Fig. 5.3c). Correspondingly, the measured quantity is a current or a voltage. The response current is commonly converted into a voltage for acquisition and digitalization, so that the impedance results as the ratio of two voltages: the stimulation voltage V_{force} (applied at the *force* electrode) divided by the circuit output voltage (linearly related to the response current measured at the *sense* electrode).
- Stimulation waveform: the excitation signal V_{force} at the force electrode can be either a single sinusoid or a more complex waveform. Examples of non-monochromatic waveforms includes square waves, sum of sinusoids at different frequencies (not harmonic), and wide-spectrum signals such as pseudo-noise signals, as illustrated in Section 5.7.3.
- Frequency range: the frequency of the stimulation signal f_{AC} is a key parameter (which is swept across a wide range

in the case of impedance spectroscopy). For measuring only resistance, a DC stimulation is enough. The most common probing range is from the Hz to the MHz region. High frequency (10–100 MHz) is less common but still achievable, both by commercial (such as model UHFLI from Zurich Instruments) and research instrument [4]. Instead, the exploration of the GHz range requires completely different circuits operating in the microwave range and often based on reflectometric approaches, which are not treated in this chapter.

- Analog vs. digital processing: in cascade to the analog circuitry representing the front-end directly interfacing with the terminals of Z_X, the measuring instrument typically includes a processing block, which can be either analog or digital. As illustrated in Fig. 5.4, the most common architecture of a modern instrument is thus composed by the analog signal conditioning bocks, ADC and DAC and a digital platform (microcontroller, DSP, FPGA, or PC) for instrument control, sampling and processing.

Other features characterizing different impedance measuring techniques include the compatibility with multi-wire schemes (three- and four-wire schemes), with differential sensing configurations, and the number of accessible terminals of the device required for the measurement (one or two). When the measurement is performed at a single terminal (by simultaneously stimulating and sensing the same node), the other electrode needs to be grounded. This configuration can be beneficial for measuring the impedance of devices in which one terminal is intrinsically grounded.

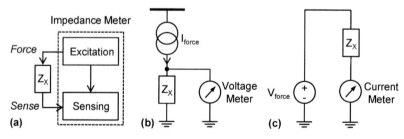

Figure 5.3 (a) Blocks of a generic impedance meter, (b) current excitation versus (c) voltage excitation scheme.

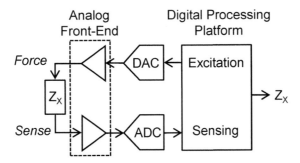

Figure 5.4 Architecture of a modern impedance-measuring instrument featuring an analog front-end and a digital back-end for signal processing and impedance calculation.

5.3 Voltage Sensing

5.3.1 Shunt Scheme

The simplest approach to measure an unknown impedance Z_X is to force a current I_{force} into it and measure the voltage $V_X = Z_X \cdot I_{force}$ that correspondingly drops across it. This scheme is the same used to measure an unknown current by inserting a shunt resistor (of a known value, typically low in order to minimize the voltage drop across it), operated in a dual mode. In this scheme only one terminal is needed and the second one can be grounded (Fig. 5.5a). However, in order to reject fluctuations of the ground potential, it is advisable to measure the voltage drop V_X by means of a differential amplifier (Fig. 5.5b). The value of I_{force} should be properly chosen in order to generate a potential V_X suitable for acquisition by the readout chain.

Note that this approach corresponds to the operative definition of the equivalent impedance of a node of an electrical network: inject a small test current and measure the corresponding potential variation.

The accuracy of the measurement lies in the accuracy of the current generator I_{force} and in the absence of stray paths in parallel to Z_X. If the value of Z_X does not exceed the MΩ range, parallel resistive paths can be typically neglected. Instead, capacitive paths need to be carefully taken into account when measuring

Z_X in frequency, since the impedance of the capacitance decreases for increasing f_{AC}, thus increasingly absorbing more and more current that no longer flows in Z_X and, thus, severely affecting the computation of the actual value of Z_X.

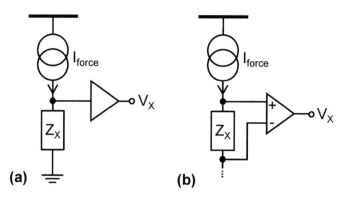

Figure 5.5 Impedance sensing scheme with single-ended (a) or differential voltage readout (b) in response to a current stimulation I_{force} injected in Z_X.

Furthermore, this solution is not suitable if Z_X is high enough to become comparable with the output impedance of a non-ideal current generator. For these two reasons (sensitivity to parallel stray impedance and suitability for relatively small impedances), this solution is typically avoided for measurement at the micro- and nano-scale where impedances tend to be very high (MΩ to GΩ ranges).

5.3.2 Ratiometric Configuration

The ratiometric configuration solves the issue of the dependence on the accuracy and stability of the current source affecting the shunt scheme. In this case, a second reference impedance Z_R of known value is put in series to Z_X so that the same excitation current I_{force} flows in both of them (Fig. 5.6a). Then, the value of Z_X is obtained measuring also the voltage drop across Z_R. The ratio between ΔV_X and ΔV_R is independent of I_{force}. A very effective way to implement the ratio is to use the V_{REF} as reference voltage of the analog-to-digital converter (ADC) used to directly sample V_X (Fig. 5.6b).

The measurement accuracy is only dependent on the accuracy and stability of Z_R which, in general, is better than that of I_{force}. This configuration grants high accuracy with a few components, at the price of doubling the acquisition channels and the voltage dynamic managed by the current source. For this reason, and being intrinsically a four-wire measurement, that cancels the parasitic series impedance of the wires as explained in Section 5.6.1, this configuration is commonly adopted for the readout of resistive temperature sensors (resistive temperature detectors, RTD). Several commercial integrated circuits embedding the current generator and two differential inputs, as well as the high-resolution Sigma-Delta ADCs [5], are available in the market.

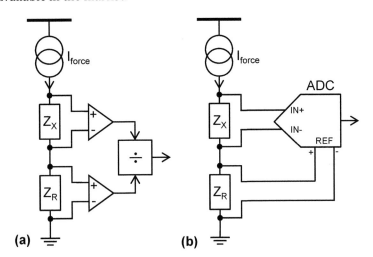

Figure 5.6 Ratiometric impedance measurement (a) and its implementation (b) with an analog-to-digital convert (ADC) with differential signal and reference inputs.

5.3.3 Half Bridge

The half bridge topology consists of a simple voltage divider (Fig. 5.7). Voltage partition takes place between the unknown impedance Z_X and a reference impedance Z_1. The series of the two impedances is energized by V_{force} and the output voltage is read in the middle node V_X, where the voltage is

$$V_X = V_{\text{force}} \frac{Z_X}{Z_X + Z_1},$$ (5.5)

from which it is straightforward to obtain Z_X, although as a non-linear function of V_X. This can be seen as a modification of the ratiometric scheme, where the series of the reference and unknown impedances is excited by a voltage source instead of a current source. Commonly, from the standpoint of circuit design, it is easier to implement a precise sinusoidal voltage source (of accurate amplitude, wide frequency tuning range and low distortion) rather than a current source. Beyond simplicity, the key advantage of this approach is that the measurement accuracy is mostly set by the accuracy of Z_1. In fact, irrespectively of the absolute value of Z_X (that can be very high, even above the MΩ range), the ratio can be limited within a desired range. However, the major drawback of this technique is that the value of Z_1 cannot be randomly chosen. In fact, if Z_1 is too low with respect to Z_X, then V_X tends to V_{force}. Analogously, if Z_1 is too high, V_X tends to zero. In both extremes of the voltage division (0% and 100%), the information about the value of Z_X is lost. The optimal value for the partition is $Z_1 = Z_X$ (50% partition). Consequently, some a priori knowledge of the approximate value of Z_X is required to properly size Z_1, significantly reducing the range of impedance values that can be explored.

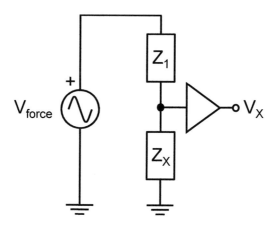

Figure 5.7 Half bridge topology for measuring Z_X in a voltage partition with Z_1.

5.3.4 Full Bridge

The full bridge represents a natural evolution of the half bridge, offering several advantages with respect to the previous methods. It is a symmetrical structure composed of two arms, i.e., of two voltage dividers, composed of four impedances, and energized by V_{force}. The unknown impedance Z_X is placed in one of the four positions (Fig. 5.8). The remaining three are composed of reference impedances Z_1, Z_2, and Z_3, of which at least one is tunable. The differential voltage ΔV_X is measured across the middle nodes of the two voltage dividers. The reference impedances are adjusted until ΔV_X is equal to zero. In this balanced condition (also called "null" condition) the value of Z_X can be easily calculated as

$$Z_X = Z_3 \frac{Z_1}{Z_2}. \tag{5.6}$$

This expression no longer depends on V_{force}, highlighting that the measurement accuracy only depends on the reference impedances. From the point of view of the readout circuit, this scheme simplifies the requirements on the differential amplifier, since, regardless of the impedance values, its differential input is close to zero (around a given common mode voltage). The symmetry and the voltage difference provide rejection of common mode disturbances, of the noise injected by V_{force} and rejection of environmental effects (such as temperature dependence) affecting all the impedances of the bridge.

Furthermore, this configuration, known as Wheatstone bridge, is particularly suitable for devices in which a pair of coupled impedances change in opposite ways, i.e., when a positive Z_{X+} increases by ΔZ, a counter impedance Z_{X-} decreases of the same amount. In this case, it is natural to connect the counter changing impedances is opposite position in the bridge, thus doubling or quadruplicating the overall effect. This is the case, for examples, of load cells realized with four resistive strain gauge (two extending, while the other are two compressing), or MEMS inertial sensors where combs interdigitated capacitors, spaced by a few microns, are used to detect the positive/negative displacement of a silicon proof mass with respect to a balanced rest position.

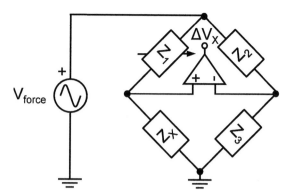

Figure 5.8 Wheatstone (full) bridge composed of a tunable arm ($Z_1 - Z_X$) to match the partition of the fixed arm reaching a null condition ($\Delta V_X = 0$)

In some implementations, the reference arm of the bridge can be replaced by a transformer with two secondary windings (balun), creating two secondary matched voltages from a single-ended V_{force}. The currents flowing into Z_X and Z_1 are compared and the latter is adjusted until the null is reached.

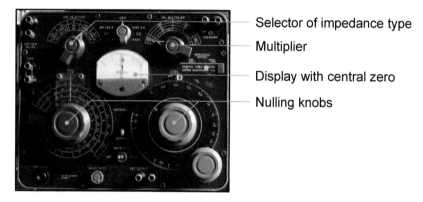

Selector of impedance type

Multiplier

Display with central zero

Nulling knobs

Figure 5.9 Historical impedance bridge (1957) with manual nulling model GR 1650-A from General Radio (photo courtesy by G. Carminati).

Despite several advantages and great popularity, the Wheatstone bridge has some limitations as well. First of all, the nulling sequence can be time consuming (particularly in manual instruments as the one shown in Fig. 5.9, but also in automatic bridges, such as the capacitive bridge model AH 2700A from Andeen-Hagerling offering the state-of-the-art capacitance resolution of 0.5 aF among commercial instruments, operating in the 50 Hz to 20 kHz range of f_{AC}). This can be a severe issue, in particular for impedance spectroscopy where several points in frequency are measured. Several switches are required to change the values of Z_1 in order to explore different decades of Z_X and some kind of a priori knowledge on the nature of Z_X (resistive, capacitive, or inductive) is required. Finally, the presence of parasitic impedances in parallel to Z_X can alter the measurement accuracy.

5.4   Current Sensing

5.4.1   Transimpedance Front-End

The principal drawback affecting all the previous schemes is the sensitivity to parasitic capacitance. Stray capacitance has typically two origins: inside the device itself (due to geometry and the materials) and due to the connection between the device and the input terminals of the measuring instrument. The latter are commonly the dominant term. For example, if coaxial cables are used to properly shield the signals from electromagnetic interferences, the presence of the outer conductive shield produces a capacitance to ground of about 50–100 pF/m depending on the cable diameter. Independently of its origin, stray capacitance to ground introduces a time constant in the voltage-sensing front-end circuit which limits the maximum probing frequency (and the response time). The higher is Z_X, the narrower is the frequency limitation. Instead, stray capacitance in parallel to Z_X provides an alternative path for the current to flow, thus introducing severe inaccuracies. An effective solution to neutralize the stray capacitance to ground is the adoption of the current sensing scheme, based on a transimpedance amplifier (TIA) illustrated in Fig. 5.10.

One terminal is excited by V_{force}, while the other terminal is connected to the TIA input. Thanks to the negative feedback, the input node is kept at ground potential (virtual ground) and the current flowing in Z_X is absorbed by the feedback branch, where it is converted into the output voltage by the feedback resistor R_F. Consequently, the output voltage is directly proportional to the current (i.e., to the admittance $Y_X = 1/Z_X$) through the value of R_F.

$$V_{OUT} = -I_X R_F = -V_{force} \frac{R_F}{Z_X}$$ (5.7)

The sizing of the best value of R_F is the result of a trade-off among current-to-voltage gain, bandwidth (properly chosen with respect to f_{AC}) and noise, as explained in the following chapter.

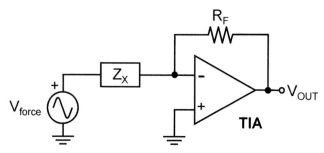

Figure 5.10 Current sensing scheme based on a transimpedance amplifier (TIA) converting the input current V_{force}/Z_X into the output voltage V_{OUT}.

In order to consider the improvement in terms of resilience to parasitic effects, let us consider and compare the schemes illustrated in Fig. 5.11. Only capacitive terms are considered, since they are the most ubiquitous kind of stray impedance. Two separate parasitic capacitance contributions are shown: C_P is in parallel to Z_X (a stray shunt capacitance), while C_S is towards ground, for instance due to a long coaxial cable connecting the Z_X sense terminal to the instrument input. In the bridge configuration C_P and C_S are not discernable and result in parallel to Z_X, altering the measurement accuracy (especially if Z_X is a capacitance too) and limiting the measurement bandwidth. Instead, in the current sensing scheme, C_S is completely neutralized thanks to the feedback which keeps at zero potential the input node, thus not allowing any current flow from the virtual ground to the actual ground (at least as long as the loop gain is high enough). Unfortunately, no beneficial effect is immediately available for C_P, that also in this configuration results unavoidably in parallel to Z_X, and additional compensation circuitry is required, as illustrated in Section 5.6.2.

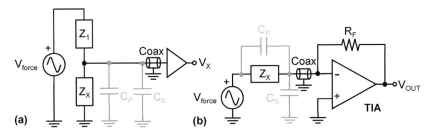

Figure 5.11 Effect of parasitic capacitances on the bridge (a) vs. the current sensing topology (b), the latter allowing complete neutralization of C_S.

Finally, it is important to underline that the current sensing configuration is also compatible with both differential sensing schemes and with single-terminal stimulation, as illustrated in Fig. 5.12. For differential sensing (Fig. 5.12a), an inverting buffer (or a fully differential drivers) is need to counter drive the differential pair (in a sort of bridge configuration). When it is balanced (i.e., $Z_{X+} = Z_{X-}$) the same current is flowing in both impedances and no current is flowing in the TIA, whose output is zero. When the pair is unbalanced, the difference current, proportional to the difference between the admittances, is amplified and sensed.

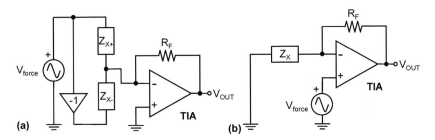

Figure 5.12 Straightforward adaptation of the current sensing scheme to differential measurement (a) and single-terminal measurement (b).

If only one terminal of Z_X is available (the other one being grounded), the current sensing scheme can still be employed, by simply applying the V_{force} signal to the non-inverting input of the operational amplifier (Fig. 5.12b). In this case, at V_{OUT} a constant contribution of V_{force} would be present and should be eliminated before further processing of the signal.

5.4.2 Summary and Comparison

It is interesting to notice that, irrespectively of the classification here introduced, all the schemes illustrated above (with the only exception of the basic shunt approach), can be seen as bridge structures. In fact, the unknown impedance Z_X is always "compared" with a reference impedance of precisely known value, in which the same current is flowing, i.e., namely: Z_R in the ratiometric configuration, Z_1 in the half and full bridge and R_F in the current sensing scheme.

Despite ultimately similar for what concerns the topology, they differ in terms of interfaces, performance, and requirements for the electronics. From the analysis previously discussed and summarized in Table 5.1, we can conclude that the preferred measurement scheme, particularly for high impedance in the micro-scale domain, is the current sensing one. In brief, it provides the following relevant advantages:

- *Precise control of the potential* applied across Z_X. In fact, differently from all the other scheme where the actual voltage drop across Z_X depends on the value of Z_X, in the current sensing scheme, V_{force} is fully dropping across Z_X. This feature is particularly important for delicate samples (such a bio-electrochemical ones) with very tight potential constraints.

- *Neutralization of parasitics to ground.*

  On the other hand, the TIA circuit, being a feedback scheme, demands more care in the design. In particular, since the loop stability depends on the value Z_X, a very robust design considering the extreme values of Z_X is required. While being very different from the point of view of parasitics, voltage and current sensing are equipollent from the standpoint of signal-to-noise ratio (SNR).

Table 5.1 Summary comparison of basic impedance measuring approaches

Technique	Stim	Out	Ter	Pro	Con
Shunt	I	V	1	Simple	Sensitive to I_{force}
Ratiometric	I	V	2	Reject I_{force} fluctuations	2 acquisition chains
Half bridge	V	V	1	Simple	Optimal value $Z_1 = Z_X$
Full bridge	V	V	1	Reject common mode	Long nulling routine
Current sense	V	I	1–2	Insensitive to parasitics	Stability of feedback

Note: Stim is the stimulation type: voltage (V) vs. current (I). Out is the readout signal (V vs. I), while Ter indicates the number of required free terminals.

5.5 Impedance Calculation

The output signal provided by the front-end stage needs to be processed in order to extract the value of impedance. If the measurement is in DC (i.e., $f_{AC} = 0$), a simple algebraic operation is required to convert the value of the DC voltage into a resistance value. If the measurement is in AC, i.e., if the impedance is complex, the output voltage has a sinusoidal shape and additional processing steps are required. In some cases the magnitude information is sufficient (for examples when measuring a capacitance), while, in other situations, both magnitude and phase need to be retrieved, as in the case of impedance spectroscopy. In order to obtain only impedance magnitude, simple circuits can be adopted, such as envelope detectors. Instead, for obtaining real and imaginary components, phase detection becomes as important as magnitude demodulation, thus requiring synchronous detection or digital processing. When tracking impedance over time at a single frequency, the control logic simply samples the impedance value with a periodic sample interval. Instead, if impedance spectroscopy is performed, the control logic increases f_{AC} by steps in order to sweep all the spectrum recoding the complex impedance value for each frequency point.

5.5.1 Envelope Detector

The simplest approach to recover the information of the sinusoid amplitude, regardless of its phase, is to rectify it. The most popular circuit to recover the envelope of a sinusoidal modulation is the peak detector based on a diode (for rectification) and a capacitor (for peak stretching) and a resistor for capacitor discharge. Figure 5.13 shows the basic peak detector featuring a single diode, i.e., performing half-wave rectification. When the input voltage exceeds the voltage sampled in the capacitor, the diode switches on and higher potentials can be stored in the capacitor, which slowly discharges through the resistor. In order to avoid the voltage loss due to the diode forward voltage (a built-in of about 0.7 V for silicon diodes), the diode can be inserted inside a feedback loop, in the so-called "superdiode" configuration (Fig. 5.13b).

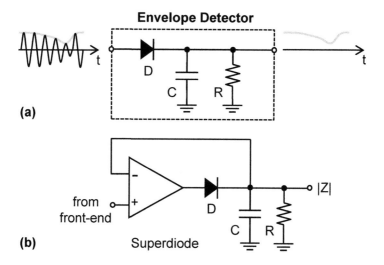

Figure 5.13 (a) Envelope detector for signal rectification (magnitude demodulation) and (b) improved implementation to avoid diode drop.

Full-wave rectification can be achieved by employing more diodes. This circuit represents a primitive demodulator (indeed used for demodulation of AM radio signals in pioneering crystal receivers), where the RC cell acts as low-pass filter of the modulated carrier and has to be properly tuned.

5.5.2 Synchronous Demodulation

Since the effect of a complex impedance Z_X is to alter in magnitude and phase the current signal with respect to the forcing voltage sinusoid, both quantities must be extracted from the output signal in order to fully quantify Z_X. A phase-sensitive circuit is required to reconstruct the phase shift between the stimulation and the readout signals. The most popular and elegant solution to realize such an operation is the *lock-in* scheme, illustrated in Fig. 5.14 in combination with the current sensing front-end. The front-end output voltage is fed in parallel to two ideal multipliers, synchronized with orthogoal reference signals. One reference is in-phase (i.e., synchronous with V_{force}) producing the real part of admittance. Instead, the in-quadrature (i.e., shifted by a quarter of period) channel demodulates the imaginary component. This operation corresponds to the projection of the complex admittance vector on two orthogonal reference phasors. In frequency domain, the lock-in operation is a shift of the original spectrum (modulated around f_{AC}) down to base-band and around $2f_{\text{AC}}$. A couple of low-pass filters (LPF) then cancel the replicas of the spectrum around $2f_{\text{AC}}$, leaving only the DC (or slowly varying) output. The non-linear operation of multiplication, i.e., frequency shifting, is very useful for different reasons. In impedance measurements, the lock-in scheme allows generating two DC signals proportional to two components of the impedance vector, modulated by slow variations of impedance during time. In other applications, the lock-in demodulator is used in combination with an external modulator, in order to shift the input spectrum to a frequency region characterized by lower noise (typically far from the flicker noise of the downstream amplifiers), amplify it and then shift it back to the original position in frequency by means of the demodulators.

In the lock-in scheme, the demodulation is always *locked* to the stimulation sinusoid, so that it behaves as a sort of intrinsically tuned band-pass filter. In fact, even if f_{AC} is swept across a wide frequency range, the averaging time is independent of f_{AC} and only set by the time constant of the LPF. This allows realizing a tunable band-pass filter with very high quality factor by means of a simple low-pass filter. For example, if the bandwidth of the LPF is BW = 1 Hz and f_{AC} is

10 MHz, then f_{AC}/BW is 10^7, hardly achievable with an actual band-pass filter tunable across decades of frequency. Consequently, the design of the LPF is not critical [6], it sets the instrument response time ($\sim 1/BW$), the noise integration bandwidth and must suitably attenuate the ripple at $2f_{AC}$. Since a constant value at the mixer input produces a modulated tone a f_{AC} at its output, it is advisable to interpose a high-pass filter between the front-end and the lock-in to remove any DC offset.

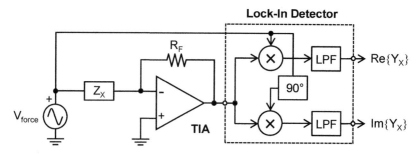

Figure 5.14 Phase-sensitive demodulation with in-phase and in-quadrature demodulators in the lock-in scheme.

The lock-in scheme can be implemented in several ways. The mixer can be an analog multiplier based on the topology of the Gilbert cell [7], adopted, for instance, in commercial integrated circuits by Analog Devices [8] such as AD835 (±1 V input and output voltage range, operating up to 250 MHz) or AD633 (with ±10 V voltage range but limited to 1 MHz) or a square-wave multiplier (much easier to implement with simple switches, in particular in CMOS implementations [9, 10] while producing folding of harmonics). The multiplication can be easily performed in the digital domain as well, either on a PC [6] or in a FPGA-embedded platform [11], paying attention to the synchronization of the acquired samples with the samples of the stimulating sinusoid, typically generated with a DAC or DDS. The low-pass filter can be either a FIR or IIR type filter.

5.5.3 Sampling-Based Techniques

As anticipated, despite the advantages of analog demodulation [8], the current trend in the design of measurement instrumentation is to digitize the front-end voltage output and perform operations in the digital domain, thanks to the abundance of low-cost, fast, and constantly improving digital processors. Once the signal is sampled with a proper sample rate, with respect to f_{AC}, a plethora of algorithms are available for signal processing and impedance reconstruction. These techniques can be grouped into two, partially overlapping, families:

- **Fourier-based algorithms**. By performing the Discrete Fourier Transform (DFT), most commonly in the form a Fast Fourier Transform (FFT) of the sampled values, it is possible to obtain the spectrum of the response sinusoid and, thus the magnitude and phase of Z_X by comparing it with the FFT of the excitation signal. More generally, Z_X is obtained in frequency domain as the ratio between the FFT of the stimulation voltage and the response current. Since the frequency resolution depends on the number of samples acquired in time, in order to achieve the same resolution of the analog lock-in demodulator, a large amount of data must be sampled and processed. For instance, if $\Delta f = 1$ Hz (easily achievable with a lock-in with BW = 1 Hz), and the max f_{AC} in impedance spectroscopy is 10 MHz, then at least 10^7 samples must be acquired. The computation of the FFT on 10 million samples can imply a high computation cost, depending on the chosen digital platform and on the constraints on the instrument response time. A single-chip impedance meter based on FFT analysis is AD5933 by Analog Devices.

- **Fitting algorithms**. Since we know that under the linearity assumption, the response signal will be a sinusoidal waveform, the acquired samples can be simply fitted with a sinusoidal function, whose amplitude and phase are free parameters (while frequency f_{AC} is fixed) optimized by a non-linear fitting routine (such as non-linear least squares or many others) [12]. In this perspective, it becomes clear that acquiring a full period of the sinusoid is no longer necessary, since fitting can be achieved even on partial domains of a sinusoid that is characterized by huge symmetries. It has been reported than only 10% of the period is sufficient to reconstruct the impedance [13].

Despite a significant reduction of the measurement time (in particular for low frequencies) achievable with these techniques, typically it is advisable to acquire several periods in order to improve the SNR. It is important to underline that with both the lock-in scheme and FFT techniques, the averaging time, set by the bandwidth of the low-pass filter; is independent of the operating frequency f_{AC} and, thus, the number of acquired periods significantly varies across the spectrum, while the resolution is constant for each frequency point and set by the LPF bandwidth integrating the noise power spectral density in proximity of f_{AC}.

5.6 Correction of Parasitics-Induced Inaccuracies

5.6.1 Multi-Wire Schemes

In order to reduce the impact of the series stray impedance, associated with the connecting cables (i.e., wire resistance and inductance), multi-wire schemes can be adopted. The most common configuration for high accuracy impedance measurement is the four-wire scheme (Fig. 5.15). The idea behind multi-wire solutions is to decouple the current sourcing function from the voltage sensing one, using different wires instead of a single one. As illustrated in Fig. 5.15a for the current forcing case, two wires (named H_{CUR} an L_{CUR}) sustain the flow of the I_{force} current, while other two wires (H_{POT} and L_{POT}) connect Z_X to the differential voltage amplifier. Since in the latter wires no current is flowing (neglecting the input bias currents of the amplifier), their series impedance is negligible, and the measured voltage V_X is the actual voltage drop across the impedance.

The same scheme can be adopted in the current sensing case (Fig. 5.15b). In fact, due to the presence of the wires parasitic resistance, the actual voltage dropping across Z_X is smaller than V_{force}. Thus, two additional wires are introduced, connecting the terminals of Z_X to a differential voltage amplifier (high input impedance), providing the correct value of V_X to be used in the computation of impedance $Z_X = V_X/I_X = R_F \times V_X/V_{OUT}$.

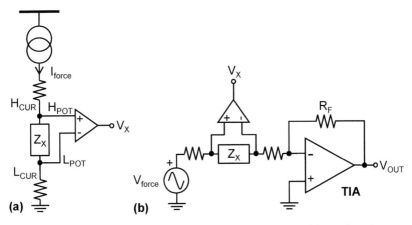

Figure 5.15 Tetrapolar measurement with current (a) and voltage (b) stimulation compensating the parasitic series resistance of wires.

In the case of electrochemical applications, it is common to use a three-electrode configuration. Although aiming at the same purpose of increasing the measurement accuracy, this configuration is slightly different from what discussed above and should not be confused with multi-wire topologies. The principal difference is that, here, active feedback is used to control the actual voltage drop V_{cell} imposed across an electrochemical interface. The three-electrode setup, shown in Fig. 5.16, is managed by a special instrument, dedicated to electrochemical measurements, among which there is electrochemical impedance spectroscopy (EIS) [14], called *potentiostat* [6]. The electrochemical reaction under observation takes place at the working electrode (WE), whose current is read by means of a TIA, biasing it a ground potential. In order to compensate the ohmic drop across the solution resistance and the non-linear voltage drop at the counter electrode (CE) interface, a third electrode called reference electrode (RE) is added in proximity of the WE. A feedback loop reads the value of the potential at the RE (characterized by particularly stable interface such as the Ag/AgCl electrode, where no current is flowing) and compares it with a set-point V_{cell} which can be the sum of a DC bias and a small sinusoidal excitation V_{force}. The CE is driven by this potentiostatic loop providing the current and the potential needed to grant that V_{cell} is dropping across the

WE interface.

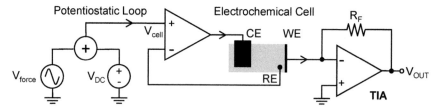

Figure 5.16 Scheme of a potentiostat employing three electrodes to control the voltage drop across the electrochemical interface at the working electrode (WE) for accurate impedance measurement in liquid, unaffected by solution resistance.

5.6.2 Calibration

Like any other kind of instrument, impedance meters must be carefully calibrated. Calibration should be performed in a correct metrological way, i.e., preserving *traceability* that is an unbroken chain of referenced calibration steps [15]. Certified calibration samples, specifically designed to minimize the impact of the setup parasitics, should be used or, at least, the measurement results should be compared with and referenced to other instruments, calibrated independently.

In order to better clarify the role of calibration, let us briefly recall three figures of merit assessing the (static) quality a measurement system:

- **Accuracy**: indicates the absolute distance (i.e., error) between the result and of the measurement and the "true" value.
- **Precision**: independently of the accuracy, it indicates the deviation (i.e., repeatability) of successive measurements of the same Z_X.
- **Resolution**: indicates the minimum variation of Z_X that can be discriminated.

Assuming that the digital section has been properly designed (i.e., suitable number of bits for the representation of numbers, suitable clock frequencies, etc), these figures of merit, i.e., the ultimate quality of the measurement, are determined by the analog front-end, handling the signals at the very beginning of the conditioning chain. Accuracy, depending on the tolerance of all the components present in the signal conditioning stages and on parasitic impedances affecting Z_X, can be significantly improved by means of the calibration. Precision is more related to the stability of the instrument, i.e., to the sensitivity of its components to environmental drifts, mainly to temperature fluctuations. Components with low thermal sensitivity, differential structures with dummy sections, active thermal stabilization, and periodic calibration can improve precision. Resolution instead, depends on the discrimination limits set by noise, as described in the next chapter.

The relative importance of these figures of merits depends on the specific application. In the characterization of the materials and devices, the absolute accuracy is very important, allowing the extraction of quantitative information. Instead, in sensors application, the absolute impedance value can be not particularly relevant, while its small variations (i.e., resolution) are of utmost importance to detect evanescent physical phenomena. Repeatability is always fundamental in order to compare measurements performed in different moments.

In addition to parasitic impedance introduced by the setup, the major source of error is due to phase delays introduced by connections and by the limited bandwidth of all the amplifiers along the signal path. Clearly, any phase error rotates the impedance vector, altering its projections on the real and imaginary axis. To better understand the importance of this effect, let us analyze a simple numerical example. As pictured in Fig. 5.17a, consider a purely capacitive impedance (C = 1 pF) measured at f_{AC} = 1 MHz. If the amplifier bandwidth is set at 100 MHz, i.e., two decades higher than f_{AC}, the phase error φ is 0.57°. Although it might appear negligible, such a phase shift produces the artifact of a real admittance (i.e., conductance) component G_{err} of 60 nS (corresponding to an apparent resistance of 16 MΩ in parallel to the capacitor).

Since the phase error change with frequency, it is important to calibrate the instrument across the whole frequency spectrum. Taking a well-characterized calibration sample, for every frequency the instrument output should be compared with the ideal output and a correction factor (for both magnitude and phase) should be stored. Then, in the case of probing a f_{AC}

for which no coefficient is present, linear or polynomial interpolation can be used between the two nearest values.

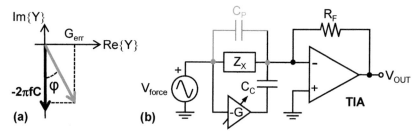

Figure 5.17 (a) Erroneous projection of a capacitive admittance on the real axis due to a phase delay φ. (b) active compensation of C_P with a tunable C_C subtracting its current.

Although numerical calibration can effectively correct the error due to the presence of a stray capacitance in parallel to Z_X (C_P in Fig. 5.17b), if the current flowing in C_P becomes large (or even dominant with respect to current in Z_X), it can severely degrade the performance of the analog chain, waste bits of the ADC and even lead to saturation. In this case, active compensation of C_P can be implemented with the scheme illustrated in Fig. 5.17b. A compensation capacitor C_C is also connected to the TIA input and is driven by a negative replica of V_{force} in order to subtract the stray current flowing in C_P, preventing it from being amplified by the TIA. Since the value of C_P is typically not known precisely, a tunable amplifier G is used. Gain tuning can be either manual or automatic [16]. Despite very effective, this differential technique (in current mode) has the only drawback of increasing the total capacitance at the TIA input node, which is detrimental from the noise point of view, as explained in the next chapter.

5.7 Other Techniques

5.7.1 Resonant Techniques

Since the ultimate resolution of the methods presented above is set by the total input impedance and by the noise performance of the readout amplifier, we wonder if there are ways to further improve the resolution above such limits. One way is to leverage on *resonance*, i.e., signal "amplification" inside a resonating passive electrical network. As illustrated in Fig. 5.18a, the unknown impedance Z_X is coupled to a resonator in such a way that variations of Z_X produce a change of the resonance frequency f_{RES}. Of course, the introduction of Z_X should preserve the resonance condition, thus excluding highly dissipative components (which reduce the quality factor Q) and basically limiting this approach to capacitive impedances.

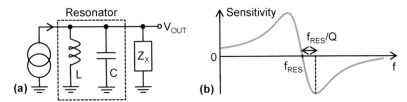

Figure 5.18 (a) Example of resonant technique in which the unknown impedance Z_X is connected to a L-C circuit. (b) Sensitivity of the output voltage to a change of the capacitance as a function of the frequency.

The change of f_{RES} can be detected in several ways. The simplest one is to measure the time between zero-crossing instants with some kind of timing circuit or time-to-digital converter (TDC). In any case, high-bandwidth readout circuitry is required. Since on the peak of resonance the sensitivity to a change of the resonator parameters is zero (Fig. 5.18b), the optimal probing frequency is on the slope of the resonance, slightly shifted with respect to f_{RES}, more precisely at $f_{RES} + f_{RES}/Q$ (which of course tends to f_{RES} high values of Q).

Beyond requiring high-frequency electronics and, thus being more sensitive to the distributed parasitics and reflections due to the connections, this technique presents the following drawbacks:

- It is only suitable for capacitive (or inductive) samples.
- It is not suitable for impedance spectroscopy since f_{RES} cannot be tuned across a wide frequency range.
- It is very difficult to extract the absolute value of Z_X whose impedance is mixed with the tank impedance.

In summary, resonant instruments are mostly suitable to detect tiny variations of reactive impedance over time [17, 18].

5.7.2 Time-Domain Techniques

If Z_X is of a single type (resistive, capacitive or inductive) so that its nature does not change over frequency (and consequently no spectroscopy is performed), simpler measurement techniques can be adopted. In particular, the diffusion of these techniques is fostered by their implementation with simple microcontrollers. In time domain, Z_X is commonly coupled to a time constant controlling the charging or discharging time of a capacitor or setting the oscillating frequency of a relaxation oscillator. By simply setting a voltage threshold and measuring the time required to exceed it, it is possible to estimate the value of Z_X.

A common solution in integrated circuits for measuring capacitance is called *Charge-Based Capacitance Measurement* (CBCM) which relies on measuring the total charge required to charge or discharge the unknown capacitor with MOSFET transistors, particularly suitable for CMOS implementation [19].

5.7.3 Non-Monochromatic Stimulation

In some applications, the measurement time of the impedance meter is important or even critical. Examples are the execution of impedance spectroscopy with a large number of points in frequency or the detection of fast events (such as the passage of micrometric objects in micro-channels [20]) by means of impedimetric sensors. In order to reduce the measurement time of a full spectrum, it is possible to apply more than a single sinusoid at a time. Excitation with simultaneous sinusoids or non-sinusoidal waveforms requires arbitrary waveform generators and more complex processing (exclusively digital in this case). Following are the most important types of spectrally rich stimulation signals:

- **Multi-sine signals** are the sum of several sinusoids at different frequencies (properly spaced in the spectrum, usually avoiding harmonics) [21].
- **Chirp signals** are sinusoids with instantaneously changing frequency. The chirp can be linear (if the frequency increases

linearly with time) or characterized by a more complex function of time.

- **Binary multi-frequency signals** are the equivalent of multi-sine discretized on two levels. A square wave is the simplest case, the equivalent of a single sinusoid, but containing all the odd harmonics.
- **Pseudorandom noise-like**, wide bandwidth signals [20].

In order to maximize the SNR, the power (i.e., the amplitude) of each sinusoidal component of the excitation signal should be maximized. However, often there are strict constraints on the maximum voltage that can be applied across Z_X (in particular in the case of bio-electrochemical samples). Thus, the phase relation among the different spectral components becomes very important. With reference to the multi-sine case (for the sake of simplicity), consider that if all the sinusoids are in phase, given a maximum rms amplitude V_X, the amplitude of each component is necessarily smaller than that of a single sinusoid. Instead, if the phases among the tones are properly arranged, a *crest factor* (defined as the ratio between the peak and the rms value of V_{force}) smaller than 1.41 (i.e., the value for a single sinusoid) can be achieved for more than seven sinusoids.

A drawback of spectrally rich stimulation is related to the linearity of the system under investigation. In fact, when linearity is only an approximation of a non-linear system, the use of single sinusoidal tone, demodulated with a lock-in detector is quite robust with respect to the generation of higher harmonics. Instead, when a full spectrum is sampled simultaneously, the generation of harmonics impacts on the measurement results.

References

1. Callegaro, L. (2012) *Electrical Impedance: Principles, Measurement, and Applications*, 1st ed. (CRC Press, USA).

2. Warwick, C. (1956). Understanding the Kramers–Kronig relation using a pictorial proof, *Physical Rev.*, **104**, pp. 1760–1770.

3. Agilent (2013). *Impedance Measurement Handbook*, 4th ed. (USA).

4. Bianchi, D., Ferrari, G., Rottigni A., Sampietro, M. (2014). CMOS impedance analyzer for nanosamples investigation operating up

to 150 MHz with sub-aF resolution, *IEEE J. Solid-State Circ.*, **49**, pp. 2748–2757.

5. Zhang, B., Buda, A. (2016) Analog front-end design considerations for RTD ratiometric temperature measurements, *Analog Dialog*, pp. 50–53.

6. Carminati, M., Ferrari, G., Sampietro, M. (2009). AttoFarad resolution potentiostat for electrochemical measurements on nanoscale biomolecular interfacial systems, *Rev. Sci. Instrum.*, **80**, pp. 124701/1–10.

7. Gilbert, B. (1968). A new wide-band amplifier technique, *IEEE J. Solid-State Circ.*, **3**, pp. 353–365.

8. Bryant J. M. (2006). Analog computation in the digital age, *Analog Dialogue* (Analog Devices).

9. Ciccarella, P., et al. (2016). Impedance-Sensing CMOS Chip for Noninvasive Light Detection in Integrated Photonics, *IEEE Trans. Circ. Sys. II: Express Briefs*, **63**, pp. 929–933.

10. Ciccarella, P., Carminati, M., Sampietro, M., Ferrari, G. (2016). Multichannel 65 zF rms resolution CMOS monolithic capacitive sensor for counting single micrometer-sized airborne particles on chip, *IEEE J. Solid-State Circ.*, **51**, pp. 2545–2553.

11. Carminati, M., Rottigni, A., Alagna, D., Ferrari G., Sampietro M. (2012) Compact FPGA-based elaboration platform for wide-bandwidth electrochemical measurements, *Proc. IEEE Int. Instrum. Meas. Technol. Conf.* pp. 264–267.

12. Doerner, S., Schneider, T., Hauptmann, P. R. (2007). Wideband impedance spectrum analyzer for process automation applications. *Rev. Sci. Instrum.* **78**, p. 105101.

13. Radil, T., Ramos, P. M., Cruz Serra, A. (2008). Impedance measurement with sine-fitting algorithms implemented in a DSP portable device, *IEEE Trans. Instrum. Meas.* **57**, pp. 197–204.

14. Orazem, M. E., Tribollet, B. (2008) *Electrochemical Impedance Spectroscopy* (Wiley and Sons, USA).

15. Callegaro, L. (2015). Traceable measurements of electrical impedance, *IEEE Instrum. Meas. Mag.*, **18**, pp. 42–46.

16. Carminati, M., Gervasoni, G., Sampietro, M., Ferrari, G. (2016). *Note*: Differential configurations for the mitigation of slow fluctuations limiting the resolution of digital lock-in amplifiers, *Rev. Sci. Instrum.*, **87**, pp. 026102.1–4.

17. Haandbæk, N., With, O., Bürgel, S. C., Heer, F., Hierlemann, A. (2014). Resonance-enhanced microfluidic impedance cytometer for detection of single bacteria, *Lab Chip*, **14**, pp. 3313–3324.

18. Tran, T., Oliver, D. R., Thomson, D. J., Bridges, G. E. (2001). 'Zeptofarad' (10–21 F) resolution capacitance sensor for scanning capacitance microscopy. *Rev. Sci. Instrum.* **72**, pp. 2618–2623.

19. Ghafar-Zadeh, E., Sawan, M. (2008). Charge-based capacitive sensor array for CMOS-based laboratory-on-chip applications, *IEEE Sens. J.*, **8**, pp. 325–332.

20. Sun, T., Holmes, A., Gawad, D., Green, N. G., Morgan, H. (2007). High speed multi-frequency impedance analysis of single particles in a microfluidic cytometer using maximum length sequences, *Lab Chip*, **7**, pp. 1034–1040.

21. Min, M., Pliquett, U., Nacke, T., Barthel, A., Annus, P., Land, R. (2008). Broadband excitation for short time impedance spectroscopy. *Physiol. Meas.* **29**, pp. 185–192.

Chapter 6

Advanced Instrumentation for High-Resolution Capacitance and Impedance Measurements

Giorgio Ferrari and Marco Carminati

Dipartimento di elettronica, informazione e bioingegneria (DEIB), Politecnico di Milano, Piazza Leonardo da Vinci 32, Milano 20133, Italy

giorgio.ferrari@polimi.it

6.1   Introduction

As discussed in the previous chapter, there are many techniques to measure the capacitance or, in general, the impedance of a sample. They are implemented in excellent commercial instruments, as the Precision LCR meter E4980 and the Impedance Analyzer E4990 by Keysight, the capacitance bridge 2700A by Andeen-Hagerling, and the lock-in amplifier-based MFIA Impedance Analyzer by Zurich Instruments just to name a few. However, despite the availability of such instruments, there are multiple motivations for the design of custom advanced instrumentation focused to a specific application. For example, electrical impedance

Capacitance Spectroscopy of Semiconductors
Edited by Jian V. Li and Giorgio Ferrari
Copyright © 2018 Pan Stanford Publishing Pte. Ltd.
ISBN 978-981-4774-54-3 (Hardcover), 978-1-315-15013-0 (eBook)
www.panstanford.com

tomography [1] and electrical capacitance tomography [2] measure the spatial map of the electrical properties of the sample under test by combining the impedance or capacitance between a multitude of electrodes properly located in the space. A multichannel instrument able to measure in parallel many electrodes greatly reduces the time required for the collection of the data.

When the measurement time for taking the impedance spectrum is of main concern, the frequency domain approach of the previously mentioned commercial instruments is not the optimal solution. Since they measure the impedance by applying a stimulus signal at a single frequency, a full impedance spectrum is acquired by sweeping the signal within the desired frequency range, thus the measurement time increases with the number of points of the impedance spectrum. A time domain approach, as discussed in the next chapter, drastically reduces the measurement time by applying a broadband signal with a multitude of frequencies. A proper acquisition and processing of the sample response to the broadband signal extracts the whole impedance spectrum in a single measurement.

In addition to provide specific functionalities for a given experiment, a properly designed custom instrument can reduce the noise of the measurement in comparison to a general-purpose instrument. For example, many micro-nano devices have a capacitance orders of magnitude smaller than the capacitance of bulky instruments making difficult to discriminate the useful signal proportional to the device under test from the instrumental noise that is proportional to the overall capacitance. A tailoring of the input amplifier to achieve a capacitance similar to the sample reduces the noise with beneficial effects on the minimum detectable signal variation or on the minimum measurement time required to obtain a certain resolution.

In this chapter, it is presented how to design custom instrumentation to improve the signal-to-noise ratio of capacitance and impedance measurements. The discussion will be mainly focused on a current sensing scheme based on a transimpedance amplifier and a lock-in amplifier for the advantages of this approach in terms of sensitivity and flexibility discussed in the previous chapter.

6.2 Integrator Stage as a Current-Sensitive Front-End Amplifier

Transimpedance amplifiers (TIA), in which the signal current made available by the Device Under Test (DUT) is converted into a voltage for further processing, are an excellent choice to perform impedance or capacitance spectroscopy. The feedback architecture (see Fig. 6.1) makes a virtual ground at the input node ideally forcing the current coming from the DUT to flow in the resistor R_F irrespective of the parasitic capacitances introduced by connections, amplifier and the sample itself. The feedback resistor sets the current-to-voltage conversion factor of the amplifier. As any electronic component, R_F adds a thermal noise that produces random fluctuations of the current flowing through the resistor. The power spectral density of these fluctuations is $4kT/R_F$, where k is the Boltzmann constant and the dimensions are A^2/Hz. The thermal noise of R_F sets a direct limit to the minimum detectable current by the amplifier[1]. Therefore, it must be chosen as large as possible to maximize the signal-to-noise ratio of the measurement. However, a large resistor R_F has a few drawbacks. Besides the limited accuracy and temperature stability of large resistors, the bandwidth of the transimpedance amplifier is given by $1/(2\pi R_F C_F)$, where C_F is the feedback capacitance added to grant the stability of the feedback loop. The minimum value of C_F is the unavoidable stray capacitance of R_F, a value around 0.1–0.2 pF that limits the bandwidth at a few kHz for a resistor of $R_F = 1$ GΩ. Thus, it is not possible to obtain simultaneously low noise (i.e., large R_F) and wide bandwidth (i.e., small R_F). In addition, the R_F resistor converts into a voltage both the DC input current and the AC input current. The maximum AC signal, that sets the dynamic range of the impedance measurement, is consequently limited by the stationary current coming from the DUT.

    In order to overcome the noise-bandwidth trade-off of the transimpedance amplifier, an effective solution is the removal of the feedback resistor maintaining only the capacitor. Figure 6.2a

---

[1]The interested reader can find an explanation of the noise analysis in many books and papers, e.g., [3–5].

shows the resulting circuit that is called integrator stage or charge preamplifier.

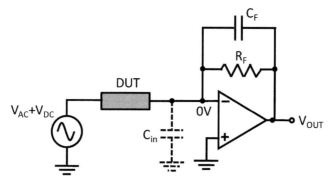

Figure 6.1 Schematic of a standard transimpedance amplifier. C_{in} is the sum of all capacitances at the input node of the amplifier (stray capacitances of the connections and of the DUT, input capacitance of the amplifier itself).

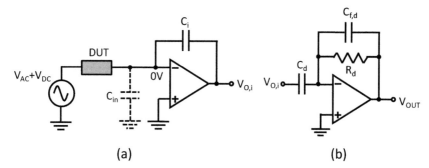

Figure 6.2 Integrator stage (a) as front-end amplifier for impedance measurements. An optional differentiator stage (b) can be connected at the output of the integrator stage to recover a flat frequency response.

This input stage offers multiple advantages. Small value capacitors in the pF range have excellent properties in terms of temperature stability, frequency response up to GHz and, notably, they are ideally noise-free components. The bandwidth of the integrator stage is limited by the operational amplifier, BW $\approx$ GBP $C_i/(C_i + C_{in})$, where GBP is the gain-bandwidth product of the operational amplifier and C_{in} the total input capacitance. Operational amplifiers with low noise and GBP exceeding 1 GHz are commercially available allowing wide bandwidth operation and high sensitivity of the impedance measurement.

In the case of a capacitive DUT, an integrator stage is also an excellent choice for a capacitance spectroscopy over a wide range of frequencies. The output voltage is proportional to the DUT capacitance C_{DUT} ($V_{0,I} = -V_{AC} \cdot C_{DUT}/C_i$) independently of the frequency. On the contrary, a transimpedance amplifier shows an output voltage proportional to the frequency requiring a tuning of V_{AC} to avoid the saturation of the amplifier at the maximum frequency of the measured spectrum.

Whenever it is relevant the measurement of the current in the time domain or the impedance spectroscopy of resistive samples, the output of the integrator stage can be connected to a differentiator stage (Fig. 6.2b) to recover a flat frequency response [6–11]. The input current is amplified by C_d/C_i and then converted into a voltage with the resistor R_d giving a gain of the full structure of $R_d \cdot C_d/C_i$. The current amplification operated by the capacitors makes negligible the noise added by the resistor R_d which, consequently, can be chosen small to enlarge the bandwidth of the differentiator stage.

An important limitation of the integrator stage is the amplifier saturation due to a stationary input current. To be operated, the schematic of Fig. 6.2a must be modified with an additional network to discharge the capacitor C_i. A switch in parallel to C_i operated when the integrator output voltage reaches a defined threshold, offers a simple solution with minimal additional noise [6]. However, it would set a limit to the time interval available to measure the DUT current. This time depends both on the DUT stationary current and on C_i. As an example, with C_i = 100 fF and I_{DC} = 10 nA, a discharge period of less than 1 ms would result, largely inferior to the time required to measure accurately the impedance in many applications. Unlimited measurement time is obtained by operating a periodic reset of the feedback capacitor and processing the charge stored in the capacitance between two reset events [12–14]. The requirement of a reset frequency higher than the bandwidth of the signal limits the maximum operating frequency of this approach.

6.2.1 Advanced Integrator Stages for Impedance Spectroscopy

A wide bandwidth integrator stage with unlimited measurement time is obtained with a continuously active reset network that removes the DC component of the input current but leaves unmodified the AC signal used for the impedance measurement. In this way, the front-end stage should behave like a pure integrator starting from very low frequencies. Figure 6.3 shows an example of active reset network using a resistor driven by a filter (H) operating a low frequency [7]. The high gain of $H(s)$ at low frequency forces the DC current through the resistor R_F. Therefore, the mean value of the integrator output is fixed around zero by the feedback loop assuring the maximum AC output swing in any bias condition of the DUT, an ideal condition for impedance spectroscopy. At higher frequency, the low gain of $H(s)$ combined to the low impedance of the feedback capacitor C_i force the input current to flow in C_i giving the frequency response of an integrator stage. The high-frequency response of the filter $H(s)$ is flat to assure the stability of the feedback loop.

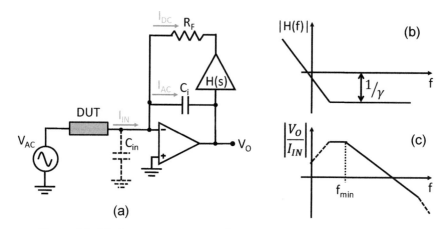

Figure 6.3 (a) Integrator stage with an active network to discharge the feedback capacitor. (b) Frequency response of the filter $H(s)$. (c) Resulting frequency response from the input current to the output voltage.

The minimum frequency f_{min} for impedance spectroscopy is set by the starting frequency of the integrator response (Fig. 6.3c). It is given by

$$f_{min} \cong \frac{1}{2\pi R_F C_i \gamma}, \tag{6.1}$$

where γ is the high-frequency attenuation of $H(s)$. The feedback elements R_F and C_i are chosen to satisfy the condition on the maximum stationary input current ($R_F = V_{sat,H}/I_{DC,max}$, where $V_{sat,H}$ is the saturation voltage of the filter H) and the maximum operating frequency (BW $\approx$ GBP $C_i/(C_i + C_{in})$) as in the case of the ideal integrator), respectively. The attenuation γ is a free parameter that can be tuned to perform the impedance measurement starting from the desired minimum frequency.

The input-referred equivalent current noise of the integrator stage with the active reset network has the same expression of a standard TIA:

$$\overline{i_{eq}^2} \approx \frac{4kT}{R_F} + \overline{e_n^2}(2\pi f)^2 (C_{in} + C_i)^2, \tag{6.2}$$

where $\overline{e_n^2}$ is the equivalent voltage noise of the operational amplifier and the noise of $H(s)$ has been neglected thanks to the typically large value of R_F. Since the bandwidth is independent of R_F, its value can be maximized to reduce the noise of the amplifier. Thus, the integrator stage with an active reset network removes the trade-off between noise and bandwidth of a standard TIA making this configuration an excellent solution for high sensitivity impedance spectroscopy over a wide range of frequencies. Figure 6.4a shows an example of experimental noise of an integrator stage with an active reset network (see Ref. [7] for details). For comparison, it is also reported the noise of a standard TIA with a feedback resistor of $R_F = 1$ MΩ that offers the same bandwidth of 1 MHz of the integrator stage but with a higher noise up to 100 kHz. The resolution of a capacitance measurement using the integrator stage connected to a lock-in amplifier is a few aF over a wide frequency range with an applied stimulus voltage of only 0.1 V, as reported in Fig. 6.4b and experimentally verified in the case of nanoscale impedance spectroscopy [15].

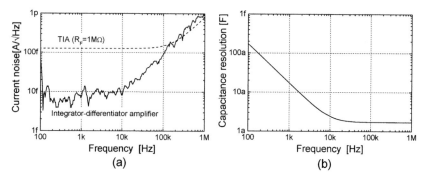

Figure 6.4 (a) Equivalent input current noise measured using an implementation of integrator stage with an active reset network [7]. The dashed line is the noise of a standard TIA with a similar bandwidth. (b) Theoretical capacitance resolution as a function of the frequency given by the noise of the integrator stage. Conditions: AC voltage applied to the DUT of 0.1 V, measurement time of 1 s.

6.3 Low-Noise Amplifiers in CMOS Technology

The small capacitance arising by submicrometric devices may impose a capacitance spectroscopy with a resolution better than 1 aF (see, for example, Chapters 11, 13, and 14 of this book). To reach such resolution, the electronic noise should be minimized over a wide frequency range, suggesting an integrator-based front-end amplifier. As shown in Eq. (6.2), the noise level is mainly limited by the effect of the voltage noise of the amplifier on the total capacitance at the input node of the amplifier. The voltage noise is limited by technological reasons to values around $\approx 1\,\text{nV}/\sqrt{\text{Hz}}$. In the case of nanodevices, the input capacitance C_{in} is largely dominated by the stray capacitance of the connections and by the input capacitance of the amplifier, being the DUT capacitance negligible. Consequently, a strategy to improve the signal-to-noise ratio (SNR) is to shorten the length of connection from the sample to the amplifier and the amplifier capacitance itself. Both these results are achieved with the integration of the amplifier in a single silicon chip. This provides the possibility of sizing the input transistors to match the input capacitance of the experimental setup. The miniaturized amplifier, a few mm^2 for an integrated solution, can be lodged very near to the sample, correspondingly reducing the stray capacitance of the connection.

In addition to a higher SNR, the integration of the electronics in an application specific integrated circuit (ASIC) can offer other advantages:

(i) multichannel capability to perform many measurements in parallel [16, 17]

(ii) high-frequency spectroscopy [18, 19]

(iii) operation in non-standard environmental conditions, for example, at cryogenic temperature [20–22]

These advantages justify the growing efforts towards integrated solutions based on CMOS technology, despite the higher design complexity with respect to standard on-the-shelf solutions. Besides the time-consuming design and the expensive fabrication costs of a custom integrated circuit, the design of current-sensitive amplifiers has to face the impossibility to integrate large value resistors. Consequently, a fully integrated standard transimpedance amplifier would suffer a high thermal noise given by the relatively small value of the feedback resistor. This limitation is overcome by specific circuits emulating large value resistors, as switched capacitors [23] for the implementation of highly linear resistances in the low-medium frequency range, pseudoresistors [24] for very high value resistances with a limited linearity and matched transistors [25] for a compromise between a good linearity and medium-high values of resistance. However, integrator stages are the preferred choice for an integrated implementation because the gain is set by a capacitor, a very well controlled component in CMOS technology, without requiring large resistances on the signal path [13]. For example, the active reset network discussed in the previous section has been successfully implemented in CMOS technology [9] improving the noise performance of more than one order of magnitude with respect to the case reported in Fig. 6.4b reaching sub-aF resolution from 10 kHz up to a few MHz.

When the integrated circuit is used to boost the performance of an existing experimental setup developed for impedance spectroscopy of nanodevices, a convenient circuit architecture is the current amplifier. Such circuit can be simply connected between the nanodevice and the low impedance input of a standard impedance analyzer, as shown in Fig. 6.5a. The integrated circuit is used to reduce the equivalent input noise of the current measurement maintaining the signal generation, signal processing and user interface of bench-top instruments.

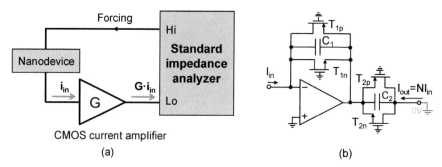

Figure 6.5 (a) Experimental setup for impedance spectroscopy on nanodevices based on a CMOS current amplifier as add-on of a bench-top instrument. (b) Simplified schematic of a CMOS current amplifier. The components T_{2p}, T_{2n}, C_2 are implemented connecting in parallel N components equal to T_{1p}, T_{1n}, C_1, respectively.

Figure 6.5b shows a CMOS current amplifier well-suited for this kind of application [26, 27]. The current amplification is obtained using matched components between the feedback path and the output path. The transistors T_{2p}, T_{2n} and the capacitor C_2 are implemented as N replicas of the components T_{1p}, T_{1n}, and C_1. Since the voltage across all the components is forced equal by the feedback loop, the output current is N time the current in the feedback components (i.e. the input current) irrespective of the non-linear behavior of the transistors. N-channel and p-channel transistors are used to amplify both positive and negative input current, whereas the capacitors are required to assure the stability of the feedback loop. The quality of the CMOS technology guarantees an excellent matching of the components with a non-linearity error less than 1% without any calibration.

The current amplification without resistors and switches allows ultra-low noise measurements from DC to a few MHz reaching a noise below 1 fA/$\sqrt{\text{Hz}}$ up to 10 kHz, as shown in

Fig. 6.6a. Correspondingly, the capacitance resolution of a lock-in amplifier coupled to the integrated circuit reaches values of a few tens of zF with an applied voltage signal of 100 mV (Fig. 6.6b), more than one order of magnitude below the solution at discrete components reported in Fig. 6.4. The best resolution reported using this approach is as low as a few zeptoFard [28].

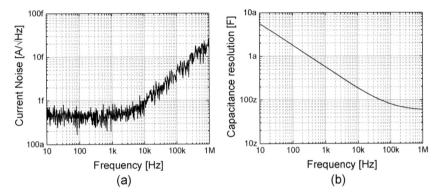

Figure 6.6 (a) Equivalent input current noise of the CMOS current amplifier (C_{in} = 1 pF, no DC input current). (b) Capacitance resolution obtainable by the integrated circuit coupled to a lock-in amplifier (AC voltage applied to the DUT of 0.1 V, measurement time of 1 s).

Figure 6.7 shows the enhancement provided by the CMOS current amplifier when used as add-on of a bench-top instrument [27]. The circuit has been connected to the low impedance input of a precision LCR meter (E4980A by Agilent) following the scheme of Fig. 6.5a. The device under test is a 0.5 pF capacitor that cannot be correctly measured by the Agilent instrument below a few kHz where the current is extremely small. By adding the CMOS amplifier, the impedance spectrum of the capacitance is properly obtained in the full frequency range of the commercial instrument. In addition to the extension of the maximum measurable impedance by the instrument, the CMOS amplifier enhances the resolution of the impedance measurement. As an example, Fig. 6.7b reports the measured impedance of the test capacitor at 100 kHz as a function of the time. Thanks to the CMOS amplifier, the fluctuations given by the instrumental noise are reduced by a factor 35 and correspondingly improving the minimum detectable impedance variation of the same factor.

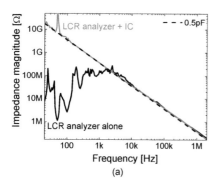
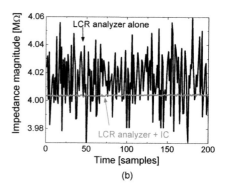

(a)

(b)

Figure 6.7 (a) Impedance spectra measured with and without the CMOS current amplifier as add-on of a precision LCR meter (E4980A by Agilent). The DUT is a capacitor of 0.5 pF and the applied voltage is V_{AC} = 3 mV. (b) Measured impedance at 100 kHz as a function of the time in the same conditions (sampling time of 0.4 s). Reprinted from [27] with permission from Elsevier.

6.4   High-Frequency Impedance Spectroscopy

There are multiple reasons to extend the frequency range of the capacitance/impedance spectroscopy up to tens of MHz or hundreds of MHz. Higher frequencies give access to the properties of faster traps or to the lifetime of the minority carries (see the first section of this book). The use of a high frequency can also be exploited for a capacitive access of the electrical properties of a sample bypassing an insulator layer on the surface (see, for example, Chapter 15 and ref. [29]).

Resonant techniques combine operation at high frequency and excellent resolution (see Chapter 5). However, they are mainly used for single frequency measurements because the resonance frequency can be tuned in a narrow range limiting the application for capacitance spectroscopy. Time-domain reflectometry can cover a few decades of operative frequency with a single measurement (see Chapter 7). However, it uses a broadband stimulation of the sample with an intrinsic reduction of the signal power for each frequency and correspondingly a decrease of the SNR.

The current sensing approach using a transimpedance amplifier with a resistive feedback (Fig. 6.1) requires a small value feedback resistor to operate at high frequency. The thermal noise of such resistor limits the sensitivity of the measurement. As discussed previously, a capacitive feedback (Fig. 6.2) combined to an operational amplifier offers wide bandwidth and low noise. In the ideal case of an operational amplifier with an infinite gain, the input current is converted into a voltage with a well-defined transfer function:

$$G_{id} = \frac{V_{0,i}}{I_{IN}} = \frac{1}{sC_i},$$ (6.3)

where s denotes the complex frequency. However, the limited gain of the operational amplifier at high frequency reduces the operative frequency range in a way dependent on the DUT and on the setup conditions. To quantify this aspect is possible to calculate the loop gain of the integrator stage of Fig. 6.2:

$$G_{loop} \approx -A(s)\frac{C_i}{C_i + C_{in} + C_{DUT}},$$ (6.4)

where $A(s)$ is the gain of the operational amplifier and for simplicity the DUT is assumed a capacitor, C_{DUT}. The actual transfer function of the amplifier is

$$G_{real} \approx G_{id}\frac{1}{1-1/G_{loop}} \approx \frac{1}{sC_i}\frac{1}{1+\dfrac{C_i + C_{in} + C_{DUT}}{A(s)C_i}},$$ (6.5)

Since the gain of the operational amplifier decreases with the frequency, the current to voltage conversion factor at high frequency becomes sensitive to the input capacitance, to the DUT impedance and to the amplifier gain itself. Thus, the accuracy of high-frequency measurements is seriously affected by the amplifier performance.

To maintain a high accuracy independent of frequency, wide bandwidth impedance analyzers use a down-conversion, amplification, up-conversion topology [30, 31]. An example of such architecture is shown in Fig. 6.8. The sinusoidal signal at the measurement frequency f_s coming from the DUT is demodulated by the multipliers $M1$ operated at the same frequency

f_s. Therefore, the DC value at the output of the multipliers is proportional to the amplitude of the input signal at the frequency f_s. The in-phase and quadrature components are separately demodulated to preserve the information of the phase of the input signal, similarly to a lock-in amplifier. The input signal translated at zero frequency is successively amplified by the stages A that provide an amplification independent of the measurement frequency f_s. The following low-pass filters (LPF) prevent the propagation of the $2f_s$ component and other unwantedharmonics given by the multipliers $M1$. Finally, the signals are up-converted by $M2$ and added together to obtain an output signal with the same frequency and phase of the input signal and a gain factor irrespective of the measurement frequency f_s.

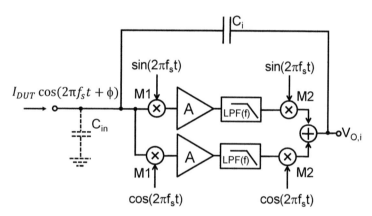

Figure 6.8 Simplified schematic of the impedance analyzer based on modulation/demodulation architecture.

By including this tuned amplifier in a capacitive feedback, as shown in Fig. 6.8, is possible to perform impedance measurements up to hundreds of MHz with a very high loop gain, thus preserving a high accuracy in the full impedance spectrum. The output voltage at the frequency f_s is related to the input current (i.e. the DUT admittance) by the transfer function of an ideal integrator (Eq. 6.3) calculated at f_s. The DC signals at the input of the multipliers $M2$ are proportional to the in-phase and quadrature terms of the output voltage at f_s, thus they are proportional to the real and imaginary parts of the DUT admittance at f_s. The architecture therefore includes the function of a lock-in amplifier without requiring external instruments, partially compensating the higher complexity of the circuit with respect to a standard amplifier.

The implementation of a demodulation-modulation architecture in 0.35 μm CMOS technology has allowed to obtain a miniaturized impedance analyzer operating from 1 kHz to 150 MHz with a sub-attoFarad resolution from 100 kHz up to 150 MHz [30].

6.5 Additional Noise Sources

The circuital solutions discussed in the previous sections reduce the noise of the current readout to extremely low values. By referring to the case of a capacitance measurement,[2] the minimum detectable capacitance is ideally limited by the equivalent input noise i_{eq}^2 of the front-end amplifier to

$$C_{min} \approx \frac{\sqrt{\overline{i_{eq}^2}(f_0)B_n}}{2\pi f_0 V_{AC}}, \tag{6.6}$$

where B_n is the equivalent noise bandwidth of the measurement, V_{AC} is the AC voltage applied to the capacitor under test and f_0 is the frequency of measurement. By adopting an integrator-based amplifier, $\overline{i_{eq}^2}$ is mainly given by the voltage noise of the operational amplifier (second term of Eq. (6.2)) producing a C_{min} of

$$C_{min} \approx (C_{in} + C_i) \frac{\sqrt{\overline{e_n^2}(f_0)B_n}}{V_{AC}}, \tag{6.7}$$

that corresponds to values of tens of zF for input capacitances of a few pF, V_{AC} of 1 V, B_n = 1 Hz and an equivalent voltage noise of a few nV$/\sqrt{\text{Hz}}$. In practical experiments, such small capacitance is not directly measured because the stray capacitance between the electrodes is commonly much larger (in the order of fF or pF). In this case, a more important parameter of a high sensitivity impedance analyzer is the resolution ΔC_{min}, that is the minimum detectable *variation* of capacitance. The resolution is also

---

[2]A similar analysis is valid for a generic impedance.

expressed in a relative form as the ratio between ΔC_{min} and the total signal itself. Assuming a linear system with a noise limited by the front-end amplifier, the resolution and the minimum detectable capacitance are the same, $\Delta C_{min} = C_{min}$, suggesting the capability of detect a few tens of zF over a total capacitance of a few pF with a relative resolution well below to one part per million (1 ppm). This high resolution is rarely achieved in a real experiment. In the following, we briefly review the main noise sources in addition to the equivalent input current noise of the front-end amplifier that can degrade the resolution to tens of ppm in many experimental setups.

6.5.1 Dielectric Noise

The input capacitance of the current-sensitive readout circuit is a key parameter for low noise measurements. Not only is it responsible of the high-frequency noise by means of the voltage noise of the amplifier, but also is an intrinsic noise source. A real capacitor has an energy dissipation due to the thermal fluctuation of dipoles in the dielectric material. The energy loss implies an admittance with a real component, i.e., an electrical model made of an ideal capacitor in parallel to a conductance. The value of the latter is proportional to the frequency and its thermal noise allows to quantify the dielectric noise of the insulator. It can be written as [32–34]

$$S_i(f) \approx 4\,kT\,2\pi fCD, \tag{6.8}$$

where C is the value of the capacitor and D is the dissipation factor of the dielectric material (also called dielectric loss and indicated with $\tan\delta$). An excellent insulator like Teflon has a dissipation factor lower than 10^{-4} making the dielectric noise negligible in practical applications. However, insulators with a lower quality have a dissipation factor order of magnitude larger. For example, FR4 epoxy laminates used for standard printed circuit boards (PCB) have a dissipation factor of $D \approx 1.7\ 10^{-2}$. Thus, a stray capacitance of 1 pF at the input node of the front-end amplifier due to the PCB adds a noise of 13 fA/$\sqrt{\text{Hz}}$ at 100 kHz, a value higher than the noise of the CMOS current amplifier reported in Fig. 6.6.

6.5.2   The Role of the Substrate

For research purposes, it is common to fabricate micro-nano devices and sensors on the top of a silicon wafer, the most used substrate in a microelectronic facility. The electrical insulation between the device and the silicon can be obtained by adding a layer of silicon dioxide, as sketched in Fig. 6.9. Although this is an excellent solution for quasi-static measurements, in the case of capacitance spectroscopy is necessary to take in account the unavoidable stray capacitance between the electrodes and the silicon substrate. The conductive strips and the bonding pads required to electrically connect the DUT are capacitively coupled with the silicon substrate (C_{sub} in Fig. 6.9) giving a spurious impedance in parallel with the DUT. For example, a bonding pad of 150 μm × 150 μm above a silicon dioxide of 100 nm thickness gives a capacitance of 8 pF. The typical low resistance of the silicon substrate connects the two bonding pads producing a total capacitance of 4 pF in parallel to the sample, thus affecting the accuracy and the noise of the capacitance measurement of the DUT [35]. The accuracy of the measurement can be improved by using a current-sensing scheme and by connecting the silicon substrate to ground (directly or via a capacitive connection by putting the silicon substrate on the top of a ground electrode). The ground connection prevents the injection of a spurious current from the driving pad to the sensing pad. However, the noise is still affected by the substrate because the C_{sub} of the sensing pad is connected to the input node of the readout circuit, thus increasing the total input capacitance and adding dielectric noise.

The effects of the substrate are augmented in presence of an electrolyte, as in the case of electrochemical experiments. The conductive liquid around to the sample may add a spurious capacitive path between the electrodes or toward ground increasing the stray capacitance up to hundreds of pF [35]. When feasible from the technological point of view, an insulating substrate, like glass or quartz, is the best solution for accurate and sensitive capacitance measurements because it strongly reduces the effects of the substrate.

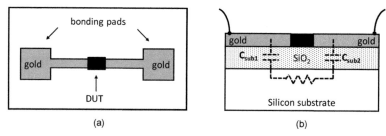

Figure 6.9 Top view (a) and cross section (b) of a two-terminal device fabricated on a silicon substrate.

6.5.3 Signal Generator

An additional noise source is the waveform generator used for stimulating the DUT. Although the concept is general, for a detailed analysis we will consider the case of an experimental setup based on a lock-in amplifier with a sinusoidal stimulus. The ideal sinusoidal voltage at frequency f_0 and amplitude V_{AC} of the signal generator is affected by three main noise sources producing a stimulus voltage of

$$V_s(t) = V_{AC} \cdot (1 + n_A(t)) \ \sin(2\pi f_0 t + n_\phi(t)) + n_w(t), \qquad (6.9)$$

where $n_A(t)$ and $n_\phi(t)$ are the amplitude and the phase noise of the sinusoidal signal, respectively, and $n_w(t)$ is an additive noise given, for example, by the output stage of the generator. The role of $n_w(t)$ is similar to the voltage noise $\overline{e_n^2}$ of the operational amplifier in the front-end amplifier: $n_w(t)$ is directly applied to the capacitance C_{DUT} of the DUT injecting a random current at the input node of the amplifier. Equation (6.7) of the minimum detectable capacitance is accordingly modified as

$$C_{min} \approx (C_{in} + C_i) \frac{\sqrt{\overline{e_n^2}(f_0) B_n}}{V_{AC}} + C_{DUT} \frac{\sqrt{\overline{n_w^2}(f_0) B_n}}{V_{AC}}, \qquad (6.10)$$

where $\overline{n_w^2}(f_0)$ is the power spectral density of $n_w(t)$ at the frequency of the measurement. Although $\overline{n_w^2}(f_0)$ is typically of 20–30 nV/$\sqrt{\text{Hz}}$, one order of magnitude higher than $\overline{e_n^2}(f_0)$, its effect can be negligible in experiments with micro-nano devices where $C_{DUT} \ll C_{in}$ due to the stray capacitance of the wire connections and the input capacitance of the front-end amplifier.

The phase noise $n_{\phi}(t)$ introduces a phase error in the synchronous demodulation operated by the LIA causing a random mixing of the real part and imaginary part of the measured admittance: the resistive component of DUT produces a random fluctuation of the measured capacitance, and vice versa. In the case of a capacitive device with a negligible resistive component, the phase noise has a limited effect on the measurement and can be totally removed by calculating the magnitude of admittance.

The amplitude noise modulates the stimulus voltage and correspondingly the demodulated signals of the lock-in amplifier are modulated by $n_A(t)$. These fluctuations give a direct limit to the minimum detectable capacitance by the instrument. Since the effect of the amplitude noise is proportional to V_{AC} and to C_{DUT}, it sets a limit to the resolution of the system as

$$\frac{\Delta C_{min}}{C_{DUT}} \approx \sqrt{\overline{n_A^2} B_n}, \tag{6.11}$$

where $\overline{n_A^2}$ is the power spectral density of the amplitude noise. Note that differently from $n_w(t)$, the amplitude noise is not translated in frequency by the demodulation of the LIA. Thus, the unavoidable $1/f$ noise of $n_A(t)$ poses a stringent limit to the best resolution of the instrument. This kind of amplitude modulation sets the resolution limit of many commercial lock-in amplifiers to tens of ppm, well above the limit of a front-end amplifier operated in the best conditions.

6.5.4 Gain Fluctuations

The amplitude noise of the signal generator is an example of gain fluctuations that play a pivotal role in high-resolution measurements. As sketched in Fig. 6.10, all the analog stages of a digital LIA-based instrument can produce gain fluctuations.

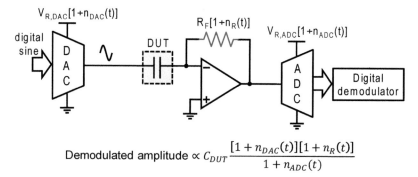

Demodulated amplitude $\propto C_{DUT} \dfrac{[1 + n_{DAC}(t)][1 + n_R(t)]}{1 + n_{ADC}(t)}$

Figure 6.10 Simplified scheme of an impedance analyzer based on a digital lock-in amplifier. The noise of the voltage references of the DAC and ADC and the fluctuations of the feedback resistor value modulate the amplitude of the signal processed by the digital demodulator.

The noise of the reference voltage used by the analog-digital converter (ADC), as well as the reference noise of the digital-to-analog converter (DAC), produce a random fluctuation of the amplitude of the digital signal processed by the instrument that limits the minimum detectable signal variation. In particular, the $1/f$ noise of the reference voltages is responsible of slow fluctuations of the LIA output that limit the performance of the instrument. The gain fluctuation of the system has an effect proportional to the processed signal, thus the relative resolution has a lower limit given by a relation similar to Eq. 6.11.

Additional sources of gain fluctuations are the amplification and conditioning stages. The value of the components that set their gain can change in time due to temperature fluctuations or to intrinsic $1/f$ noise. This is exemplified in the experimental results shown in Fig. 6.11 [36] where the same DUT is measured with three different instruments. The upper curve, obtained with a commercial lock-in amplifier, shows slow $1/f$-like fluctuations imputable to gain and amplitude noise. They set the resolution of the instrument to 15 ppm. The center and lower curves are the output of a custom LIA [36] with a gain fixed by standard resistors (center line) or by resistors with a low temperature coefficient (5 ppm/K instead of 50 ppm/K of standard resistors). The better stability of the gain in the last case is essential to achieve a resolution of 0.7 ppm.

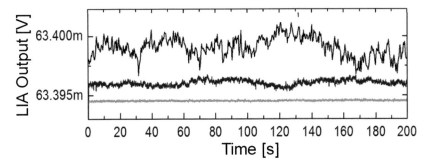

Figure 6.11 Experimental comparison of the instrumental noise of a commercial LIA (SR830 by Stanford Research System, upper curve) and of a custom LIA with the gain set by standard resistors (center curve) or by LTC resistors (lower curve). © 2014 IEEE. Reprinted, with permission, from [36].

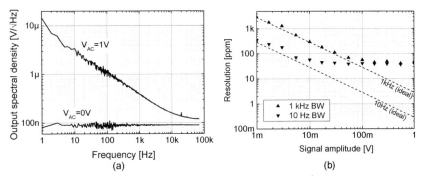

Figure 6.12 (a) Power spectral densities of the LIA output with $V_{AC} = 0$ V and $V_{AC} = 1$ V. They are measured connecting the stimulus signal directly to the input of the HF2LI instrument. (b) Resolution experimentally achieved as a function of the signal amplitude for two filtering bandwidths. The dashed lines are the theoretical resolution assuming the noise measured with $V_{AC} = 0$ V.

The power spectral density of the LIA output reflects the spectrum of the gain fluctuations, thus it is commonly affected by $1/f$-like noise irrespective of the frequency of the impedance measurement. As a consequence, its effect is not effectively reduced by averaging the output of the LIA. Figure 6.12 reports the noise of a state-of-art commercial LIA (HF2LI by Zurich Instruments) with the stimulus signal directly connected to the input. In absence of the voltage signal, the measured noise of the LIA output is white and limited by the front-end amplifier (Fig. 6.12a). The ideal resolution achievable by such instrumental noise as a function of the stimulus voltage is shown in Fig. 6.12b for two bandwidths of the LIA low-pass filters. By applying the voltage signal, the gain fluctuations add a significant $1/f$ noise (Fig. 6.12a) setting a maximum resolution of about 40 ppm, independently of the bandwidth set by the low-pass filter. Similar resolution limits have been measured for other fast digital lock-in amplifiers.

6.6 Techniques for the Detection of Small Admittance Variations

High-resolution measurements require a full control of the experimental setup. Sections 6.2–6.4 have discussed how to design the front-end amplifier to reduce its noise contribution. Section 6.5 has reviewed some additional sources of noise and has pointed out the role of slow fluctuations of the acquired signal amplitude in setting the ultimate resolution limit of an impedance measurement. Such amplitude fluctuations are mainly given by (i) gain fluctuations of the amplifying stages; (ii) amplitude noise of the signal generator; (iii) gain noise of the analog-to-digital converter. Although gain fluctuations of the amplifiers can be limited by using very stable components (as resistors with a low temperature coefficient) along the signal path, the reduction of the amplitude noise of the generation and of the acquisition stages is not an easy task. In the following sub-sections, we briefly discuss two techniques for the reduction of the effects due to the gain fluctuations that enable measurements with a resolution better than 1 ppm.

6.6.1 Differential Approach

The effect of gain fluctuations and amplitude noise on the measurement is proportional to the signal amplitude. In experiments where it is important to detect a small variation of a relatively large capacitance, a way to reduce the effect of gain fluctuations is to avoid the processing of the large constant term of the capacitance by keeping just the variation of the signal with respect to a reference value. This can be accomplished using a differential approach [37], as the example shown in Fig. 6.13.

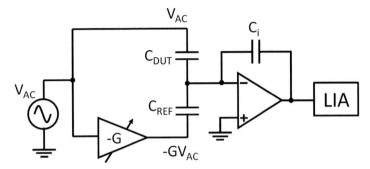

Figure 6.13 Example of differential setup for high-resolution capacitance measurements.

A reference capacitor C_{REF} is used to cancel the current injected by C_{DUT} at the input node of the amplifier. The two capacitors are driven by sinusoidal signals in counter-phase. In the case of a perfectly matched condition, $V_{AC} \cdot C_{DUT} = G \cdot V_{AC} \cdot C_{REF}$, the front-end amplifier and the LIA have no signal at their input, thus the effect of gain fluctuations is null. The amplitude noise of the V_{AC} generator is nulled as well. A variation of C_{DUT} unbalances the differential structure producing an output signal proportional to $\Delta C = C_{DUT}-C_{REF}$. The resolution limit due to the slow gain fluctuations is then related to ΔC with an overall improvement of the performance when $\Delta C \ll C_{DUT}$. A resolution better than 1 ppm is reported in the case of capacitance measurements [37].

A key parameter of the differential configuration is the matching of reference path with respect to the DUT path. This can be achieved by designing the DUT with a differential structure, as in the case of many MEMS sensors [38, 39], or by adding a reference capacitor to the experimental setup. A calibration procedure to balance the differential structure at the begin of the experiment is required in many cases. To simplify the calibration, a variable gain amplifier is commonly used to generate the counter-phase signal ($-GV_{AC}$) with the correct amplitude, as sketched in Fig. 6.13.

Note that the DUT and the reference path must be balanced in magnitude and in phase to compensate effectively the gain fluctuations. Such severe condition makes difficult to implement the differential approach in the case of a DUT with a complex impedance or when the experiment requires a change of the operating conditions (frequency, bias, temperature).

6.6.2 Enhanced Lock-In Amplifier for High-Resolution Measurements

A different approach to enhance the resolution of an impedance measurement is the switched ratiometric lock-in amplifier [40]. The idea is to cancel the spurious fluctuations of the signal amplitude by means of a ratio between the amplitudes of the DUT signal and a reference signal. Differently from the differential approach that requires a balanced condition at each instant time, the ratiometric approach operates on the mean value of the amplitudes without demanding a precise control of amplitude and phase of the reference signal. The only requirement on the reference signal is to experience on the average the same gain fluctuations of the DUT signal. An example of implementation of this technique, called Enhanced LIA (ELIA), is shown in Fig. 6.14 [40].

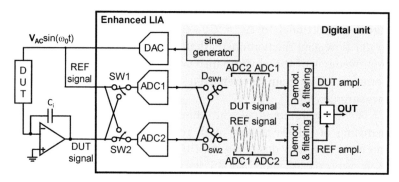

Figure 6.14 Simplified architecture of a switched ratiometric dual-channel LIA for high resolution measurements.

The dual-channel LIA simultaneously acquires the DUT signal and the reference signal that in this case is the stimulus signal.

The digital processor demodulates and filters the signals as a standard LIA. Finally, the instrument calculates the ratio between the amplitudes of the DUT and REF signals. The amplitude noise of the stimulus signal produces the same random fluctuations of the DUT and REF amplitudes, thus its effect is cancelled-out by the ratio operation. In order to remove the gain fluctuations of the two ADCs, the inputs of the converters are switched periodically between the DUT and the REF signals. The digitized samples are processed in real time using the digital switches D_{SW1}, D_{SW2} (synchronous with the hardware switches SW_1, SW_2) in order to reconstruct continuously in time the DUT and REF signals in the digital domain. By choosing the switching frequency f_{SW} faster than the slow gain fluctuations of the ADCs, the DUT and REF signals in a period $1/f_{SW}$ are acquired with the same gain of the ADC. Therefore, the reconstructed digital DUT and REF signals show the same gain fluctuations and the final ratio operation cancel the gain noise of the ADCs.

An example of effectiveness of the switched ratiometric technique is reported in Fig. 6.15 [40]. A 5 ppm variation of the DUT signal has been obtained by applying the stimulus signal to a voltage divider with an attenuation factor changed every 10 s (Fig. 6.15a). A resistor of 48 MΩ is periodically connected in parallel to a resistor of 250 Ω to produce a variation of 1.25 mΩ and correspondingly changes the DUT signal. Figure 6.15b reports the experimental results using three different instruments. All measurements have V_{AC} = 0.3 V, f_0 = 3 kHz and LIA bandwidth of 1 Hz. The upper curve is the results obtained with a commercial LIA (HF2LI by Zurich Instruments) and shows the impossibility to sense ppm variation being the instrumental limit of 39 ppm. The custom ELIA instrument has a resolution of 9 ppm when used as a standard lock-in amplifier without the ratiometric technique. The resolution is boosted to 0.6 ppm by the switching ratiometric technique (switching frequency of 1 kHz) allowing a clear detection of the DUT variations. The ELIA instrument provides such sub-ppm resolution for frequencies up to 6 MHz without any dependence by the phase between the DUT and REF signals. Therefore, the ELIA can be used to enhance the resolution of the impedance measurements without calibration neither setup changes.

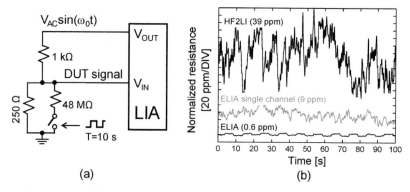

(a)   (b)

Figure 6.15 (a) Experimental setup to produce well-controlled DUT variation of 5 ppm; (b) assessment of the resolution capability of ELIA (adapted from [40]).

6.7 Summary

In the quest for higher sensitivity in capacitance and impedance spectroscopy, it is necessary to reduce the noise added by the instrumentation. In this chapter, we have shown that the design of custom amplifiers can reduce the equivalent input noise at extremely low levels. Such result is obtained by following two main strategies. First, the amplification at the frequency of measurement is operated by capacitors in a suitable feedback architecture. Small value capacitors offer a lower noise compared to resistors and have an excellent frequency behavior with negligible parasitic effects up to hundreds of MHz. Secondly, the capacitance at the input node of the instrument should be minimized for an effective reduction of the high-frequency noise. Both the instrument and experimental setup must be carefully designed to reach capacitances of a few pF at the input node. To achieve this goal, the realization of the front-end amplifier as a custom integrated circuit is particularly advantageous. The capacitance of the integrated amplifier can be tailored to maximize the performance of the experiment and the small size of the silicon chip allows to lodge it very near to the sample under investigation minimizing the stray capacitance of the connections.

In the second part of the chapter (Sections 6.4 and 6.5), we have shown that to fully profit of a low noise front-end amplifier is important the optimization of all parts of the experiment: signal generator, device under test, connections, amplifiers and the conversions between analog and digital domains. In particular, very high-resolution measurements in the range of a few ppm require a special care for the reduction of slow gain fluctuations of the analog and analog/digital components. A differential approach or an enhanced lock-in amplifier architecture specifically conceived for high-resolution measurements are useful to cope with gain fluctuations of fast digital lock-in amplifiers.

Acknowledgments

The authors would like to acknowledge M. Sampietro for the constant support and many fruitful discussions. D. Bianchi, P. Ciccarella, and G. Gervasoni are acknowledged for contributing to the development of the circuits and techniques reported in this chapter. This work was supported by Fondazione Cariplo through the project DRINK-ABLE (No. 2014-1285), and by the European Union under grant agreement no. 688172 (H2020-ICT-STREAMS project).

References

1. Holder, D. S. (2004). Electrical impedance tomography: Methods, history and applications. *CRC Press* pp. 456 doi:10.1118/1.1995712.

2. Yang, W. Q., and York, T. A. (1999). New AC-based capacitance tomography system. *IEE Proc.–Sci. Meas. Technol.* **146**, p. 47.

3. Gray, P. R., Hurst, P. J., Lewis, S. H., and Meyer, R. G. (2009). *Analysis and Design of Analog Integrated Circuits.* 5th ed. (John Wiley & Sons, USA).

4. Horowitz, P., and Hill, W. (1989). *The Art of Electronics.* (Cambridge University).

5. Leach, W. M. (1994). Fundamentals of low-noise analog circuit design. *Proc. IEEE* **82**, pp. 1515–1538.

6. Sherman-Gold, R. (2008). *The Axon Guide for Electrophysiology and Biophysics Laboratory Techniques.* (Axon Instruments).

7. Ferrari, G., and Sampietro, M. (2007). Wide bandwidth transimpedance amplifier for extremely high sensitivity continuous measurements. *Rev. Sci. Instrum.* **78**, p. 94703.

8. Ciofi, C., Crupi, F., Pace, C., Scandurra, G., and Patanè, M. (2007). A new circuit topology for the realization of very low-noise wide-bandwidth transimpedance amplifier. *Instrum. Meas. IEEE Trans.* **56**, pp. 1626–1631.

9. Ferrari, G., Gozzini, F., Molari, A., and Sampietro, M. (2009). Transimpedance amplifier for high sensitivity current measurements on nanodevices. *IEEE J. Solid-State Circuits* **44**, pp. 1609–1616.

10. Goldstein, B., Kim, D., Xu, J., Vanderlick, T. K., and Culurciello, E. (2012). CMOS low current measurement system for biomedical applications. *IEEE Trans. Biomed. Circuits Syst.* **6**, pp. 111–119.

11. Rosenstein, J. K., Wanunu, M., Merchant, C. A., Drndic, M., and Shepard, K. L. (2012). Integrated nanopore sensing platform with sub-microsecond temporal resolution. *Nat. Methods* **9**, pp. 487–492.

12. Bennati, M., et al. (2009). A sub-pA $\Delta\Sigma$ current amplifier for single-molecule nanosensors. *Dig. Tech. Pap.–IEEE Int. Solid-State Circuits Conf.* pp. 348–350 doi:10.1109/ISSCC.2009.4977451.

13. Crescentini, M., Bennati, M., Carminati, M., and Tartagni, M. (2014). Noise limits of CMOS current interfaces for biosensors: A review. *IEEE Trans. Biomed. Circuits Syst.* **8**, pp. 278–292.

14. Gore, A., Chakrabartty, S., Pal, S., and Alocilja, E. C. (2006). A multichannel femtoampere-sensitivity potentiostat array for biosensing applications. *IEEE Trans. Circuits Syst. I Regul. Pap.* **53**, pp. 2357–2363.

15. Fumagalli, L., et al. (2006). Nanoscale capacitance imaging with attofarad resolution using ac current sensing atomic force microscopy. *Nanotechnology* **17**, pp. 4581–4587.

16. Ciccarella, P., Carminati, M., Sampietro, M., and Ferrari, G. (2016). Multichannel 65 zF rms resolution CMOS monolithic capacitive sensor for counting single micrometer-sized airborne particles on chip. *IEEE J. Solid-State Circuits* **51**, pp. 2545–2553.

17. Laborde, C., et al. (2015). Real-time imaging of microparticles and living cells with CMOS nanocapacitor arrays. *Nat. Nanotechnol.* **10**, pp. 791–795.

18. Bakhshiani, M., Suster, M. A., and Mohseni, P. (2014). A broadband sensor interface IC for miniaturized dielectric spectroscopy from MHz to GHz. *IEEE J. Solid-State Circuits* **49**, pp. 1669–1681.

19. Ferrari, G., Bianchi, D., Rottigni, A., and Sampietro, M. (2014). CMOS impedance analyzer for nanosamples investigation operating up to 150 MHz with Sub-aF resolution. in *2014 IEEE International*

Solid-State Circuits Conference Digest of Technical Papers (ISSCC) pp. 292–293. doi:10.1109/ISSCC.2014.6757439.

20. Antonio, D., Pastoriza, H., Julián, P., and Mandolesi, P. (2008). Cryogenic transimpedance amplifier for micromechanical capacitive sensors. *Rev. Sci. Instrum.* **79**, p. 84703.

21. Guagliardo, F., and Ferrari, G. (2013). in *Single Atom Nanoelectronics*, eds. Prati, E., and Shinada, T., pp. 187–210 (Pan Stanford Publishing).

22. Tagliaferri, M. L. V., et al. (2016). Modular printed circuit boards for broadband characterization of nanoelectronic quantum devices. *IEEE Trans. Instrum. Meas.* **65**, pp. 1827–1835.

23. Caves, J. T., Rosenbaum, S. D., Copeland, M. A., and Rahim, C. F. (1977). Sampled analog filtering using switched capacitors as resistor equivalents. *IEEE J. Solid-State Circuits* **12**, pp. 592–599.

24. Harrison, R., and Charles, C. (2003). A low-power low-noise CMOS amplifier for neural recording applications. *IEEE J. Solid-State Circuits* **38**, pp. 958–965.

25. Gozzini, F., Ferrari, G., and Sampietro, M. (2006). Linear transconductor with rail-to-rail input swing for very large time constant applications. *Electron. Lett.* **42**, p. 1069.

26. Ferrari, G., Farina, M., Guagliardo, F., Carminati, M., and Sampietro, M. (2009). Ultra-low-noise CMOS current preamplifier from DC to 1 MHz. *Electron. Lett.* **45**, p. 1278.

27. Carminati, M., Ferrari, G., Bianchi, D., and Sampietro, M. (2013). Femtoampere integrated current preamplifier for low noise and wide bandwidth electrochemistry with nanoelectrodes. *Electrochim. Acta* **112**, pp. 950–956.

28. Carminati, M., Ferrari, G., Guagliardo, F., and Sampietro, M. (2011). ZeptoFarad capacitance detection with a miniaturized CMOS current front-end for nanoscale sensors. *Sens. Actuators A Phys.* **172**, pp. 117–123.

29. Morichetti, F., et al. (2014). Non-invasive on-chip light observation by contactless waveguide conductivity monitoring. *IEEE J. Sel. Top. Quantum Electron.* **20**, p. 8201710.

30. Bianchi, D., Ferrari, G., Rottigni, A., and Sampietro, M. (2014). CMOS impedance analyzer for nanosamples investigation operating up to 150 MHz with sub-aF resolution. *IEEE J. Solid-State Circuits* **49**, pp. 2748–2757.

31. Agilent Technologies (2013). *Impedance Measurement Handbook: A Guide to Measurement Technology and Techniques.* (4th edition).

32. Akiba, M. (1997). 1/f dielectric polarization noise in silicon P-N junctions. *Appl. Phys. Lett.* **71**, p. 3236.

33. Clément, N., Nishiguchi, K., Fujiwara, A., and Vuillaume, D. (2011). Evaluation of a gate capacitance in the sub-aF range for a chemical field-effect transistor with a Si nanowire channel. *IEEE Trans. Nanotechnol.* **10**, pp. 1172–1179.

34. Uram, J. D., Ke, K., and Mayer, M. (2008). Noise and bandwidth of current recordings from submicrometer pores and nanopores. *ACS Nano* **2**, pp. 857–872.

35. Carminati, M., et al. (2012). Accuracy and resolution limits in quartz and silicon substrates with microelectrodes for electrochemical biosensors. *Sens. Actuators B Chem.* **174**, pp. 168–175.

36. Gervasoni, G., et al. (2014). A 12-channel dual-lock-in platform for magneto-resistive DNA detection with ppm resolution. in *2014 IEEE Biomedical Circuits and Systems Conference (BioCAS) Proceedings* pp. 316–319. doi:10.1109/BioCAS.2014.6981726.

37. Carminati, M., Gervasoni, G., Sampietro, M., and Ferrari, G. (2016). Note: Differential configurations for the mitigation of slow fluctuations limiting the resolution of digital lock-in amplifiers. *Rev. Sci. Instrum.* **87**, p. 26102.

38. Barlian, A. A., Park, W.-T., Mallon, J. R., Rastegar, A. J., and Pruitt, B. L. (2009). Review: Semiconductor piezoresistance for microsystems. *Proc. IEEE* **97**, pp. 513–552.

39. Benmessaoud, M., and Nasreddine, M. M. (2013). Optimization of MEMS capacitive accelerometer. *Microsyst. Technol.* **19**, pp. 713–720.

40. Gervasoni, G., Carminati, M., and Ferrari, G. (2017). Switched ratiometric lock-in amplifier enabling sub-ppm measurements in a wide frequency range. *Rev. Sci. Instrum.* **88**(10), p. 104704.

Chapter 7

Time Domain–Based Impedance Detection

Uwe Pliquett

Institut für Bioprozess- und Analysenmesstechnik e.V.,
Heilbad Heiligenstadt, D-37308, Germany

uwe.pliquett@iba-heiligenstadt.de

The electrical impedance is useful for material characterization. Especially, process relevant material characterization requires fast but precise measurement. Particularly, for multichannel systems or distributed systems, probably powered with harvested energy, minimal resources are available. The assessment of the electrical impedance based on measurements in time domain is a fast and robust technique. Post-processing of time functions as an answer to a broad bandwidth stimulus like step function yield either impedance spectra or material characteristics like relaxation times.

Capacitance Spectroscopy of Semiconductors
Edited by Jian V. Li and Giorgio Ferrari
Copyright © 2018 Pan Stanford Publishing Pte. Ltd.
ISBN 978-981-4774-54-3 (Hardcover), 978-1-315-15013-0 (eBook)
www.panstanford.com

7.1 Introduction

Dielectric measurements are widely used for material characterization in terms of storage capabilities of electric energy [1, 2]. The physical quantity directly measured between the clamps of the instrumentation is the capacity. Basic processing with elimination of the electrode geometry yields the permittivity. However, although calibration yields practically useful results, for deeper insight into material behavior, model based assessment of molecular properties like polarization or dipole moment is required.

The directly assessable quantity for the energy storage capabilities, the permittivity, is a complex number due to the dielectric loss.

$$\underline{\varepsilon} = \varepsilon' + j\varepsilon'' = \varepsilon' + \frac{\sigma'}{j\omega} \tag{7.1}$$

where σ' is the real part of the electrical conductivity and ω is the angular frequency $\omega = 2\pi f$. Other than the conservative real part of the permittivity, the conductivity is associated with dissipative behavior, e.g., electrical energy is converted into other forms of energy like for instance heat.

The complex conductivity is related to the permittivity by

$$\underline{\sigma} = j\omega\underline{\varepsilon}_r\varepsilon_0. \tag{7.2}$$

The result of a measurement is always related to the geometry of the electrode system and yields the admittance, $\underline{Y} = G + jB$, rather than the conductivity. The real part G is the conductance while B is the susceptance. In the simple case of plan parallel electrodes

$$\underline{Y} = \underline{\sigma}A/d = \underline{\sigma}/k, \tag{7.3}$$

where k is the geometry factor with the usual unit cm^{-1}. Since most electrode systems are more complicate, k is determined by calibration using material of known conductivity. In electrochemistry potassium chloride (e.g., 100 mM KCl at 25°C) is regarded as standard calibration electrolyte.

The impedance, $\underline{Z}$, is the complex resistance and simply the reciprocal of the admittance

$$\underline{Z} = 1/\underline{Y} = R + jX, \tag{7.4}$$

where R is the real part of the impedance and X the imaginary part [3] also termed reactance. It should be noted that the real part is associated with energy dissipation while the imaginary part is conservative. This is the opposite of the permittivity, where the real part is conservative.

Capacitive elements yield negative X while positive X are associated with inductivity. While Ohmic resistances are frequency independent, reactive parts (capacitors and inductors) exhibit frequency-dependent resistance.

$$X_{\text{C}} = \frac{1}{\omega C} = \frac{k}{\omega \varepsilon'_r \varepsilon_0} \quad \text{and} \quad X_{\text{L}} = \omega L \tag{7.5}$$

In electrochemistry and in material science, the inductive behavior is commonly negligible for frequencies lower than hundreds of MHz. Therefore, inductive behavior is not further considered.

Due to the reactive part, the impedance is frequency dependent. Most practical applications benefit from spectroscopic information rather than the sole measurement at dc or a single frequency.

Since impedance spectroscopy relates to frequency domain, time domain–based measurements need further processing to extract spectroscopic information that in the case of time-invariant and electrically linear systems yields the same information.

Alternatively, direct processing in time domain is a promising approach, especially if resources are limited. The resulting physical quantities like relaxation time and relaxation strength correspond directly to quantities in frequency domain like characteristic frequency and circumference of frequency dispersion.

The great advantage of measurements in time domain is the short measurement time making this approach suitable for fast measurements, which is especially useful for time-varying impedances. Having small excitation amplitude, dynamic impedance, also termed small signal impedance or incremental impedance, can be measured at multiple voltage offsets.

This chapter describes the assessment of electrical impedance based on measurement in time domain.

7.2 Frequency Dispersion and Relaxation

Electrical impedance measurement is a method for nondestructive and label-free characterization of any kind of material. Spectral information is useful for characterization of frequency dependent polarization behavior. At low frequency, lateral movement of charge carriers can polarize for instance large supra-molecular structures in electrolytes. Interfacial polarization is much faster with characteristic frequencies in the RF-range. Debye polarization, which is the orientation of dipoles in an outer electric field, is much faster and is mostly found in GHz-range. Electronic polarization predominates in the visible frequency range (THz) but is behind the scope of this chapter.

7.2.1 Electrical Properties in Frequency Domain

A key feature of most material is the considerable drop in impedance magnitude by several orders of magnitude over a broad frequency range. A region with pronounced frequency dependence of the impedance is called frequency dispersion [4]. If the underlying process relates to a single polarization mechanism, it can be characterized by the circumference of the dispersion, e.g., the difference of the real part of the impedance at low and high frequency. The second characteristic feature is the characteristic frequency, which is the frequency where the imaginary part approaches a minimum.

The representation in frequency domain is either the Bode- or the Nyquist diagram.

Although not always advantageous, an equivalent circuit can help to understand the impedance spectrum [5]. A simple circuit for the spectrum shown in Fig. 7.1, at this point independent of the physical background, would be a lossy capacitor (parallel circuit of capacitor and resistor) in series to a second resistor (Fig. 7.2) [6]. The values of the elements for the spectrum in Fig. 7.1 are as follows: $C_p = 1$ nF, $R_p = 10\ \Omega$, $R_b = 30\ \Omega$.

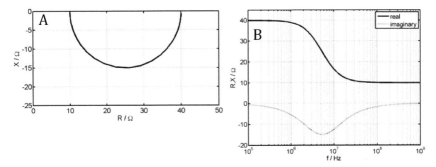

Figure 7.1 Representation of impedance as (A) Nyquist diagram and (B) as Bode plot.

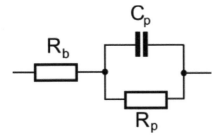

Figure 7.2 Equivalent circuit for a polarization process (C_p, R_p) in series to a bulk resistor R_b. In this case, the polarization process is modeled as lossy capacitor.

The impedance $\underline{Z}$ of the circuit in Fig. 7.2 is

$$\underline{Z} = R_b + \frac{\dfrac{R_p}{j\omega C_p}}{R_p + \dfrac{1}{j\omega C_p}} = \frac{R_b + R_p + j\omega C_p R_p R_b}{1 + j\omega C_p R_p} \tag{7.6}$$

In order to find the real and the imaginary part, the equation is expanded by the conjugate of the denominator.

$$\underline{Z} = \frac{R_b + R_p + j\omega C_p R_p R_b}{1 + j\omega C_p R_p} \cdot \frac{(1 - j\omega C_p R_p)}{(1 - j\omega C_p R_p)} = \frac{R_b + R_p + \omega^2 C_p^2 R_p^2 R_b - j\omega C_p R_p^2}{1 + \omega^2 C_p^2 R_p^2} \tag{7.7}$$

The real part is

$$Z' = \frac{R_b + R_p + \omega^2 C_p^2 R_p^2 R_b}{1 + \omega^2 C_p^2 R_p^2} \tag{7.8}$$

and the imaginary part

$$Z'' = \frac{-\omega C_p R_p^2}{1 + \omega^2 C_p^2 R_p^2} \tag{7.9}$$

The characteristic behavior of the polarization is described by the circumference of the frequency dispersion that is given by the difference between the real part of the impedance at high and low frequencies:

$$\Delta Z_p' = \Delta Z_{\omega=0}' - \Delta Z_{\omega \to \infty}' = R_b + R_p + R_b = R_p \tag{7.10}$$

The characteristic frequency $f_p = \omega_p/2\pi$ is the frequency where the imaginary part reaches its minimum. In order to find this, Z'' is differentiated with respect to ω and the zero point calculated.

$$\frac{dZ''}{d\omega} = \frac{-C_p R_p^2 (1 + \omega^2 C_p^2 R_p^2) + \omega C_p R_p^2 (2\omega C_p^2 R_p^2)}{(1 + \omega^2 C_p^2 R_p^2)^2} = \frac{-C_p R_p^2 + \omega^2 C_p^3 R_p^4}{(1 + \omega^2 C_p^2 R_p^2)^2} = 0 \tag{7.11}$$

It follows that $f_p = \dfrac{1}{2\pi C_p R_p}$ is the characteristic frequency of the polarization.

7.2.2 Electrical Properties in Time Domain

When measuring the electrical properties in frequency domain, the system should be in steady state. In time domain, a system is excited with a transient function and the relaxation into a new state is observed. In this case, the frequency, also regarded as differential operator, becomes complex

$$\frac{d}{dt} \circ\!\!-\!\!\bullet\ s = \sigma + j\omega, \tag{7.12}$$

where σ is a real constant[1]. σ is a decay factor of the exponential function e^{st}. The imaginary part of s represents the real frequency.

The response of the system can be calculated by convolution of the excitation signal and a proper weight function. The latter is the equivalent in the time domain of the impedance in the frequency domain. The weight function can be found by inverse Laplacian transformation of the impedance multiplied with the Laplace transformed Dirac function $g(t) = \mathcal{L}^{-1}\{Z \cdot \mathcal{L}[\delta(t)]\}$.

Since the Dirac function $\delta(t)$ corresponds to 1 in frequency domain the weight function as a function of frequency $G(s)$ is simply

$$G(s) = 1 \cdot \underline{Z} = \frac{R_b + R_p + sC_pR_pR_b}{1 + sC_pR_p} \tag{7.13}$$

The transformation into time domain is possible by inverse Laplace transformation using a correspondence table or the residue theorem. Therefore, it is advantageous to rewrite Eq. 7.13 as

$$G(s) = \frac{1}{C_pR_p}\left((R_b + R_p)\frac{1}{\frac{1}{C_pR_p} + s} + C_pR_pR_b \frac{s}{\frac{1}{C_pR_p} + s} \right) \tag{7.14}$$

The correspondence table for the both s-terms is shown in Table 7.1.

Table 7.1 Correspondence table for the terms used in Eq. 7.14

$F(s)$ (frequency domain)	$f(t)$ (time domain)
$\dfrac{1}{a+s}$	e^{-at}
$\dfrac{s}{a+s}$	$-ae^{-at}$

---

[1]It should be noted that the constant σ is not related to the complex conductivity σ defined in Eq. (7.2).

This yields the weight function $g(t)$ for $t > 0$. Note that the stimulus is a current with a Dirac function behavior. Therefore the current flowing in R_b is zero for $t > 0$ and no voltage appears across R_b. Therefore, the resistor R_b does not appear in the weight function.

$$g(t) = \frac{1}{C_p R_p}\left(R_b + R_p - \frac{C_p R_p R_b}{C_p R_p}\right)e^{-\frac{t}{C_p R_p}} = \frac{1}{C_p}e^{-\frac{t}{\tau}}, \tag{7.15}$$

where τ is the time constant of the relaxation. The response of an LTI (linear and time-invariant) system to a step function (integral of the Dirac function) is the unit-step response $h(t)$

An elegant way for inverse Laplace transformation uses the residue theorem. The time function is simply the sum of all residues (Res) which is for single poles:

$$g(t) = \sum_{i=1}^{N}\underset{S \to S_i}{\mathrm{Res}}(G(s)) = \lim_{S \to S_i}\sum_{i=1}^{N}G(s)(s - s_i)e^{s_i t}, \tag{7.16}$$

where N is the number of poles. Applying the residue theorem to

$$G(s) = \frac{1}{C_p R_p}\left(\frac{R_b + R_p + sC_p R_p R_b}{\frac{1}{C_p R_p} + s}\right) \tag{7.17}$$

with the sole zero point of the nominator (pole) $s_1 = -\dfrac{1}{C_p R_p}$-yields

$$g(t) = \frac{1}{C_p R_p}\left(R_b + R_p - \frac{C_p R_p R_b}{C_p R_p}\right)e^{-\frac{t}{C_p R_p}} = \frac{1}{C_p}e^{-\frac{t}{\tau}} \tag{7.18}$$

which is exactly the same solution as in Eq. 7.15.

The meaning of the weight function is a function of time after application of a Dirac function. With respect to Ohm's law, $U = IR$, a current pulse with infinite height but an area $\int i\,dt = 1As = 1\,C$ is applied which yields the voltage across the capacitor C_p of 1 As/C_p and an exponentially decaying curve (Fig. 7.3) where the decay constant equals τ.

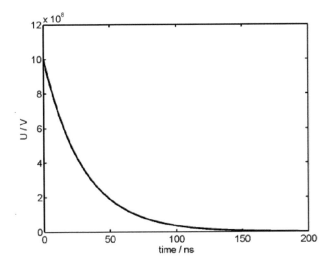

Figure 7.3 Voltage as function of time after a current impulse (Dirac function) with 1 As of delivered charge for the circuit shown in Fig. 7.2. The characteristic parameters are the time constant and the relaxation strength (amplitude). Note the high voltage which is due to the charge of 1 C at a capacity of 10 nF!

7.3 Comparison between Time and Frequency Domain

The information contained in the representation in both domains is exactly the same for a LTI system. While the material is described by impedance as function of frequency, the corresponding measure in time domain is the weight function. The characteristic properties are the circumference of the dispersion in frequency domain and the relaxation strength in time domain. The time constant is directly related to the characteristic frequency by $f_p = \frac{1}{2\pi\tau_p}$. Figure 7.4 shows different presentations for three well-distinguished polarization mechanisms. The polarizations are ideal with discrete lines in the time spectrum. This corresponds to exact half circles in frequency domain. Most materials, however, show a distribution of time constants which broadens the lines in the time spectrum and depresses the semicircle in the Nyquist diagram.

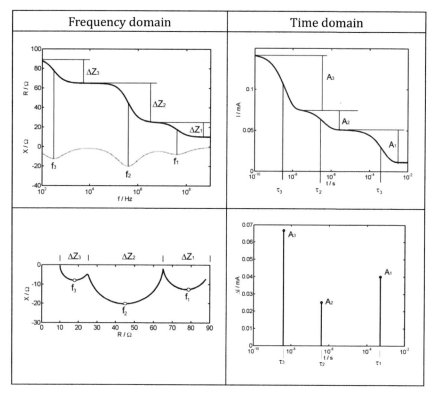

Figure 7.4 Representation of passive electrical properties in frequency and time domain for three well-distinguished polarization mechanisms. In frequency domain, either the Bode- or the Nyquist diagram is used while in time domain either a time function (here the current response to a voltage step) or the spectrum of time constants represents the material properties. The figures are created in Matlab using three serial combinations RC ($R_x C_x$) in parallel. A further parallel resistor R_p is responsible for the dc-part. The values of the elements are: R_p = 10 Ω, R_1 = 15 Ω, C_1 = 260 pF, R_2 = 40 Ω, C_2 = 10 nF, R_3 = 25 Ω, C_3 = 20 µF.

7.4 Assessment of Impedance in Time Domain

The application of Dirac functions would be favorable from theoretical point of view but creates major problems for practical applications. A transient function, much simpler to generate, is the step function $s(t)$. The product of the impedance and the Laplace transformed step function

$$\mathcal{L}(s(t)) = \frac{1}{s} \qquad (7.19)$$

yields the transfer function while the result of the inverse Laplace transformation is the step answer $h(t)$. Practically, a current step is applied while the voltage across the material is monitored (or vice versa). The voltage in frequency domain due to a current step at the circuit in Fig. 7.2. is

$$U(s) = \frac{I_0}{s} Z(s) = \frac{I_0}{s} \cdot \frac{R_b R_p + s C_p R_p R_b}{1 + s C_p R_p} = \frac{I_0}{C_p R_p} \left(\frac{R_b + R_p + s C_p R_p R_b}{s \left(\frac{1}{C_p R_p} + s \right)} \right) \qquad (7.20)$$

The time function is

$$U(t) = \frac{I_0}{C_p R_p} \left(\frac{R_b + R_p}{\frac{1}{C_p R_p}} e^{0 \cdot <} + \frac{R_b + R_p - \frac{1}{C_p R_p} \cdot C_p R_p R_b}{-\frac{1}{C_p R_p}} e^{\frac{t}{C_p R_p}} \right) \qquad (7.21)$$

$$= I_0 \left(R_b + R_p - R_p e^{\frac{t}{\tau}} \right)$$

When comparing the characteristic measures in time and frequency domain, it is obvious that the time constant $\tau_p = C_p R_p$ is the reciprocal of the characteristic angular frequency ω_p or related to the characteristic frequency f_p as

$$f_p = \frac{1}{2\pi\tau_p} = \frac{1}{2\pi C_p R_p}, \qquad (7.22)$$

which was derived previously from the impedance.

The relaxation strength is the amplitude at $t = 0$ and is $I_0 R_p$. The impedance

$$\underline{Z} = \frac{R_b + R_p + j\omega C_p R_p R_b}{1 + j\omega C_p R_p} \qquad (7.23)$$

at dc is $R_0 = R_p + R_b$ while it is just $R_\alpha = R_b$ when the frequency approaches infinite which yields a difference $\Delta Z_p = R_p$. Obviously, it is besides the current amplitude the same as in time domain.

7.4.1 Current and Voltage Excitation (Galvanostatic and Potentiostatic Measurement)

Potentiostatic (i.e., forcing the voltage) or galvanostatic (i.e., forcing the current) excitation does not yield any difference for pure resistors. However, the measured time constant (or characteristic frequency) for complex impedances depends on the source impedance of the instrumentation. In the example above, galvanostatic excitation was used. It can be shown that both, relaxation strength and time constant is different in potentiostatic regime. In this case, the current is the quotient of voltage and impedance:

$$I(s) = \frac{U_0}{sZ(s)} = \frac{U_0}{s} \cdot \frac{1 + sC_p R_p}{R_b + R_p + sC_p R_p R_b} = \frac{U_0}{C_p R_p R_b} \left(\frac{1 + sC_p R_p}{s\left(\dfrac{R_b + R_p}{C_p R_p R_b} + s \right)} \right) \quad (7.24)$$

The poles are the zero points of the denominator:

$$s_1 = 0 \quad \text{and} \quad s_2 = -\frac{R_b + R_p}{C_p R_p R_b}$$

This yields the time function

$$I(t) = U_0 \left(\frac{1}{R_b + R_p} + \frac{1 - \dfrac{R_b + R_p}{R_b}}{-(R_b + R_p)} e^{-\frac{t(R_b + R_p)}{C_p R_p R_b}} \right) = U_0 \left(\frac{1}{R_b + R_p} + \frac{R_p}{R_b(R_b + R_p)} e^{-\frac{t}{\tau}} \right)$$

$$(7.25)$$

The relaxation strength is finally $I(0) - I(\infty) = \dfrac{R_p}{R_b(R_b + R_p)}$ and the time constant is $\tau = \dfrac{C_p R_b R_p}{R_b + R_p}$.

While a current source with high internal impedance leaves the circuit open, it is shortened by connecting a voltage source with low internal impedance. Therefore, the time constant is the product of the capacitor and both resistors in parallel. This can be found for the electrical properties in frequency domain as well when calculating the characteristic frequency for the same object from impedance Z and admittance $Y = 1/Z$.

7.4.2 Periodic Signals

The previous examples of transformation use only a single step or Dirac function. If a square wave is applied, it should be regarded as infinite sequence of equally spaced steps with alternating polarity [7]. The result is the sum of all step responses with time delay. If the nth step response $g(t, n)$ is considered, it should become

$$g(t, n) = \sum_{i=-\infty}^{n} -1^n g(t - nt). \qquad (7.26)$$

-1^n creates the alternating polarity and T is the time delay between two steps. Practically, about 10 periods are sufficient. The calculation of the steady periodic signal becomes simple if the period is chosen sufficiently long ensuring a full decay to zero before starting the next period. The advantage of this approach is the possibility of employing efficient algorithms like fast Fourier transformation (FFT, see below) for signal processing. Moreover, if measurement time is not critical, averaging over several periods can tremendously increase signal quality due to suppression of uncorrelated noise.

7.4.3 Transformation from Time- to Frequency Domain

Although processing in time domain is quite sufficient, the presentation of the impedance spectrum is often favored. The typical approach is the measurement of the response to a broad bandwidth signal with subsequent transformation into frequency domain.

Transient functions such as Dirac functions or single steps are transformed into frequency domain by Laplace transformation using the complex frequency $s = \sigma + j\omega$.

$$F(s) = \mathcal{L}(f(t)) = \int_{0}^{\infty} f(t)e^{-st}dt \qquad (7.27)$$

This, however, is only possible for analytical functions. Sampled signals are represented as vector $f(kT_s)$, $k = 1...N$, where T_s is the sampling interval and N the number of elements. For such signals, z-transformation which can be regarded as discrete Laplace transformation is required.

$$F(z) = Z(f(t)) = \sum_{k=1}^{N} f(k)z^{-k}$$ (7.28)

A popular algorithm for solving this equation is the Bluestein–algorithm which requires extensive computing power for large vectors.

If the time function is periodic and the system in steady state, s reduces to $j\omega$. In this case, the simpler Fourier transformation is adequate.

$$F(j\omega) = \mathcal{F}(f(t)) = \frac{1}{\sqrt{2\pi}} \int_{-\infty}^{\infty} f(t)e^{-j\omega t} dt$$ (7.29)

As in case of transients, sampled signals need a discrete transformation which is termed DFT (discrete Fourier transformation). This results in a complex coefficient, $c_n = a_n + jb_n$, for each frequency line (harmonic frequency) where a_n are the real coefficients and b_n are imaginary.

$$c_n = \frac{1}{N} \sum_{i=0}^{N-1} f(i)e^{-j\frac{2\pi n_i}{N}},$$ (7.30)

with $n = 0 ... N - 1$. N is the length of the vector and i is the index of the discrete vector f. The value $2\pi/N$ is the fundamental frequency and $2\pi n/N$ are the harmonics.

The fastest algorithm for discrete Fourier transformation is the fast Fourier transformation (FFT) which is always applicable if the vector has a length of power of two. In practice, zero padding is used in order to trim input vectors for taking advantage of this fast algorithm.

Sampling of a function not only cuts the upper frequency (half of the sample frequency) but also comprises the accuracy of the Fourier coefficients (Fig. 7.5). If the sample intervals reach zero, discrete Fourier transformation approaches the performance of the analytical solution.

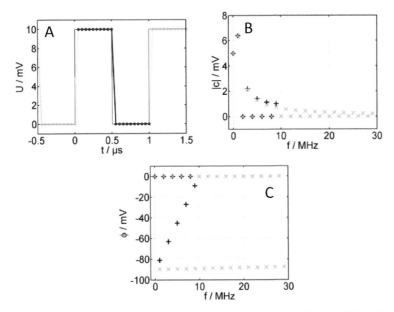

Figure 7.5 (A) Sampled rectangular wave with low sample rate (dotted), (B) magnitude and (C) the phase of the Fourier transformed signal (one period used, bold black in A). The thin black line shows a square wave sampled with high resolution. The (+) in (B) and (C) are from low sample rate while the (x) arise from the densely sampled signal. Note the marked difference in the phase but also the uncertainties in the magnitude.

The 0th coefficient is the dc-component and the first line is the fundamental frequency (1 MHz). Due to symmetry, all even harmonics are zero. Especially time functions with strong transients (Dirac function, step) are very sensitive to sampling. Low sampling rate always compromises signal quality.

Impedance measurements based on assessment of functions in time are done by application of a broad bandwidth stimulus to the material under test (MUT) in either potentiostatic or galvanostatic configuration. The basic frontend circuits are shown in Fig. 7.6.

The MUT in Fig. 7.6A is excited between the amplifier with low output impedance and the transimpedance amplifier with a virtual ground as input. The voltage monitor can use either a separate pair of electrodes (tetrapolar interface) or is connected to the current electrodes [8]. The output of the transimpedance is used as current monitor. The core element of the galvanostatic frontend in Fig. 7.6B is the controlled current source which injects the current into the MUT. In order to keep signal distortion minimal, current and voltage monitor use the same circuitry. The voltage monitor directly measures the voltage drop across the MUT whereas the current monitor uses a sensing resistor R with a differential amplifier.

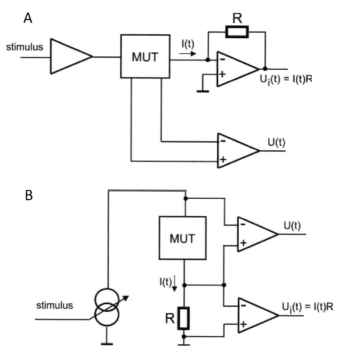

Figure 7.6 Principle frontend circuitry for (A) potentiostatic and (B) galvanostatic impedance measurement.

The typical approach is digitizing both voltage and current and transformation into the frequency domain. The impedance is then calculated by

$$Z(\omega) = \frac{\mathcal{F}(U)}{\mathcal{F}(I)}.$$ (7.31)

Due to the symmetry of rectangular waves, each even harmonic should be removed, since a division by zero or at least near zero in real signals yields non-interpretable results. Moreover, the spectrum should be cut at half of the sampling rate since calculations of harmonics higher than half of the sampling frequency are meaningless.

Repeated sequences of broad bandwidth signals are suitable for very fast impedance detection with high repetition rate. The fastest approach uses exactly one complete period. Although for potentiostatic excitation only the current and for galvanostatic measurements only the voltage should be measured, monitoring both is recommended for precision measurements.

Especially for square wave functions with the low energy at the higher harmonics, the spectrum becomes noisy at high frequency (higher than ≈100th harmonic frequency) and further processing is required for obtaining suitable results. If the time for averaging over several periods is not available [9], averaging over increasing intervals within one period is a suitable way for noise suppression. This is mathematically correct for the usually dominating stochastic noise. Practically, the impedance at odd harmonics within the desired frequency range is averaged. Although not required, logarithmic spacing is chosen due to its simple algorithm. $Z(j\omega)$ spectrum calculated in this way coincides well with precise measurements in frequency domain. Noticeably, the practical approach of step sinus measurements (frequency domain) bases often on logarithmic spacing of the frequency vector, especially for broad frequency ranges over several decades.

For averaging, a logarithmic vector over the desired frequency range is created and the mean of the impedance of all lines between the elements of the frequency vector is calculated (Fig. 7.7). Since the frequencies corresponding to the averaged values differ from those in the frequency vector, a new one is created as the means between two adjacent frequencies.

Because the spacing of the frequencies in the low frequency region is rather large with respect to the frequency and the noise is acceptable, usually no averaging is used up to the 100th harmonic frequency.

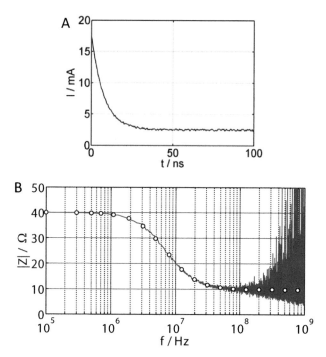

Figure 7.7 (A) Response of the circuit in Fig. 7.2 to a voltage step of 100 mV without averaging. For the measurement, a square wave without offset and a voltage of 200 mV was used. (B) magnitude of the impedance calculated from one complete period of the response (current) and the applied voltage square wave without averaging (line) and with logarithmic averaging (circles).

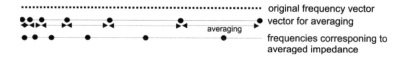

Figure 7.8 Schematic for logarithmic averaging.

7.5 Excitation Signals for Time Domain Measurements

Although any broad bandwidth signal would be suitable for measurements in time domain [10, 11], only few signals with special features are commonly in use (Table 7.2).

Table 7.2 Broad band width excitation signals

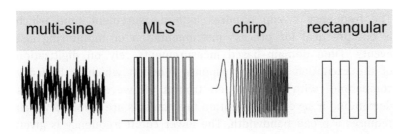

Today, multi-sine excitation (i.e., a signal obtained by summing a finite number of sine waves at different frequency) is very popular in impedance measurement practice since the energy of each frequency component can be tailored individually making this excitation signal favorable for applications with equidistant sampling. Chirp signals have a flat frequency spectrum over a wide frequency range with fast falloff at high and low frequencies. This makes them interesting for applications where with lowest stimulation energy the best signal to noise ratio is needed. Moreover, this avoids energy falling out of the frequency range of interest which is important for application to living matter like for instance monitoring heart function with implanted electrodes [12]. While the analog signals need a digital-to-analog converter (DAC) for creation, digital signals like square wave or maximum length sequence (MLS, also termed pseudo random noise or M-sequence) are generated by switching between two values. MLS is a series of alternating, non-equidistant steps where all lengths of high and low levels are present [13]. The great advantage of MLS is their simple generation using a shift register with feedback and noise suppression due to synchronous sampling of excitation and response signal. The advantage of steps and square wave functions is the ease of generation and excellent suitability for relaxation measurements. They are less suited for applications with uniform sampling because of the low energy at high frequency and therefore high noise. To solve this problem, averaging over increasingly more lines with higher frequency such as logarithmic averaging is recommended.

7.5.1 Huge Data Volume for Continuous Measurements

The frequency sweep requires long measurement time which is not practical for process instrumentation or monitoring fast events. Time domain–based measurements rely on digitization of the excitation and the response function which is basically compatible with modern electronics. However, it becomes demanding if several polarization mechanisms are involved which requires a broad bandwidth. The lower cutoff frequency is given by the duration of the excitation signal or in the case of periodic signals by the period. This means, if the lowest frequency is for instance 100 Hz, the duration of the excitation should be at least 10 ms. In order to not violate the sampling criteria, the sampling frequency should be at least twice the highest frequency component in the signal. Measuring in a frequency band between 100 Hz and 10 MHz (5 decades) requires a minimal sampling rate of 20 MS/s over 10 ms. The resulting vector length of 200,000 samples for each channel (excitation and response) already exceeds the capabilities of most microcontrollers but is still compatible with instrumentation based on digital oscilloscopes and, for instance, personal computers. Additionally, a higher sampling frequency is often chosen due to noise constraints. This is still feasible for single measurements but exceeds the capabilities of most instrumentation for high-throughput monitoring (e.g., 100 measurements/s). A widely accepted compromise is the use of narrow bandwidth, for instance between 10 kHz and 1 MHz where a time requirement for the measurement can be reduced down to 1 ms while the data volume is theoretically only 200 samples (one channel) for the single measurement. With data processing based on FFT, real time monitoring even with making decisions is feasible.

The narrow frequency range may be a major limitation since it compromises selectivity. A way out is adaptive sampling with small sampling intervals during fast changes of the signal but slowing down the sample rate for parts of the signal where changes are slow [14]. A typical feature of step responses for capacitive objects is the sum of exponential functions, changing fast immediately after the step but much slower with time.

Therefore, together with the simple generation, this excitation signal is favored for adaptive sampling. Other signals like multi-sine or chirp are not suitable for sampling based data reduction to an extent as it is possible for step response.

A single step has a continuous frequency spectrum. Although steps are simple to generate and processing is straight forward, in practice square wave functions without dc-offset, i.e., a series of equidistant steps with alternating polarity, are often used. This has the advantage of charge neutrality, which is essential in electrochemical measurements, but yields a discrete spectrum. The harmonics are multiples of the fundamental frequency equaling the repetition frequency of the square wave. For a symmetric wave, all even harmonics are zero and the amplitude of all odd harmonics declines by $1/n$ where n is the number of the harmonic.

A major problem associated with simple sampling is the violation of the sampling criteria because the slow-changing part of the signal exhibits noise and therefore higher frequency compounds as well. Using an anti-aliasing filter is not a solution because it would compromise the high frequencies immediately after the step. Basically, an adaptive anti-aliasing filter with high cutoff frequency at the beginning and decreasing cutoff frequency with proceeding time would be required. A practical solution is the partial integration and subsequent sampling of the signal (Fig. 7.9). This greatly reduces uncorrelated noise and yields a short vector with highly significant samples. Given the nature of the response for capacitive objects, which is generally the sum of exponential functions, short integration times immediately after the step and increasingly longer integration time toward the end of a half period are adequate. For practical reasons, logarithmic spacing is favored. The exact boundary times for the integration are not important but should be exactly determined for subsequent data processing.

A favored length of the sample vector is between 4 and 10 samples per time decade. As practical example, for a frequency range between 100 Hz and 10 MHz only 50 samples are necessary compared to a vector length of 200000 for uniform sampling with the same resolution.

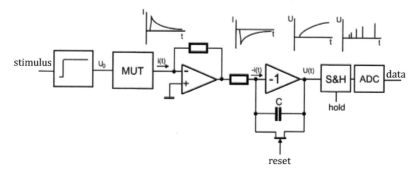

Figure 7.9 Schematic for the instrumentation with signal integration. A controller generates both, the stimulus (rectangular wave) and the schedule for sampling. Here, voltage steps are applied yielding an exponentially decaying current. The capacitor is discharged during the negative half period. The integrated curve is sampled at predefined times $t_{s,n}$.

7.5.2 Transformation into Frequency Domain

Because of non-uniform spacing, discrete Fourier transformation (e.g., fast Fourier transformation) as developed for uniform vectors is not feasible. This requires an analytical solution for obtaining the Fourier coefficients using partial integration between adjacent sampling points. It is important to compensate the dc-offset for periodic signals, either by hardware or during post-processing.

The signals of voltage and current are transformed piecewise between sample points. First, the sampled vector should be differentiated accounting for the integration by the instrumentation.

As an example, a sampled current vector will be used. For simplification, a small piece of an exponential function is regarded to be linear which has been proven reliable for practical applications. The differentiated vector I_n based on the sampled vector $I_{s,n}$ is therefore

$$I_n = \frac{I_{s,n} - I_{s,n-1}}{t_{s,n} - t_{s,n-1}}. \tag{7.32}$$

Based on linearity, the elements of the time vector t_n are the means between two sample times.

$$t_n = \frac{t_{s,n} + t_{s,n-1}}{2} \qquad (7.33)$$

The further calculation involves analytical integration of the product between current and the cosine-function (real part, sine for imaginary) between two sample points with summarizing over all sample intervals.

A further simplification is due to the symmetry of the rectangular wave. Since the even harmonics become zero, only odd harmonics should be calculated. Moreover, based on symmetry between the half periods, only integration from $t = 0$ s up to $t = T/2$ is required. Therefore, for obtaining the right value for the coefficients, they have to be doubled.

A linear function between the sample points t_n and t_{n-1} is calculated as $I_i(t) = at + b$. The coefficients a_i and b_i are determined for all adjacent time points.

$$a_i = \frac{I_i - I_{i-1}}{t_i - t_{i-1}} \qquad b_i = I_i - a_i t_i \qquad (7.34)$$

The integrated product of current and cosine between t_{i-1} and t_i yields the real part of A, A_a for the k-th harmonic with m partial integrations (number of time points). The 0th value is zero due to the reset immediately before the step.

$$
\begin{aligned}
A_{a,k} &= \sum_{i=1}^{m} \int_{t_{i-1}}^{t_i} (a_i t + b_i) \cos(k\omega_0 t) dt \\
&= \sum_{i=1}^{m} \left[\frac{a_i[\cos(k\omega_0 t) + k\omega_0 t \sin(k\omega_0 t)]}{(k\omega_0)^2} + \frac{b_i \sin(k\omega_0 t)}{k\omega_0} \right]_{t_{i-1}}^{t_i}
\end{aligned}
$$

$$(7.35)$$

The imaginary part A_b can be found as

$$
\begin{aligned}
A_{b,k} &= \sum_{i=1}^{m} \int_{t_{i-1}}^{t_i} (a_i t + b_i) \sin(k\omega_0 t) dt \\
&= \sum_{i=1}^{m} \left[\frac{-a_i[\sin(k\omega_0 t) - k\omega_0 t \cos(k\omega_0 t)]}{(k\omega_0)^2} + \frac{b_i \cos(k\omega_0 t)}{k\omega_0} \right]_{t_{i-1}}^{t_i}.
\end{aligned}
$$

$$(7.36)$$

Other than FFT, not all frequency components ($k\omega_0$) need to be calculated. For $k = 0 \ldots N$, where N is the upper cutoff of the spectrum, the result is the same as calculated by FFT. The highest useful value of N is $T/(t_2 - t_1)$ which arises from the highest sample rate immediately after the occurrence of the step. T is the period of the signal. The Fourier transformed current is then

$$\mathcal{F}(I(t)) = 2(A_{a,k} + jA_{a,k}) \tag{7.37}$$

with either $k = 0 \ldots N$ or k equals to the selected harmonics. Although not required, it is favorable to have a logarithmically spaced frequency vector. The multiplication with 2 accounts for the fact that only one half period was used for calculation. In this case, a fully symmetric response to positive and negative steps is assumed.

Finally, the impedance will be calculated as using the Fourier transformed current and voltage:

$$\underline{Z}(\omega) = \frac{\mathcal{F}(U(t))}{\mathcal{F}(I(t))}. \tag{7.38}$$

7.5.3 Direct Processing in Time Domain

A relatively fast algorithm lies in the partial fit of relaxations yielding a time spectrum (relaxation strength a vs. time constant τ) as shown in Fig. 7.10 for KCl solution contacted by microelectrodes (5 µm distance, 1 mm length).

The normalized relaxation strengths are with respect to the sum of the relaxation strength of all polarization mechanisms. For micro-electrodes in electrolyte, one dominating relaxation arising mostly from the electrode-electrolyte interface was evident.

Common data processing involves nonlinear fitting which is problematic because of the behavior of the data. Most changes occur within about 1%–2% of the time where several time constants and therefore the corresponding polarizations are located. A simple fit gives good coincidence with the slow part of the function but neglects almost completely the information from the immediate time after the applied step. A suitable and successful algorithm is partial fitting, starting at the end with large time intervals and fitting only the slowest time constant. After recalculation of this particular time function over the entire time window, it is subtracted from the original curve. This procedure is repeated with moving the time window toward the time immediately after the step until all data are processed. A recursive algorithm can test for a minimum of relaxation strength, A, and break the procedure, independent of the number of time constants found. This approach is universal but lacks a basis for further interpretation.

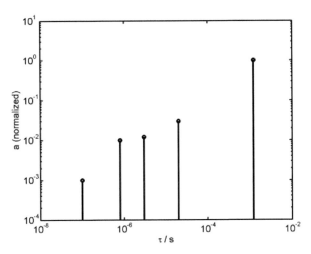

Figure 7.10 Time-spectrum calculated from the voltage across gold microelectrodes electrodes in 100 mM KCl solution.

Having an electrochemical or physical model with given number of time constants requires a fit as described above but with as many iterations as there are time constants within the model.

For well-separated time constants a single step fit becomes feasible after preprocessing of the data. It has been found that logarithmic averaging of the equidistantly sampled data together with the logarithmically spaced time vector gives a reliable fit through several (up to 3) polarization events. Based on practical experience, it should be avoided by biasing the fitting procedure that negative relaxation strength appears in the result.

Due to the fit procedure described, only ideal relaxations were calculated. More sophisticated algorithms use distribution of relaxation times (DRT) which can also account for constant phase elements (CPE) describing for instance the impedance of the electrode interface [15]. In most cases, relaxations are calculated from impedance spectrum rather directly from time response. This, however, is beyond the scope of this chapter.

Non-linear data fitting is not suitable for fast processing in microcontrollers. However, together with a special sampling regime, very fast processing using an analytical method enables real time monitoring and also data logging with high time resolution with just a microcontroller. This can be shown using data from body impedance analysis (BIA) [16] where a square wave voltage was applied between leg and arm and the voltage between a second pair of electrodes between the applicator electrodes was measured. The total current through the body is shown in Fig. 7.11A together with the integrated signal as voltage across the integration capacitor (B).

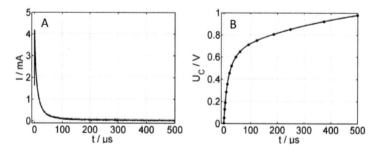

Figure 7.11 (A) Total current passing though the human body during body impedance analysis (BIA) and (B) voltage after signal integration (voltage at the integration capacitor). The circles show the non-uniformly spaced sampling points.

Starting from the end of the signal U_C (0.5 ms in Fig. 7.11B), the largest time constant ($i = 1$) and the relaxation strength can be calculated as

$$\tau_i = \frac{t_i - t_{i-1}}{\ln\left(\dfrac{U_{i-1} - U_i}{U_i - U_{i+1}}\right)} \qquad a_i = \frac{U_{i-1} - U_i}{e^{\frac{-t_{i-1}}{\tau_i}} - e^{\frac{-t}{\tau}}} \tag{7.39}$$

As obvious from the figure, this segment has a dc-component:

$$c_i = U_{i-1} - a_i e^{\frac{-t_{i-1}}{\tau_i}} \qquad (7.40)$$

After subtraction of the recalculated voltage at all time points from the entire vector, the next points are used for calculation of the next relaxation. This will be repeated for a given number of relaxations or until the relaxation strength falls behind a pre-defined minimum. A possible result, here for microelectrodes, is shown in Fig. 7.10.

7.5.4 Impedance Calculation

For systems with galvanostatic excitation with the step height of I_0, this algorithm opens the door for very simple calculation of the impedance spectrum. Although acquired on quite different basis ($U(t)$ in Fig. 7.6B) the time function looks similar to Fig. 7.11B and the same mathematical procedure as described before can be used for data processing. This yields a vector a (relaxation strength, voltage amplitude) and another vector for t (time constant). A further linear factor is the dc-part ($\Sigma R_{pn} + R_b$).

A basic equivalent circuit where each polarization is represented as RC-combination is given in Fig. 7.12A.

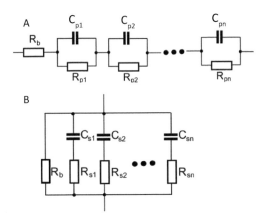

Figure 7.12 Universal electrical equivalent circuit for n polarization mechanisms. Each polarization is modeled as one RC-combination. (A) Equivalent circuit favored for excitation with current step and (B) parallel circuit for simple processing of current response due to a voltage step.

The impedance based on a and τ is

$$Z = \frac{1}{I_0}\left(a_0 + \sum_{i=1}^{n}\frac{a_i}{1+j\omega t_i}\right) = R_b + \sum_{i=1}^{n}\frac{R_{p,i}}{1+j\omega R_{p,i}C_{p,i}}, \qquad (7.41)$$

where a_0 is the dc-part and is calculated as sum of all c-coefficients $a_0 = \sum_{i=1}^{n}c_i$. I_0 is the magnitude of the current step. It should be noted that this simple approach works only with nearly ideal current step, e.g., fast rising edge, no overshot and stable current without decay during data acquisition.

Using the data from microelectrode (time spectrum in Fig. 7.10), the impedance determined directly from current step response was compared to a measurement using multi-sine excitation [17] (Fig. 7.13).

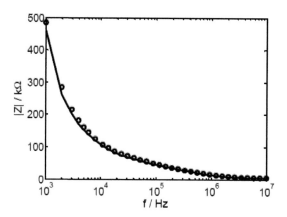

Figure 7.13 Reference measurement based on multi-sine excitation (dots) and impedance magnitude directly calculated from time domain (line) for an electrode distance of 5 μm.

A similar algorithm is possible for voltage-controlled excitation. In this case, it is advantageous but not required to represent the polarizations as parallel circuit of serial RC-combinations (Fig. 7.12B). The current due to a voltage step is

$$I = U_0\left(\frac{1}{R_b} + \sum_{i=1}^{n}\frac{1}{R_{s,i}}\cdot\frac{j\omega R_{s,i}C_{s,i}}{1+j\omega R_{s,i}C_{s,i}}\right) \qquad (7.42)$$

The terms U_0/R_x correspond to the amplitude factors a_i and R_xC_x are the time constants τ_i of the sampled signal.

Using transformation of both, current and the voltage, signals of less quality (e.g., overshot, decay during data acquisition) can be used but require more computing. Other approaches, mostly related to handling diffusive phenomena with distributed time constants are handled using constant phase elements.

7.5.5 Time Domain Transmissometry

The configuration for transmission was already introduced in Fig. 7.6 and is probably most utilized for measurement in the low- and medium-frequency range. The material under test is clamped between electrodes preferentially but not necessarily parallel plate electrodes and a current is passed through. Due to Kirchhoff's law, at low frequency the current of the incident signal equals the transmitted one (Fig. 7.14). The voltage dropping across the electrodes (separate voltage monitoring electrodes are possible as well) depends on the MUT.

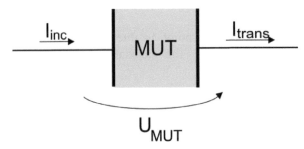

Figure 7.14 Basic configuration for transmission measurement. Although it should be independent of potentiostatic and galvanostatic excitation, the internal impedance of the excitation source matters, especially for the time constants measured.

The impedance is simply $\underline{Z}(\omega) = \underline{U}(\omega)/\underline{I}(\omega)$ which implies, that if measured in time domain, the current and voltage should be transformed into frequency domain.

Although not relevant for most impedance measurements of technical objects like semiconductors, tetrapolar electrode systems should be used for measurements using metal electrodes in electrolyte due to the unavoidable electrode polarization.

7.5.6 Time Domain Reflectometry

Impedance measurements in reflection mode are not limited to high-frequency measurements but in most cases used for measurements above 500 MHz up to the 100 GHz. While transmission measurement collects information from the bulk of the object, reflectometry assesses mostly the surface behavior. In principle, an electromagnetic wave is applied to the surface of a material and the reflected wave is monitored.

As seen in Fig. 7.15B, the voltage of the incident wave and the reflected one adds while the current of the reflected wave has the opposite direction of the incident wave. Therefore, the voltage U at the interface between transmission line and MUT, which is at the distance $d = 0$, is

$$U = U_{inc} + U_{ref} \tag{7.43}$$

while the current is the difference

$$I = I_{inc} - I_{ref} \tag{7.44}$$

The current at this point is due to the voltage and the wave impedance Z_0 of the coaxial cable which is often 50 Ω.

Therefore

$$I = \frac{U_{inc} - U_{ref}}{Z_0}. \tag{7.45}$$

With U and I, the impedance of the MUT is found as

$$Z = \frac{U_{inc} + U_{ref}}{U_{inc} - U_{ref}} Z_0. \tag{7.46}$$

The reflection factor Γ is

$$\Gamma = \frac{U_{ref}}{U_{inc}} = \frac{Z - Z_0}{Z + Z_0}. \tag{7.47}$$

This means, the reflected voltage U_{ref} is positive for $Z > Z_0$ but negative for $Z < Z_0$. If the MUT-impedance is matched to the wave impedance of the cable, $U_{ref} = 0$. For nonreactive MUT, i.e., no transmission, Γ equals the scattering parameter S_{11}.

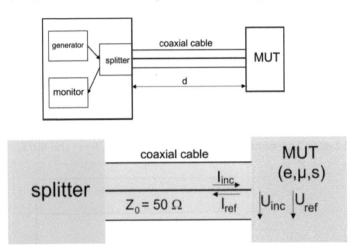

Figure 7.15 (A) Principle of reflectometry. A generator, monitor and the electrode system with the material under test are interconnected with a power splitter. In general, the connection between beam splitter and MUT uses wave guides like coaxial cable. An often used electrode system is a coaxial electrode, i.e., a cylinder with center pin, but also open ended coaxial probes or antennas are in use. (B) current and voltage in a coaxial probe. The subscript "inc" denotes the incident wave while "ref" stays for "reflected."

The time delay between the splitter and the MUT depends on the length and the behavior of the cable. The signal speed is governed by the telegraph equation and is

$$v_{signal} = \frac{1}{\sqrt{LC}}, \tag{7.48}$$

where L is the inductance and C is the capacitance of the cable with respect to the length. At high frequency follows

$$v_{signal} = \frac{c_0}{\sqrt{\varepsilon_r \mu_r}}, \tag{7.49}$$

where c_0 is the speed of light in vacuum, ε_r the relative permittivity and μ_r the permeability between inner and outer conductor. Since μ_r is nearly one for typical coaxial cable, the signal speed is

$$V_{signal} = \frac{c_0}{\sqrt{\varepsilon_r}}$$

(7.50)

Despite small dispersion of the permittivity of the dielectric used for the cable, it can be regarded as constant for the specified frequency range of the cable. Therefore, the delay time between the incident and the reflected wave is $t_d = 2d/v_{signal}$; d is the cable length between the MUT and the splitter (Fig. 7.15A). The factor 2 accounts for the travel to and back from the MUT.

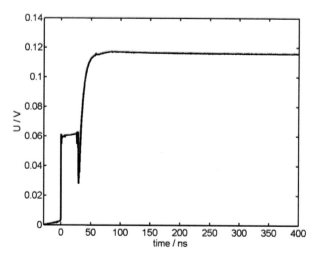

Figure 7.16 Typical TDR (time domain reflectometry) signal arising from di-water (distilled water) at an open-end coaxial probe for a rectangular excitation monitored by a digital oscilloscope. The cable length (RG-58) was 3 m.

The signal at the monitor, as shown in Fig. 7.16 for di-water jumps at $t = 0$ s (incident wave). After the traveling time to the MUT and back, the reflected wave superposes the incident one. The reflected voltage is first negative, because the high-frequency impedance of water is less than the wave impedance of the cable but becomes positive with proceeding stimulus corresponding to lower frequency where the water shows a higher impedance than Z_0.

The superposed voltage at the splitter, e.g., the monitored signal is

$$U = U_{inc,t_0} + U_{ref,t_d}. \tag{7.51}$$

The current at this point is then

$$I = I_{inc,t_0} - I_{ref,t_d} = \frac{U_{inc,t_0} - U_{ref,t_d}}{Z_0}. \tag{7.52}$$

These time functions are not suitable for direct calculation of the impedance. First, the voltage and current needs to be calculated from the voltage monitored at the point of the splitter. The incident signal can be found with matched probe (50 Ω termination) where the reflected energy is zero.

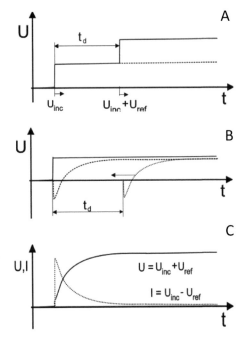

Figure 7.17 (A) Incident and reflected signal with open probe; (B) incident signal (blue) and reflected signal (red) before and after time shift and (C) sum (voltage) and difference (current * Z_0).

In order to find the delay time t_d, a measurement with open cable yields a steep rising response (Fig. 7.17A). By subtracting the incident signal shifted by t_d from the measured signal yields the reflected signal which needs a time shift by t_d in order to have timely coincidence between the incident and reflected signal (Fig. 7.17B). The correction of the signal for t_d is critical and in practical applications often needs a fit to the most probable time. Since the incident and reflected voltages at open cable are theoretically the same but time shifted, a suitable target function for optimization is the area of their difference which should be exact zero for t_d. Finally, the voltage and current can be calculated (Fig. 7.17C). Further processing can be done as described, directly in time domain or by using Fourier transformation. In the latter case, a complete period is necessary, which can be created by assembling the vectors for voltage and current with the negative one. The exact length of the period is not important but needs to be taken into account for calculation of the harmonic frequencies. The impedance is just the ratio of voltage and current as described above. In practice, considerable noise appears in the high-frequency region which can be suppressed by logarithmic averaging.

Especially for high-frequency applications for MUT with low conductivity, a representation as permittivity is often favored. Calculation of the permittivity needs a geometry factor for the electrode system. This can be found using the capacitance of the electrode system filled with reference material like distilled water $\varepsilon_r = 78$ at 25°C). The geometry factor is then

$$k = \frac{\varepsilon_r \varepsilon_0}{c}.$$

(7.53)

The permittivity is calculated from impedance as

$$\varepsilon_r = \frac{k}{j\omega\varepsilon_0 Z}.$$

(7.54)

The assessment of the permittivity spectrum at high frequency is also termed TDDS (time domain dielectric spectroscopy).

7.6   Dynamic Impedance Measurement

In general, impedance measurement and especially the transformation between time and frequency domain is limited to linear and time invariant materials. However, time–varying impedance is often of primary interest, for instance when monitoring body functions (heartbeat, respiration). Characterization of nonlinear objects like pn-junctions, electrode-electrolyte interfaces or biological material [18] uses techniques, such as, assessment of current–voltage characteristics by application of a voltage ramp and measuring the resulting current (voltammetry) [19]. Using different scan rates gives the opportunity to distinguish between reversible and non-reversible reactions and diffusion processes.

A suitable method for characterization of time-varying and non-linear material is the measurement of the dynamic impedance [20]. The measurement time is short with respect to the characteristic time of changes (sub-millisecond) and the excitation signal is small enough falling into a quasi-linear range of the current–voltage characteristics when superposing a dc-voltage.

As an example, dynamic impedance measurement was used for characterization of an electrode-electrolyte interface. The electrode material was boron-doped diamond on a niobium support (Fraunhofer Institut für Schicht- und Oberflächentechnik IST, Braunschweig, Germany). The electrolyte bathing the electrodes was a 100 mM HCl titrated with NaCl to pH 5.

A basic arrangement for characterization of electrodes in electrolytes is the stepwise increase of the voltage offset (Fig. 7.18) up to a maximum value and the subsequent decrease to the starting point. At each step an impedance measurement is performed.

The voltammogram (Fig. 7.18C) does not give advantage over established techniques like square wave voltammetry or ordinary cyclic voltammetry. However, the possibility to assess spectral information between 20 Hz and 200 kHz at each voltage step opens the door for assessment of transfer resistance, diffusive transport within the Gouy–Chapman layer and additionally about electrode reaction like protonation and de-protonation of HCl. It should be noted that no additional redox-system was used.

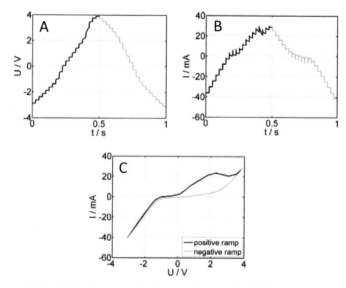

Figure 7.18 (A) Voltage applied to a diamond electrode (bore doped diamond on niobium, $A = 1$ cm²) in contact with electrolyte (100 mM HCl with NaCl at pH 9). A three-electrode system with an Ag/AgCl electrode (in vivo metric, Carson, Ca, USA) was used with an electrode of same type for voltage monitoring. The scheme was similar to Fig. 7.6 but with one current injection and voltage monitoring electrode connected to the diamond electrode. Each step applied had voltage of 350 mV which was distorted due to the behavior of the counter electrode. (B) Total current through the electrode system. (C) Voltammogram calculated from voltage and current at the end of each step. The black portion corresponds always to the positive (rising) ramp while the negative ramp is gray.

A marked point is the impedance where the C/V-characteristic has a negative slope, which is due to the full turnover and further limitation of electron transfer by diffusion at 2.2 V at the positive slope. The phase reaches a minimum hinting to overwhelming influence of diffusion capacitance. The lowest impedance was found at the highest magnitude of the offset because the decomposition voltage was exceeded. The increased impedance at low offset voltage with small imaginary part is due to diffusion limited charge transfer. More sophisticated modeling and data mining will be found in literature about electrochemistry [21].

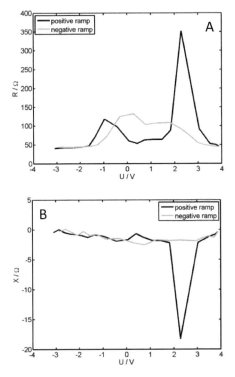

Figure 7.19 Real (A) and imaginary part (B) of the impedance at 20 kHz depending on the offset voltage.

7.7 Summary

Electrical impedance measurement is a method for electrical characterization of material very similar to capacitance measurements or dielectric characterization. For process measurement, fast and robust methods are required. Especially for applications where distributed sensors in large quantity are used, affordable instrumentation which is compatible with modern electronics is the basis for the applicability of this technique. A feasible concept is the measurement of the response of a broad bandwidth signal in time domain and transformation of the answer into the frequency domain for calculation of the electrical impedance. This approach requires extensive resources for broad bandwidth measurements. The processing of relaxation phenomena after step functions gives the same information as the measurement in frequency domain or with other broad-bandwidth signals in time domain but allows comfortable data reduction by non-uniform sampling. In order to avoid undersampling, partial integration of the response signal acts like a dynamic anti-aliasing filter.

While an analytical method based on partial integration is suitable for transformation of the signal into the frequency domain, direct processing in the time domain by relaxation spectroscopy will give comparable results using a fraction of the otherwise required resources.

When measuring dynamic impedances of time-varying and non-linear objects, processing the relaxation after step functions gives the opportunity of continuous sampling of an entire frequency spectrum in time but also with voltage offset.

References

1. Agilent T, *Agilent Impedance Measurement Handbook: A guide to Measurement and Techniques*. 2009.

2. Basoukov EE, MacDonald JR, *Impedance Spectroscopy: Theory, Experiment and Applications*: Wiley-Interscience; 2010.

3. MacDonald JR, *Impedance Spectroscopy*. New York: John Wiley & Sons; 1987.

4. Grimnes S, Martinsen OG, *Bioimpedance and Bioelectricity Basics*: Academic Press; 2014.

5. Choy TT, Ye J, Circuit model-based analysis of the impedance rheopneumogram. *Med Prog Technol* 1993, 19(4): 179–185.

6. Shoar Abouzari MR, Berkemeier F, Schmitz G, Wilmer D, On the physical interpretation of constant phase elements. *Solid State Ionics* 2009, 180(14–16): 922–927.

7. Itkis M, Ghajar JB, Hariri RJ, The square-wave approach to impedance measurement of brain tissue water dynamics. *J Neurosci Methods* 1995, 59(2): 237–244.

8. Bragos R, Sarro E, Fontova A, Soley A, Cairo J, Bayes-Genis A, Rosell J, Four versus two-electrode measurement strategies for cell

growing and differentiation monitoring using electrical impedance spectroscopy. *Conf Proc IEEE Eng Med Biol Soc* 2006, 1: 2106–2109.

9. Farre R, Rotger M, Navajas D, Optimized estimation of respiratory impedance by signal averaging in the time domain. *J Appl Physiol* 1992, 73(3): 1181–1189.

10. Kaczka DW, Barnas GM, Suki B, Lutchen KR, Assessment of time-domain analyses for estimation of low-frequency respiratory mechanical properties and impedance spectra. *Ann Biomed Eng* 1995, 23(2): 135–151.

11. Min M, Pliquett U, Nacke T, Barthel A, Annus P, Land R, Broadband excitation for short-time impedance spectroscopy. *Physiol Meas* 2008, 29(6): S185–S192.

12. Min M, Ollmar S, Gersing E, Electrical impedance and cardiac monitoring-technology, potential and applications. *Int J Bioelectromagnetism* 2003, 5(1): 53–56.

13. Sachs J, Peyerl P, Wöckel S, Kmec M, Herrmann R, Zetik R, Liquid and moisture sensing by ultra-wideband pseudo-noise sequence signals. *Meas Sci Technol* 2007, 18(4): 1074.

14. Pliquett U, Gersing E, Pliquett F, Evaluation of fast timedomain based impedance measurements on biological tissue. *Biomed Tech* 2000, 45: 6–13.

15. Ciucci F, Chen C, Analysis of electrochemical impedance spectroscopy data using the distribution of relaxation times: A Bayesian and Hierarchical Bayesian approach. *Electrochim Acta* 2015, 167: 439–454.

16. Heymsfield SB, Wang Z, Visser M, Gallagher D, Pierson RN, Jr., Techniques used in the measurement of body composition: an overview with emphasis on bioelectrical impedance analysis. *Am J Clin Nutr* 1996, 64: 478–484.

17. Nacke T, Barthel A, Friedrich J, Helbig M, Sachs J, Peyerl P, Pliquett U, *A New Hard and Software Concept for Impedance Spectroscopy Analyzers for Broadband Process Measurements.* In: 2007; Berlin-Heidelberg. Springer Verlag, pp. 194–197.

18. Schwan HP, Linear and nonlinear electrode polarization and biological materials. *Ann Biomed Eng* 1992, 20(3): 269–288.

19. Hamann CH, Vielstich W, *Elektrochemie*. Weinheim: Wiley-VCH Verlag GmbH; 1998.

20. Pliquett U, Fast impedance measurements and non-linear behaviour. In: *Proceedings of the XII International Conference on Electrical Bio-Impedance, Gdansk, 2004*. 2004, pp. 739–742.

21. Bondarenko A, Ragoisha G, *Inverse Problem in Potentiodynamic Electrochemical Impedance.* Nova Science Publishers: New York, NY, USA 2005: pp. 89–102.

SECTION III: APPLICATIONS

Chapter 8

Comparison of Capacitance Spectroscopy for PV Semiconductors

Adam Halverson

Photonics Lab,
GE Global Research, Niskayuna, NY 12309, USA

halverso@ge.com

Admittance measurements are a simple and straightforward technique for measuring electronic properties of contacted semiconductor devices. Typical devices to which admittance spectroscopy can be applied include pn-diodes, photodiodes, transistors, and solar cells. This chapter focuses on admittance measurements applied to thin-film solar cell devices. Admittance measurements are a facile way to access the electronic properties of majority carriers in a completed device, that is they are a non-destructive way to look inside a completed device to determine the electronic properties of the main semiconductor material. Material properties that can be determined with these measurements include the free carrier density, energy depth and capture cross section of the electronic states that produce these majority carriers, the size of energy barriers induced due

Capacitance Spectroscopy of Semiconductors
Edited by Jian V. Li and Giorgio Ferrari
Copyright © 2018 Pan Stanford Publishing Pte. Ltd.
ISBN 978-981-4774-54-3 (Hardcover), 978-1-315-15013-0 (eBook)
www.panstanford.com

to interfaces and contact potentials, or any other contribution to the measured capacitance signal.

8.1 Measurement and Instrumentation

Admittance measurements entail the application of a DC voltage and a small perturbative AC voltage to a semiconductor device and measurement of the subsequent current response of the device. By comparison of the phase and magnitude of the AC current response to the AC voltage perturbation, the real and complex components of the current response can be determined. Many commercial instruments exist that can measure the admittance signals as a function of DC bias, AC bias, and AC frequency, generally referred to as LCR meters. Typical examples often encountered in labs include the Agilent 4284A, the more recent Agilent E4980A, and the Quadtech 1920, all precision LCR meters.

An alternative approach to measure the complex current response of a device is to use a lock-in amplification system. A lock-in amplifier is a phase-sensitive detector that can discern a signal oscillating at a reference frequency from a large background of other noise sources. Thus, the AC voltage perturbation is input to both the device being studied and as the reference signal for the lock-in amplifier. The lock-in, using various phase-sensitive detection techniques, is able to separate the amplitude and phase shift of the current signal gathered from the device.

A very simple method to calibrate the response of a lock-in amplifier to a known capacitance is to substitute a capacitor with a known value into the measurement system. Then, for a given set of perturbation and detection parameters (frequency, AC amplitude, preamplifier gain), the phase angle can be nulled, thus transferring all lock-in signal into the X channel and cancelling any parasitic contributions from self-impedance or lead resistance. Thus forth, the X channel of the lock-in will be proportional to the calibrated capacitance for which the phase was nulled. For measurements over a range of frequencies, this process can be repeated at each frequency setting and a calibration file accumulated that can translate complex impedance measurements at any frequency into capacitance and conductance components.

Lock-in amplifiers have several advantages over LCR meters in detection of small AC perturbative signals: they have excellent noise-rejection properties (often able to find ppm signals within a noisy background), and their transient response is fast enough to perform time-dependent measurements such as deep-level transient spectroscopy (DLTS), or transient photocapacitance spectroscopy (TPC).

8.2 Measurement Frequency, Temperature, and the Emission Energy

The relationship between the measurement parameters frequency and temperature and the physical parameters of energy depth and capture cross section are described by thermionic emission theory and detailed balance [1]. The relationship, referred to as the Richardson equation or Arrhenius equation, is of the form

$$E_\mathrm{e} = -kT\ln\left(\frac{\omega}{f}\right) = -kT\ln\left(\frac{\omega}{N_\mathrm{v}(T)\langle v\rangle\sigma_\mathrm{h}}\right),$$

where k is Boltzmann's constant, T is the sample temperature, ω is the measurement frequency, f is the attempt-to-escape frequency, and E_e is the emission energy or activation energy (E_act) to be measured. The attempt-to-escape frequency v is related to underlying physical parameters of the level contributing to the capacitance by detailed balance arguments yielding

$$f = N_\mathrm{v}(T)\langle v\rangle\sigma_\mathrm{h},$$

where $N_\mathrm{v}(T)$ is the effective density of states in the valence band, $\langle v\rangle$ is the average thermal velocity, and σ_h is the capture cross section of the state involved.

The emission energy depends linearly on temperature and logarithmically on measurement frequency, which informs the range and steps over which temperature and frequency must be varied to gather an accurate and efficient picture of the electronic processes being measured. Thus, temperature can be sampled in linearly uniform steps or with step sizes that vary based on temperature regime. Data collected for this report is often collected at 2K steps at very low temperatures and 10K steps at higher temperatures. The frequency steps must be sampled logarithmically so as to have enough data points to sample of many orders of magnitude of frequency, but not so many that the measurement is inefficient. Typically, a starting and stopping frequency are specified, and then a number n of frequency points to measure are specified. To calculate the frequency points, one must compute the ratio $r = 10^{\wedge}(\log 10(f_{final}/f_{initial})/n)$. The frequencies can then be generated by $f_n = f_{start} \times r^n$. This function is generally captured in most programming languages as the log space function although with varying implementations.

8.3 Temperature Control

There are several best practices for temperature control and measurement, especially as regards the temperature of thin films on comparatively large, insulating substrates. Because of the linear dependence of the emission energy on the semiconductor temperature, small errors in temperature measurement can lead to large errors in calculation of the resulting emission energy. For this reason, immersion cryostats provide the best conditions for rapid, uniform, and stable temperature control. In an immersion cryostat, the sample sits in a carrier gas (e.g., helium, nitrogen) that is able to cool both the sample and substrate passively via radiation, and actively via convection. This is opposed to an evacuated cryostat where the sample sits in an evacuated chamber on a cold-finger that is in contact with a refrigerated gas, but little or no atmosphere is available in the chamber with which to convectively cool. Many thin-film samples are on thick electrically and thermally insulating substrates and for practical electrical contacts must be mounted with the film side away from the cold finger. In an evacuated chamber, this can lead to severe thermal gradients between the film and the cold-finger and can wreak havoc on either the temperature control routine (very long and slow feedback in the control loop if the measurement sensor is placed on the film) or the temperature measurement itself. High-quality immersion cryostats are available on the market by companies such as Linkam Scientific Instruments Ltd. [2] and Instec Inc. [3], and several simple flow-through cryostat designs are available in the scientific literature [4].

8.4   Area Measurement

Another significant source of error is in the measurement of the active area of the device. Capacitance and conductance measurements are typically presented in area normalized units, e.g., nF/cm^2, facilitating the comparison to other devices. The device area also enters into depletion width calculations linearly and into the carrier density calculation in Mott–Schottky analysis as a squared dependence [5].

If a cell does not have a well-defined area or shape, such as through lithographic or laser scribing processes, then accurate area measurement can be tricky. The most cost-effective and reliable measurements for millimeter scale and larger devices have been flatbed scanners. One can simply scan the device itself along with two perpendicular length standards. The standards can be used to apply a pixels/mm calibration in each direction in an image analysis program and the area can be determined either by tracing the outline of the active area of the part manually or using a thresholding algorithm.

8.5   Device Measurements

Several thin-film solar cell devices were obtained from research labs and universities. The devices have different semiconductor absorbers and span the range from production scale technology to new materials under development. Important parameters for devices detailed in this chapter are shown in Table 8.1.

All measurements were performed in a two-wire configuration with a Quadtech 1920 Precision LCR meter. The cryostat was a modified Linkam Scientific Instruments LTS420 with an LNP95 liquid nitrogen pump. Samples were mounted to a home-built jig constructed out of machined aluminum nitride (AlN) and electrically functionalized with a 50 ohm 2D coaxial pattern of gold terminating in Yokowo J-2307P-1-00-000 pogo pins for contacting the cell front and back contact and U.FL ultra small surface mount coaxial connectors mounted on the substrate connect to the LCR meter. An adaptor plate of AlN with through-holes for the pogo pins sits between the contacting jig and the samples under test. The entire assembly is held on to the cold-finger in the Linkam stage using spring-loaded screws and a frame for even pressure. Thus, the film is in intimate contact with the cold finger through two thermally conducting but electrically insulating layers, with 50 ohm coaxial contacts extending from the LCR meter to the contact points of the cells. A custom LabView data acquisition system acquires and saves the data through a GPIB connection to the LCR meter and a serial connection to the Linkam temperature controller. Temperature was typically stepped in 10°C increments from the range of –180°C to 80°C. Oftentimes for better accuracy of in the measurement of a shallow acceptor energy depth, 2°C steps can be made from –190°C to –170°C, although this was not performed in this study. An equilibration time of three minutes was allowed after each temperature ramp to ensure thermal stability and uniformity. Transient measurements of capacitance, low-light intensity open-circuit voltage, dark current, and photoluminescence indicate that the sample temperature is in equilibrium with the stage temperature after this dwell time.

Table 8.1 List of PV devices used in this chapter and their specifications

Name	Absorber	Source	Efficiency (%)	V_{oc} (V)
CdS/CdTe	CdTe	GE Global Research	14%	0.822
CAIGS	$CuAgInGaSe_2$	IEC, UDelaware	17.4%	0.707
CZTS	CuZnSnS	Purdue University	9.1%	0.390
Perovskite	$FA_{0.3}MA_{0.7}PbI_3$	University of Toledo	11.9%	1.070
GaAs	Epi GaAs	GE Global Research	12.7%	0.879

8.6 CdS/CdTe

Figure 8.1 shows the capacitance and conductance/ω curves for a CdS/CdTe thin-film solar cell. Two steps in capacitance are observed with activation energies of 89 and 260 meV (see the next section for determination of activation energies from these data). Previous studies have indicated that the lower of the two activation energies is related to the shallow acceptor responsible for doping the CdTe absorber layer p-type. Generally, an activation energy with a range of values between 60 and 90 meV is observed for this level [6].

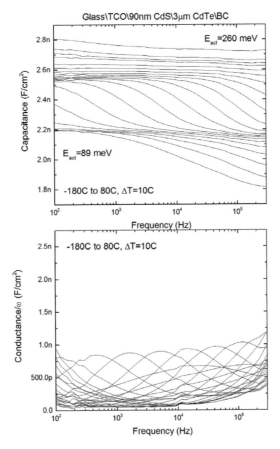

Figure 8.1 Capacitance and conductance/ω data for a 90 nm CdS/3 µm CdTe thin-film photovoltaic device.

The larger energy step has been determined to be related to a barrier to hole collection at the back contact. Bias-dependent admittance sweeps show no change to the apparent activation energy, ruling out a signal from a broad interface state. Devices made with different back contact metals show a linear dependence on this activation energy with the metal work functions [7].

8.7 CuAgInGaSe$_2$

In Fig. 8.2, the capacitance and conductance/ω curves for a CuAgInGaSe$_2$ device from the Institute of Energy conversion are shown. This device has a layer stack of soda lime glass/700 nm molybdenum/2.5 μm CAIGS absorber/50 nm CdS/50 nm ZnO/150 nm ITO/3 μm NiAl-grid. There are several notable features of these admittance curves. In the capacitance graph, there are clearly a few curves that show an abrupt capacitance increase at modest temperatures, that then relax at higher temperatures, thus the capacitance does not monotonically increase with temperature at a fixed frequency. This is likely an indication of the relaxation of a metastable state as the sample is warmed up. Transposing the data and selecting a few frequencies can illuminate this idea, as shown in Fig. 8.3, although the temperature steps are too granular to perform a direct fitting. Many discussions of the source of this metastable state have been made and are outside the scope of this chapter [8].

In the conductance graph in Fig. 8.2, only very faint signatures of peaks are observed. Frequency-dependent capacitance and conductance/ω curves are connected by the Kramer–Kronig relationship, which connects the real and imaginary parts of complex functions [9]. Thus, where there are steps in the capacitance signal there should be peaks in the conductance signal. In the case of this CuAgInGaSe$_2$ device, there is a parasitic leakage current that is exponentially activated in temperature and frequency, and this signal dominates the conductance/ω peaks that should be observed. Careful subtraction of the exponential component can reveal the conductance/ω peaks, or more advanced analysis of the capacitance steps can be undertaken.

CuAgInGaSe₂ Institute of Energy Conversion, University of Delaware

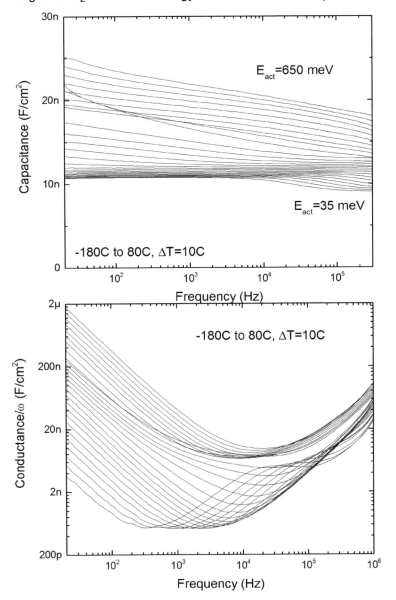

Figure 8.2 Capacitance and conductance/ω data for a CuAgInGaSe₂ with a thickness of $\sim$2.5 µm.

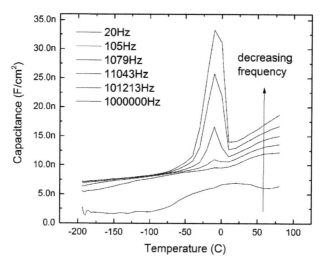

Figure 8.3 Transposed capacitance data at select frequencies demonstrating the relaxation of an unknown metastable state in the CuAgInGaSe$_2$ film.

8.8 CuZnSnSe$_2$

CuZnSnSe$_2$ devices are known for their optimal band gap match to the solar spectrum and their reliance only on abundant materials. While they crystallize in a kesterite crystal structure, there are secondary phases which can condense out of the system as well, leading to defective phases and interfaces that interfere with their efficient operation and stability [10, 11]. Figure 8.4 shows the admittance spectra for a CZTS device.

The capacitance spectra shown in Fig. 8.4 show two indistinct step-like features. The very broad features indicate that the electronic defects that give rise to these steps have a very broad energy distribution [12]. In fact, the steps are not fully resolved within a single frequency sweep of the system, which does not hinder the determination of a distinct activation energy, but does hinder accurate determination of the attempt-to-escape frequency or capture cross section of the defect. The activation energy is determined by the spacing of the curves, which can still be determined as described in detail later, whereas the attempt-to-escape frequency is determined by the inflection point of the curves, which for these broad curves is incompletely resolved. Again, a strong power law leakage signal obscures the conductance/ω peaks.

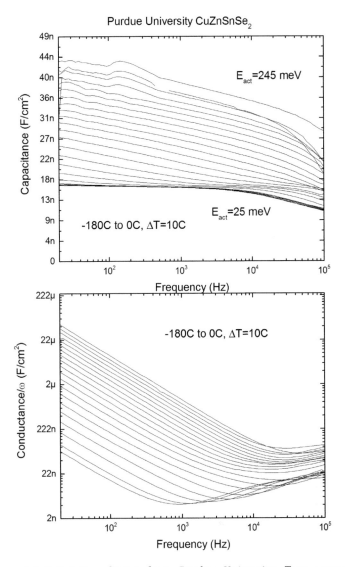

Figure 8.4 CuZnSnSe$_2$ device from Purdue University. Two apparent activation energies are shown from incompletely resolved capacitance steps.

8.9 Perovskite

Methylammonium lead triiodide (MAPI) has seen enormous developments in peak research cell efficiency and a significant amount of attention as a low cost, solution processed, high quality material for solar energy harvesting [13, 14]. However, the material is lead-based and water soluble, raising stability issues due to poor encapsulation and environmental issues should the encapsulation become compromised [15]. However, the materials themselves are known for long carrier lifetimes and the devices have very high voltages, with imputed defect densities less than 10^{10} cm^{-3} [16]. Their band gaps are broadly tunable and ideally match the solar spectrum [17]. These qualities make them very promising materials for solar conversion, hence the great interest from research groups. Figure 8.5 shows the capacitance and conductance/ω curves for a methylammonium formamidinium lead triiodide perovskite solar cell obtained from the University of Toledo. The first feature seen in these capacitance curves is the large magnitude of the capacitance signal. Room temperature capacitance over 100 nF/cm^2 is nearly an order of magnitude larger than the capacitance for competing CAIGS and CZTS and two orders of magnitude larger than the CdTe devices previously discussed in this chapter. CdTe, CIGS, and CZTS are all compensated semiconductors doped by native defects, the net doping resulting from large densities of both native acceptors and native donors. Perovskites are known to be acceptor doped by metal vacancies due to non-stoichiometry in the lattice, and appear to be robust against deviations in stoichiometry creating compensating donors.

The next interesting observation with the perovskite is that there are two activated steps observed, but with nearly identical activation energies. The entire activation step is not observed for the higher capacitance step, which limits the ability to determine the attempt-to-escape frequency, but the uniformly spaced capacitance curves can be analyzed to provide an estimate activation energy. Due to the nature of the device, higher temperature measurements where a complete step could be measured were not possible. Lower frequency measurements could be attempted although at lower frequencies $1/f$ noise becomes an issue and can obscure true signals from the device. Device simulations indicate that a signal of this sort can arise from a double-acceptor state with similar activation energies for each of the occupation states but different capture cross sections.

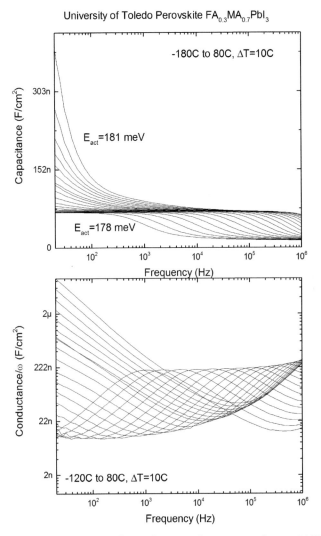

Figure 8.5 Capacitance and conductance/ω curves for a MAFAPbI$_3$ perovskite solar cell device from the University of Toledo.

8.10 GaAs Device

Admittance data from an experimental planar GaAs device is shown in Figs. 8.6 and 8.7. A single, prominent step with an activation energy of 281 meV is easily seen, as well as the associated conductance peaks. This experimental device was lower efficiency than a typical GaAs device, and the peak arises due to a barrier between the active layers of the device and the substrate upon which is was deposited. This is apparent from the very large magnitude of the capacitance, which rules out a deep level response, and from subsequent optimization of these devices which corrected this issue. However, the quality of the capacitance and conductance data for this device are striking, with the Kramers–Kronig relationship that relates the data immediately apparent [9]. Admittance measurements performed on a commercial GaAs device revealed a shallow p-type acceptor 64 meV above the valence band (not observed in this device) and much lower overall capacitance.

8.11 Determination of the Activation Energy: Arrhenius Plots

Once measurements of the capacitance as a function of frequency and temperature are complete, further analysis is required to determine the activation energy and attempt-to-escape frequency [1]. These are the exponential factor and prefactor to the Arrhenius thermal activation equation

$$\omega = v_0 \mathrm{Exp}\left(-\frac{E_{act}}{kT}\right),$$

which is simply a solution of equation the Arrhenius equation shown earlier for the measurement frequency ω. Thus, a plot of the loge of the characteristic frequency versus $1/kT$ yields a downward sloping line with a slope equal to the activation energy in meV and an infinite temperature ($1/kT = 0$) intercept equal to the attempt-to-escape frequency. The task then is to determine the characteristic frequency for each curve.

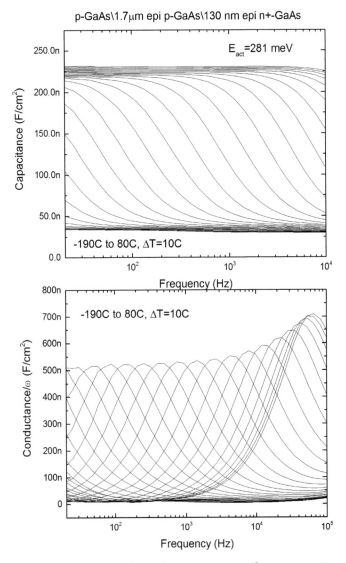

Figure 8.6 Capacitance and conductance curves for an experimental planar junction GaAs device. The admittance response originates from a poor interface to a contacting layer.

Due to lack of large effects from external resistances and other parasitic effects, the capacitance signal is often the cleanest signal from which to extract the activation energy [18]. Any series of capacitance steps that are linearly spaced with temperature will give a linear Arrhenius plot. Thus, determination of the frequencies at each temperature that give a fixed capacitance value is sufficient to determine the activation energy. However, this does not uniquely determine the attempt-to-escape frequency. The most robust technique for determination of the inflection frequencies is by differentiating the capacitance data versus the logarithmic frequency

$$\frac{dC}{d(\log \omega)} = -\omega \frac{dC}{d\omega} \propto \frac{G}{\omega},$$

where the last proportionality is a consequence of the Kramers–Kronig relationship. Because real-world data is being differentiated, oftentimes noise-low frequency data must be neglected as the differentiation significantly amplifies the noise. The peak position from this differentiated data can be determined, and careful comparison with the capacitance data reveals that the peak location is coincident with the halfway-point of the height of the capacitance step. This is why the attempt-to-escape frequency cannot be determined when the entire step is not resolved.

Figure 8.7 shows a series of conductance curves derived from the earlier discussed planar GaAs device. Each curve is individually run through an off-the-shelf peak detection algorithm with a threshold and minimum width as the inputs and peak locations as the outputs. The locations of the peaks that are found are overlaid on the data for verification. These peak positions are then plotted in an Arrhenius fashion and an exponential fit to the data is determined. Because of the x-axis is plotted as $1/kT$, the negative of the slope of the fit is equivalent to the activation energy for that data set. This process can be repeated on differentiated capacitance data (as detailed above) in situations where the capacitance data is either less noisy than the conductance data, or where the conductance data is overwhelmed by other parasitic responses such as shunt conductance.

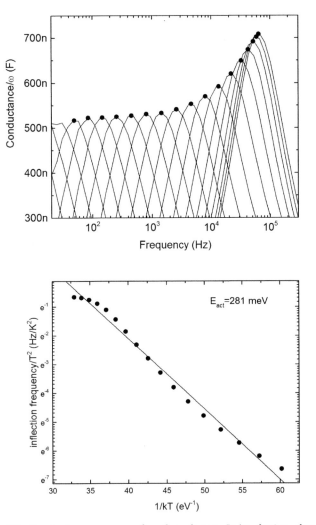

Figure 7.6 Conductance curves for the planar GaAs device shown in figure 6 along with peak positions from a peak detection algorithm. These peaks are then plotted in an Arrhenius plot with an exponential fit with a slope of 281 meV.

8.12 Conclusions

The technique of admittance spectroscopy has been applied to a diverse range of solar cell devices, demonstrating its utility as a probe of the electronic responses of the device. Capacitance and conductance data gathered as a function of temperature and probe frequency are able to elucidate key information about these signals, including activation energies. Techniques for gathering these data in a robust fashion are discussed as well as methods for analyzing the data to determine physical parameters of the signals.

Acknowledgments

The author would like to thank Jian Li for the invitation to contribute to this book, as well as Loucas Tsakalakos at GE Global Research (GaAs), Dewei Zhao, Changlei Wang, and Yanfa Yan at the University of Toledo (perovskite), Bill Shafarman at the Institute of Energy Conversion (CAIGS), and Rakesh Agarwal at Purdue University (CZTS) for providing samples for measurement.

References

1. D. Losee, Admittance spectroscopy of impurity levels in Schottky barriers, *J. Appl. Phys.* 46, 2204 (1975).

2. http://www.linkam.co.uk/.

3. http://www.instec.com/.

4. S. S. Andrews and S. G. Boxer, A liquid nitrogen immersion cryostat for optical measurements, *Rev. Sci. Instrum.* 71, 3567 (2000); doi: 10.1063/1.1287343.

5. J. T. Heath, J. D. Cohen, and W. N. Shafarman, Bulk and metastable defects in $CuIn_{1-x}Ga_xSe_2$ thin films using drive-level capacitance profiling, *J. Appl. Phys.* 95, 1000 (2004); http://dx.doi.org/10.1063/1.1633982.

6. F. H. Seymour, V. Kaydanov, T. R. Ohno, and D. Albin, Cu and $CdCl_2$ influence on defects detected in CdTe solar cells with admittance spectroscopy, *Appl. Phys. Lett.* 87, 153507 (2005); http://dx.doi.org/10.1063/1.2099515.

7. J. V. Li, A. F. Halverson, O. V. Sulima, S. Bansal, J. M. Burst, T. M. Barnes, T. A. Gessert, and D. H. Levi, Theoretical analysis of effects of deep levels, back contact, and absorber thickness on capacitance-voltage profiling of CdTe thin-film solar cells, *Solar Energy Mat. Solar Cells*, 100, 126 (2012).

8. T. Eisenbarth, T. Unold, R. Caballero, C. A. Kaufmann, and H.-W. Schock, Interpretation of admittance, capacitance-voltage, and current-voltage signatures in CuInGaSe$_2$ thin film solar cells, *J. Appl. Phys.* 107, 034509 (2010).

9. C. León, J. M. Martin, J. Santamaría, J. Skarp, G. González Díaz, et al., Use of Kramers–Kronig transforms for the treatment of admittance spectroscopy data of pn junctions containing traps, *J. Appl. Phys.* 79, 7830 (1996); http://dx.doi.org/10.1063/1.362391.

10. A. Nagoya, R. Asahi, R. Wahl, and G. Kresse, Defect formation and phase stability of Cu$_2$ZnSnS$_4$ photovoltaic material, *Phys. Rev. B* 81, 113202 (2010).

11. J. J. Scragg, T. Ericson, T. Kubart, M. Edoff, and C. Platzer-Björkman, Chemical insights into the instability of Cu$_2$ZnSnS$_4$ films during annealing, *Chem. Mater.* 2011, 23 (20), 4625–4633, DOI: 10.1021/cm202379s.

12. T. Walter, R. Herberholz, C. Müller, and H. W. Schock, Determination of defect distributions from admittance measurements and application to Cu(In,Ga)Se$_2$ based heterojunctions, *J. Appl. Phys.* 80, 4411 (1996); http://dx.doi.org/10.1063/1.363401.

13. M. A. Green, A. Ho-Baillie, and H. J. Snaith, The emergence of perovskite solar cells, *Nat. Photonics* 8, 506–514 (2014), doi:10.1038/nphoton.2014.134.

14. M. Liu, M. B. Johnston, and H. J. Snaith, Efficient planar heterojunction perovskite solar cells by vapour deposition, *Nature* 501, 395–398 (19 September 2013) doi:10.1038/nature12509.

15. G. Niu, X. Guoa, and L. Wang, Review of recent progress in chemical stability of perovskite solar cells, *J. Mater. Chem. A* 2015, 3, 8970–8980, DOI: 10.1039/C4TA04994B.

16. D. Shi, V. Adinolfi, R. Comin, M. Yuan, E. Alarousu, A. Buin, et al., Low trap-state density and long carrier diffusion in organolead trihalide perovskite single crystals, *Science* 30 January 2015: 347(6221), 519–522, DOI: 10.1126/science.aaa2725.

17. J. H. Noh, S. H. Im, J. H. Heo, T. N. Mandal, and S. I. Seok, Chemical management for colorful, efficient, and stable inorganic–organic hybrid nanostructured solar cells, *Nano Lett.*, 2013, 13 (4), 1764–1769, DOI: 10.1021/nl400349b.

18. J. H. Scofield, Effects of series resistance and inductance on solar cell admittance measurements, *Solar Energy Mater. Solar Cells* 37 (1995) 217–233.

Chapter 9

Capacitive Techniques for the Characterization of Organic Semiconductors

Dario Natali[a,b] and Mario Caironi[b]

[a]*Dipartimento di Elettronica, Informazione e Bioingegneria,*
Politecnico di Milano, Piazza L. da Vinci 32, 20133, Milano, Italy
[b]*Center for Nano Science and Technology @PoliMi,*
Istituto Italiano di Tecnologia, via Pascoli 70/3, 20133 Milano, Italy

dario.natali@polimi.it, mario.caironi@iit.it

Organic semiconductors (OSCs) are an appealing class of materials because they are solution processable and can be deposited by means of cost-effective printing methods, thus having the possibility to address large areas and flexible substrates (Ogawa, 2015; Ishii et al., 2015; Caironi and Noh, 2015). In addition, their optoelectronic properties can be tuned by means of chemical tailoring. All the basic building blocks of electronics can be developed in the framework of organic electronics: organic light-emitting diodes (OLEDs) (Reineke et al., 2013; Thejokalyani and Dhoble, 2014), field-effect transistors (OTFT) and circuits (Sirringhaus, 2014; Heremans et al., 2016; Bucella et al., 2015;

Capacitance Spectroscopy of Semiconductors
Edited by Jian V. Li and Giorgio Ferrari
Copyright © 2018 Pan Stanford Publishing Pte. Ltd.
ISBN 978-981-4774-54-3 (Hardcover), 978-1-315-15013-0 (eBook)
www.panstanford.com

Baeg et al., 2013b), photovoltaics (OPV) (Kang et al., 2016), memories (Zhang et al., 2016) and sensors (Someya et al., 2015), most notably photodetectors (Pace et al., 2015; Baeg et al., 2013a; Natali and Caironi, 2016). The integration of these devices can give rise to lightweight and even conformable optoelectronic systems. While demonstrators and first products are already being proposed, some of the fundamental properties of organic materials are not fully understood yet, and this implies the lack of generalized quantitative models. Reliable techniques for investigating material properties and for comparing experimental results to theoretical predictions are highly desirable: this chapter outlines how the analysis of the capacitance (or admittance) of organic devices as function of applied bias, frequency, and modulation intensity can be used to obtain useful material parameters.

9.1 Organic Semiconductors: A Brief Introduction

OSCs are materials based on sp^2-hybridized carbon atoms. Hybridized orbitals give rise to three strongly localized covalent σ-bonds which constitute the molecular backbone; the remaining half-filled p orbitals give rise to π-orbitals delocalized over the entire molecule (π-conjugation), constituting the frontier orbitals. Oversimplifying, the highest occupied molecular orbital (HOMO) and the lowest unoccupied molecular orbital (LUMO) play somehow the role of the valence and conduction band edges respectively of inorganic semiconductors.

OSCs can be classified as single small molecules, oligomers (given by the repetition of few monomer units) and polymers (given by the repetition of many monomer units). In the solid-state they give rise to molecular solids, because intermolecular bonds, which are usually of Van der Waals type, are characterized by a relatively low energy, below 40 kJ/mol, compared to 220 kJ/mol of the average energy of the covalent Si-Si bond. This has two consequences: on the one hand the tendency towards the formation of ordered structures is weak because the gain in enthalpy is low, while, on the other hand, the processing thermal budget is relatively low (less than few hundreds of degree Celsius). In addition, by means of suitable functionalization, many OSCs can be made soluble in appropriate solvents, so that solution-based deposition protocols can be adopted and high-throughput, low-cost printing techniques borrowed from graphical arts can be exploited. Depending on the nature of the molecule and on the deposition technique, aggregation and packing in the solid-state can result in amorphous materials, inhomogeneous materials with nano- and micro-crystalline regions interspersed in an amorphous matrix, up to single crystals (by means of dedicated vapor-phase techniques or, in exceptional cases also directly from solution).

The lack of long-range order, together with a strong electron–phonon coupling, causes charge carrier localization on a single molecule or on very few neighboring molecules at best, to be compared to delocalized Bloch waves typical of crystalline inorganic semiconductors. Carrier transport can be envisioned as a — relatively ineffective — thermally activated tunneling from molecule to molecule, usually referred to as hopping. The resulting mobility is low, often below 10^{-1} cm^2/Vs. Only in the latest generation of high-performing polymers remarkable carrier mobility values in the 1–10 cm^2/Vs range have been recorded, but only at the high carrier density of about 10^{19} cm^{-3} reached in the accumulation channel of transistors.

The disordered nature of OSCs causes a spreading of the HOMO and LUMO levels: gaussian, exponential and various other models have been proposed in the literature to describe the resulting density of states (DOS). For a large class of DOS decaying faster than E^{-4}, where E is the energy, transport can be described as an excitation of carriers from the DOS tail up to a so called transport level, which is the same irrespective of the starting excitation level. The energetically distributed states below the transport energy do not equally contribute to transport: the deeper a state lies in energy, the smaller the probability per unit time of being excited to the transport energy. This is formally similar to a multiple trapping and release scenario often adopted in inorganic semiconductor theory, with the transport energy playing the role of the mobility edge, the deep portions of the DOS acting as deep traps and the intermediate portions of the DOS acting as shallow traps.

It has to be reminded that OSCs are mainly employed as nominally intrinsic semiconductors, even though an unintentional doping with a density up to 10^{15} cm^{-3} is often present. Doping is not a common processing technique yet: even though much progress has been made in the field of OLEDs, doping of OTFTs is still relatively rare (Lüssem et al., 2013, 2016). Issues related to doping are the following: dopant ionization is far from being effective, due to the relatively low dielectric constant of OSCs ($\varepsilon_r \simeq 3$–4); dopants being interstitial suffer from a certain thermodynamic instability; n-type doping has to rely on molecules with low ionization potential, which tend to be environmentally unstable (even though schemes where stable dopant intermediates are employed have emerged recently). Molecularly doped polymers such as poly(3,4-ethylendioxythiophene) polystyrene sulfonate (PEDOT:PSS) are a notable exception.

Finally, considering that thermal generation is negligible in most cases (the energy gap of many materials is around 2 eV), this implies that excess carriers are always a large perturbation with respect to thermodynamic equilibrium.

9.2 Capacitive Measurements in the Space Charge Limited Regime

A large number of works in the literature has been devoted to the application of capacitance (and admittance) measurements to two-terminal, metal/semiconductor/metal devices operating in the space charge limited regime (SCL) (Lampert and Mark, 1970). In the absence of traps and with a constant mobility, the static current can be expressed as $J = (9/8)\varepsilon\mu V^2/L^3$, where L denotes the interelectrode spacing and μ the carrier mobility. When a small sinusoidal signal is superimposed to a voltage bias, the space charge accumulated into the device is probed. The impedance reads (Montero et al., 2009)

$$Z(\omega) = \frac{6}{g_0(i\omega\tau_0)^3}\left[1 - i\omega\tau_0 + \frac{1}{2}(i\omega\tau_0)^2 - \exp(-i\omega\tau_0)\right], \qquad (9.1)$$

where $\tau_0 = (4/3)\,L^2/\mu V$ is the carrier transit time, $C_g = \varepsilon/L$ is the parallel plate geometrical capacitance and $g_0 = \dfrac{dJ}{dV} = \dfrac{3}{\tau_0}C_g$ is the static conductance. Modeling the impedance as the parallel connection of an equivalent resistor and of an equivalent capacitance, the latter is equal to $(3/4)C_g$ in the low frequency limit, where carriers are able to respond to the small signal, while it tends to the geometrical capacitance in the high frequency limit, where the space charge is not able to follow the sinusoidal perturbation. The capacitance spectrum shows a stepwise increase at a frequency which is proportional to the reciprocal transit time, from which it is possible to deduce the carrier mobility. The mobility is more clearly resolved looking at the differential susceptance $\Delta B(\omega) = \omega[C(\omega) - C_g]$, which shows a maximum located at $0.72\tau_0^{-1}$.

The condition of constant mobility is highly idealized and is rarely encountered in OSCs: carrier mobility is very often found to depend both on the carrier density and on the electric field. For a dependence in the form of a power law of the carrier density, a model often adopted in OSCs (Ramachandhran et al., 2006), the low-frequency value of the capacitance is altered with respect to the constant-mobility case, but still the transition from the low frequency to the high frequency regime can be exploited to extract the carrier mobility. The picture gets further complicated by the presence of trapping centers. The factors affecting the capacitance spectrum are the trap density of states and the trap kinetics. To gain insight, we briefly discuss the case of a single trap level (Montero et al., 2009). A shallow, fast single trap level (where fast means that trapped carriers are in quasi-equilibrium with carriers at the transport level) diminishes the carrier mobility due to trapping/detrapping phenomena, hence the position of the maximum for ΔB is shifted towards lower frequencies according to the ratio between trapped carrier density and transport carrier density. A shallow, slow, single trap level (viz. a trap that is not in quasi-equilibrium with transport states) results in an increase of the low-frequency capacitance well above $(3/4)C_g$. However, for sufficiently high frequencies, traps are not able to capture and emit carriers and they cease to be effective. As a consequence, the ΔB maximum is the same as in the trap-free case. For a deep, single trap level the low frequency capacitance spectrum is well below $(3/4)C_g$ and the ΔB maximum is shifted to lower frequency with respect to the trap-free case. The simulations reported in Figs. 9.1 and 9.2 summarize the situation; note that reported spectra oscillations are expected to be damped by including the contribution from diffusion currents (Ramachandhran et al., 2006).

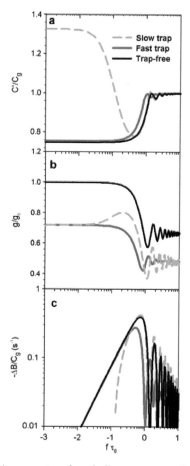

Figure 9.1 Simulation spectra for shallow traps at 2 V. Fast traps are plotted by pink lines, slow traps by orange dashed lines and trap-free spectrum by black lines. Panel (a) reports capacitance spectra normalized to C_g for fast and slow traps. For slow traps capacitance increases at low frequency. Panel (b) shows conductance spectra normalized to g_0. Panel (c) shows the negative differential susceptance $-\Delta B(\omega)$ normalized to C_g, used to extract transit times. In the case of fast trapping, the peak shifts at lower frequency, whereas for slow trapping no deviation from the trap free case is observed. Reproduced with permission from Montero et al. (2009). Copyright 2009, Elsevier.

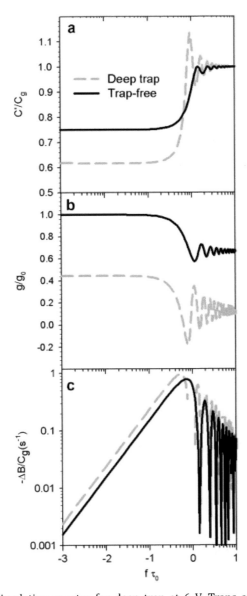

Figure 9.2 Simulation spectra for deep trap at 6 V. Traps are plotted by maroon dashed lines and trap-free spectrum by black lines. Panel (a) reports capacitance spectra normalized to C_g for fast and slow traps. Panel (b) shows conductance spectra normalized to g_0. Panel (c) shows the negative differential susceptance $-\Delta B(\omega)$ normalized to C_g. Reproduced with permission from Montero et al. (2009). Copyright 2009, Elsevier.

In the more realistic case of a distribution of traps, both shallow and deep traps coexist according to their energy depth, so that both the low frequency capacitance and the mobility are affected. For a general treatment of the effect of traps the interested reader is referred to the works by Dascalu (Dascalu, 1966, 1968) and Kassing (Kassing and Kähler, 1974; Kassing, 1975) and to the more recent approach by Takagi and co-workers (Takagi et al., 2015).

Another approach, instead of markedly distinguishing between carriers at transport level and trapped ones, considers the slow and widely time-distributed hopping processes occurring in the DOS between high energy states and states lying in the DOS tail (Germs et al., 2011): the mobility turns out to be frequency dependent. The advantage of this approach is that the frequency dependent mobility can be calculated from the DOS characteristics (assumed to be gaussian) and from the steady state mobility, with no need for introducing phenomenolgical parameters as in the work of Martens et al. (1999).

For the case of bipolar injection into organic semiconductors, the interested reader is referred to the paper by Schmeits (2007). The region where both electrons and holes are present and recombine has been correlated with the insurgence of a negative component of the capacitance at low frequency (Lungenschmied et al., 2009; Ehrenfreund et al., 2007), even though alternative explanations have been given as well in terms of hot electrons (Tripathi and Mohapatra, 2010) or self-heating effects (Okumoto and Tsutsui, 2014).

The analyses so far rely on the assumption of an ideally ohmic injecting contact, as it is usual in the framework of space charge limited conduction regime. Indeed carrier injection into organic semiconductors can be far from ideal and hampered by energetic injection barrier given by the mismatch between the metal Fermi level and the semiconductor HOMO/LUMO levels (Natali and Caironi, 2012). As far as the mobility determination, if injection issues are not properly accounted for, they lead to underestimating the actual carrier mobility (Tang et al., 2015). It has been reported that injection effects are negligible for barriers below 0.2 eV (Takagi et al., 2015).

9.3 Capacitive Measurements in Schottky-Contacted Single-Carrier Organic Devices

In this section we consider the case of a Schottky contact between a metal and a semiconductor. We assume that a depleted space charge region exists close to the metal/semiconductor interface. This can be the case when the semiconductor is unintentionally doped, for example due to the interaction with ambient oxygen and moisture. For a recent account on progresses in the field of intentional doping the interested reader is referred to the work of Leo et al. (Lüssem et al., 2013, 2016).

We start considering an ideal, Schottky-contacted semiconductor with a N^- density of p-type dopants biased in moderate reverse bias. We call $W = \sqrt{2\varepsilon(V_{bi} - V_{DC})/(qN^-)}$ the extent of the space charge region, where V_{bi} is the built-in voltage and V_{DC} the DC bias voltage. When a small sinusoidal signal V_{ac} is superimposed to V_{DC}, the depleted region acts as a capacitance: to a first degree of approximation, carriers are added and removed at the boundary between the depleted and the neutral region, so that a differential, specific capacitance $C = \varepsilon/W$ can be measured. The dependence of C on V_{DC} can be exploited to extract V_{bi} by plotting C^{-2} versus V_{DC}. In addition, the classic profiler equation can be used (and is valid also in case of non-uniform doping): $N^-(x) = C^3(q\varepsilon A^3)^{-1} \, dV/dC$, where x is the distance from the metal–semiconductor junction.

The situation changes if a discrete deep trap, located at E_T is present (Fig. 9.3). We consider for simplicity the stepwise approximation for the Fermi–Dirac occupation statistics, and we define X_T as the coordinate where E_T equals the Fermi level E_F : for $0 < x < X_T$ deep traps are negatively charged, whereas for $X_T < x < W$ deep traps are neutral. Therefore the presence of a deep trap modifies the charge in the depleted region and its extension W for a given applied bias V_{DC}. If a sinusoidal signal is superimposed to the DC bias, in addition to the free carrier response (located at $x = W$), the contribution due to charging and discharging of deep traps has to be taken into account.

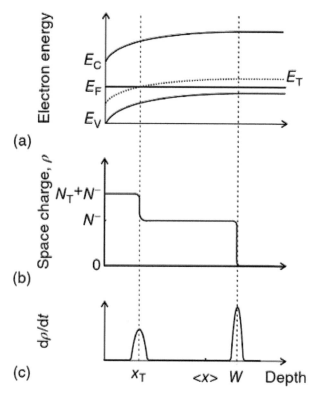

Figure 9.3 Schottky contact with one deep trap state at E_T. Panels (a), (b), and (c) show the spatial dependence of: energy levels, space charge and space charge variation upon a small signal, respectively. Reproduced with permission from Heath and Zabierowski (2011). Copyright 2011, J. Wiley and sons.

Only deep traps in resonance with the Fermi level, viz. those located around $x = X_T$, can alter their occupancy state and contribute to the capacitance. Whether this occurs or not, depends on the frequency of V_{ac}. Only if it is sufficiently low, then the trapping state is in quasi-equilibrium with the transport state and can quasi-statically follow the applied V_{ac}. For a quantitative analysis, we introduce the trap characteristic emission time, which can be reasonably modeled as an exponential function of the energy difference between E_T and the transport level E_{tr}: $\tau_p = v_{esc}^{-1} \exp[(E_T - E_{tr})/kT]$, where v_{esc} is the attempt to escape frequency. Thanks to the fact the τ_p is a rapidly varying function of energy, we can say that only states whose emission time satisfies $\omega < 1/\tau_p$, where ω is the angular frequency of applied signal, can alter their occupancy state. During a CV measurement, V_{DC} is usually swept slowly, so that trap states are expected to respond. As a consequence, even if V_{ac} angular frequency is high, viz. $\omega > 1/\tau_p$, the measured capacitance is affected by traps: only free carriers respond at $x = W$, but W depends on the static charge accumulated in the junction, which comprises both ionized dopants and traps. To a first approximation, the application of the profiler equation yields an apparent doping $N_{app} = N_T (x_T) (x_T/W) + N^-(W)$ (Kimerling, 1974). If ω is relatively low, then the measured capacitance is given by the first moment of the charge variation due to both free carriers at $x = W$ and traps at $x = X_T$.

Now we move from the case of a discrete deep trap to the case of distributed deep traps. In principle, all along the depleted region the Fermi level can be at resonance with a portion of the DOS, but again the emission time has to be taken into account. It is then useful to define a coordinate x_ω which acts as a cut-off abscissa (Fig. 9.4): for a given modulation angular frequency ω, only the region $x_\omega < x < W$ can follow V_{ac}, whereas for $0 < x < x_\omega$ the DOS portion in resonance with the Fermi level is lying energetically too deep to follow V_{ac}. Going quantitatively in this case requires making some approximations. An often adopted strategy is due to Walter (Walter et al., 1996): assuming a constant electric field, the DOS can be derived as it follows: $N_T (E_\omega) = V_{bi}/(qW)(dC/d\omega) (\omega/k_B T)$, where $E_\omega = k_B T \ln(2\pi\nu_{esc}/\omega)$.

A *caveat* has to be given: the analysis so far assumes that applied frequencies are in any case lower than the dielectric relaxation frequency, or in other words that carrier transport across the device occurs on a time scale which is faster than the applied signal modulation period. As organic semiconductors are notoriously characterized by relatively low mobility, this limitation has to be taken into account (Li et al., 2011).

Finally, we recall that in case of organic photovoltaic blends the DOS population can be altered by means of illumination. By biasing the solar cell at open circuit, recombination equals photogeneration and a chemical capacitance, due to the accumulation of photocarriers, can be measured (Garcia-Belmonte et al., 2010). By varying the illumination intensity, the separation between the hole and electron quasi-Fermi levels is varied, which is measured by the open circuit voltage V_{oc}. In the specific case of the widely adopted P3HT:PCBM photovoltaic blend, the P3HT phase is environmentally doped, hence its quasi Fermi level is not much altered by illumination. The consequence is that V_{oc} is proportional to the electron quasi-Fermi level. It can be shown that considering the 0 K approximation for the Fermi–Dirac statistics, the chemical capacitance is directly proportional to the electron DOS at a given position of the quasi-Fermi level, viz. $C_\mu = Lq^2g(E_{Fn})$, where L is the sample length. At low frequency, the overall impedance can be modeled as the parallel connection of C_μ and a resistance embodying recombination mechanisms. By varying the illumination intensity different regions of the DOS can be probed. In addition to the electron DOS, the charge carrier density and recombination kinetics can be addressed with this technique (Garcia-Belmonte et al., 2010).

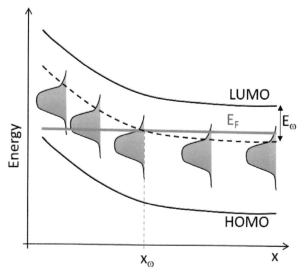

Figure 9.4 Schottky contact with a distribution of deep trap states. The Fermi level all across the space charge is always in resonance with some part of the DOS, but — for a given angular frequency ω of V_{ac} — only states whose energy distance from the transport level (here for simplicity embodied by the LUMO) is beyond a critical energy value E_ω can respond. The energetical constraint translates into a space constraint: only states located at $x > x_\omega$ can respond.

9.4   *C–V* Measurements on Metal–Insulator– Semiconductor Capacitors

Recently it has been shown that capacitance–voltage (*C–V*) measurements performed on metal–insulator–semiconductor capacitors combined with numerical simulation enable the extraction of the semiconductor DOS (Maddalena et al., 2015; Sung et al., 2016). In fact, the shape of the *C–V* curve depends on the width of the DOS: the larger the DOS width, the smoother the *C–V* curve. Going from depletion to accumulation the center of gravity of the charge variation induced by the signal V_{ac} is shifted from the metal/semiconductor interface towards the semiconductor insulator/interface. As a consequence, the measured capacitance grows from a value given by the series connection of the depleted semiconductor and of the insulator up to a value close to the insulator capacitance in deep accumulation (see Fig. 9.5). This transition occurs more abruptly when the semiconductor DOS width is smaller. In this technique, the assessment of the DOS width is performed by measuring the capacitance at a suitably low frequency to ensure the device is in quasi-equilibrium. Thanks to the quasi-equilibrium condition, carrier mobility is not involved. On the one hand, the disentanglement between the DOS and the carrier mobility is a great advantage of this technique over those involving carrier transport, because (i) carrier transport theory is not well assessed yet and is still a subject of debate in the scientific community and (ii) it depends on the DOS itself. On the other hand the quasi-equilibrium condition simplifies numerical simulations, which require the solution of a non-linear Poisson equation: compared to a drift-diffusion based approach, this is less complex and numerically more accurate. To verify the capacitor is in quasi-equilibrium, a frequency scan of the capacitance (*C–F* measurement) is performed: the low frequency, quasi-equilibrium region is identified as a plateau, followed by a decrease of the capacitance at higher frequency occurring when carriers, due to the finite transit time across the semiconductor start lagging behind V_{ac} (note that organic insulators can have a frequency-dependent permittivity that might obscure the identification of the plateaux). In addition, the semiconductor has to be patterned or selectively deposited so that its area is smaller than the area of both the bottom and of the top electrode: this suppresses the slow lateral spreading of accumulated carriers which would impede the identification of the *C–F* plateau. The method has been successfully applied to n-type and p-type semiconducting polymers under the hypothesis of a Gaussian DOS (Maddalena et al., 2015) or of a DOS given by the superposition of multiple gaussians (Sung et al., 2016). As an example, Fig. 9.6 reports the fitting of simulated to experimental *C–V* curves and their first derivative for the good electron transporting polymer P(NDI2OD-T2). Furthermore, with the extracted DOS it was possible to reproduce linear transfer characteristic curves of transistors assuming the Extended Gaussian Disorder Model for the carrier mobility (Coehoorn and Bobbert, 2012), with no fitting parameters apart from a multiplicative constant representing the low field, low density carrier mobility (Fig. 9.7). Recently, this technique has been extended in order to properly model also MIS C-F curves in out-of-equilibrium conditions. More details can be found in Ref. Africa et al. (2017). Doing so, a more detailed modeling of the metal/semiconductor contact was introduced.

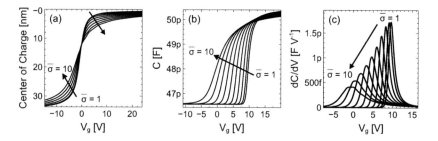

Figure 9.5 Simulations for MIS capacitor based on a n-type organic semiconductor characterized by a gaussian DOS $g(E) = N_0/(\sigma\sqrt{2\pi})$ $\exp[-E^{-2}/2\sigma^2]$. Panel (a): position of the center of gravity of the charge variation with respect to the semiconductor/insulator interface as a function of V_g, calculated for various values of disorder parameter $\bar{\sigma} = \sigma/k_B T$. In panels (b) and (c), the *C–V* curves and their first derivatives are shown: the larger the disorder parameter, the smoother the *C–V* curve and the broader the d*C*/d*V* peak. Simulation parameters: semiconductor thickness 35 nm, insulator thickness 445 nm, semiconductor relative dielectric constant 2.9, insulator relative dielectric constant 2.82, energy difference between metal Fermi level and semiconductor LUMO 1 eV, $N_0 = 10^{21}$ cm^{-3}, $T = 295$ K. Adapted with permissions from Maddalena et al. (2015). Copyright 2015, Elsevier.

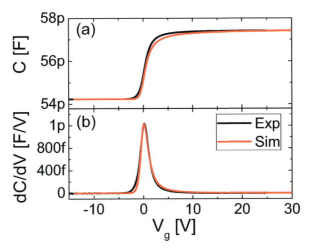

Figure 9.6 Fitting of simulated to experimental *C–V* curves of a MIS capacitor based on a prototypical n-type semiconducting polymer P(NDI2OD-T2): (a) *C–V* curve; (b) first derivative of *C–V* curve. Measurement frequency 9283 Hz. Numerical fitting is obtained with a gaussian DOS width of 78 meV. Reproduced with permissions from Maddalena et al. (2015). Copyright 2015, Elsevier.

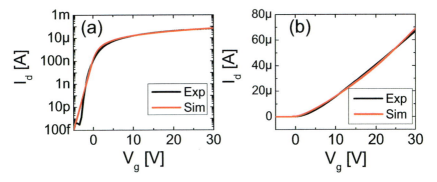

Figure 9.7 Fitting of simulated to experimental linear transfer characteristic curves of a OTFT based on P(NDI2OD-T2) on (a) semilogarithmic and (b) linear scale. Numerical fitting is obtained with the gaussian DOS width of 78 meV extracted from *C–V* curve and with a low-field, low-density mobility prefactor of about 3×10^{-2} cm^2 V^{-1} s^{-1}. The transistor drain to source voltage was set at 5 V. $W/L = 10^3$, Reproduced with permissions from Maddalena et al. (2015). Copyright 2015, Elsevier.

9.5   Conclusions

We have reviewed the main applications of capacitive techniques in the field of organic semiconductors.

In the space charge limited regime, the mobility can be extracted exploiting the crossover from the low frequency region, where carriers can follow the sinusoidal small signal, to the high frequency region where the space charge is not able to respond any longer. The main issue here is due to the presence of trapping centers that can alter and/or obscure the low to high frequency crossover. In addition, an issue specific to organic semiconductors, the resistance introduced by the injecting contact, which is rarely an ideal ohmic contact, should be taken into account properly.

As to Schottky diodes, the DOS of deep trapping centers can be extracted, but it should be underlined that the techniques are borrowed from inorganic semiconductors and applied with little adaption to the peculiarities of organic semiconductors: only recently the issue of carrier mobility, which limits the maximum frequency that can be applied, has been taken into account.

Finally, we have resumed a technique specifically developed for organic semiconductors which exploits capacitance–voltage measurements on metal–insulator–semiconductor capacitors to extract the density of states.

As a conclusive remark, it has to be underlined that it is true that capacitive-based technique are very powerful and can help to discover and investigate many characteristic of organic semiconductors and devices. Nevertheless the limitations and hypothesis behind each technique should be set clear, in order to properly validate and critically evaluate obtained results.

Acknowledgments

The authors are grateful to Dr. C. de Falco for careful reading of the manuscript and for his fruitful suggestions.

References

Africa, P. C., de Falco, C., Maddalena, F., Caironi, M., and Natali, D. (2017). Simultaneous extraction of density of states width, carrier mobility and injection barriers in organic semiconductors, *Scientific Reports* 7, article number 3803, doi 10.1038/s41598-017-03882-8.

Baeg, K.-J., Binda, M., Natali, D., Caironi, M., and Noh, Y.-Y. (2013a). Organic light detectors: Photodiodes and phototransistors, *Advanced Materials* 25, 31, pp. 4267–4295, doi:10.1002/adma.201204979.

Baeg, K.-J., Caironi, M., and Noh, Y.-Y. (2013b). Toward printed integrated circuits based on unipolar or ambipolar polymer semiconductors, *Advanced Materials* 25, 31, pp. 4210–4244, doi:10.1002/adma.201205361, URL http://doi.wiley.com/10.1002/adma.201205361.

Bucella, S. G., Luzio, A., Gann, E., Thomsen, L., McNeill, C. R., Pace, G., Perinot, A., Chen, Z., Facchetti, A., and Caironi, M. (2015). Macroscopic and high-throughput printing of aligned nanostructured polymer semiconductors for MHz large-area electronics, *Nature Communications* 6, p. 8394, doi:10.1038/ncomms9394, URL http://www.nature.com/doifinder/10.1038/ncomms9394.

Caironi, M., and Noh, Y.-Y. (2015). *Large Area and Flexible Electronics* (Wiley-VCH), ISBN 9783527336395.

Coehoorn, R., and Bobbert, P. A. (2012). Effects of Gaussian disorder on charge carrier transport and recombination in organic semiconductors, *Physica Status Solidi (A)* 209(12), pp. 2354–2377, doi:10.1002/pssa.201228387, URL http://doi.wiley.com/10.1002/pssa.201228387.

Dascalu, D. (1966). Small-signal theory of space-charge-limited diodes, *International Journal of Electronics* doi:10.1080/00207216608937906.

Dascalu, D. (1968). Trapping and transit-time effects in high-frequency operation of space-charge-limited dielectric diodes. Frequency characteristics, *Solid State Electronics* 11(4), 491–499.

Ehrenfreund, E., Lungenschmied, C., Dennler, G., Neugebauer, H., and Sariciftci, N. S. (2007). Negative capacitance in organic semiconductor devices: Bipolar injection and charge recombination mechanism, *Applied Physics Letters* 91(1), p. 012112, doi:10.1063/1.2752024, URL http://scitation.aip.org/content/aip/journal/apl/91/1/10.1063/1.2752024.

Garcia-Belmonte, G., Boix, P. P., Bisquert, J., Sessolo, M., and Bolink, H. J. (2010). Simultaneous determination of carrier lifetime and electron density-of-states in P3HT:PCBM organic solar cells under illumination by impedance spectroscopy, *Solar Energy Materials and Solar Cells* 94(2), pp. 366–375, doi:10.1016/j.solmat.2009.10.015.

Germs, W. C., van der Holst, J., van Mensfoort, S., Bobbert, P., and Coehoorn, R. (2011). Modeling of the transient mobility in disordered organic semiconductors with a Gaussian density of states, *Physical Review B* 84(16), p. 165210, doi:10.1103/PhysRevB.84.165210, URL http://link.aps.org/doi/10.1103/PhysRevB.84.165210.

Heath, J., and Zabierowski, P. (2011). *Capacitance Spectroscopy of Thin-Film Solar Cells* (Wiley-VCH Verlag GmbH and Co. KGaA), ISBN 9783527636280, pp. 81–105, doi:10.1002/9783527636280.ch4, URL http://dx.doi.org/10.1002/9783527636280.ch4.

Heremans, P., Tripathi, A. K., de Jamblinne de Meux, A., Smits, E. C. P., Hou, B., Pourtois, G., and Gelinck, G. H. (2016). Mechanical and electronic properties of thin-film transistors on plastic, and their integration in flexible electronic applications, *Advanced Materials* 28(22), pp. 4266–4282, doi:10.1002/adma.201504360, URL http://doi.wiley.com/10.1002/adma.201504360.

Ishii, H., Kudo, K., Nakayama, T., and Ueno, N. (eds.) (2015). *Electronic Processes in Organic Electronics*, Springer Series in Materials Science, vol. 209 (Springer Japan, Tokyo), ISBN 978-4-431-55205-5, doi:10.1007/978-4-431-55206-2, URL http://link.springer.com/10.1007/978-4-431-55206-2.

Kang, H., Kim, G., Kim, J., Kwon, S., Kim, H., and Lee, K. (2016). Bulk-heterojunction organic solar cells: Five core technologies for their commercialization, *Advanced Materials* 28(36), pp. 7821–7861, doi:10.1002/adma.201601197, URL http://doi.wiley.com/10.1002/adma.201601197.

Kassing, R. (1975). Calculation of the frequency dependence of the admittance of SCLC diodes, *Physica Status Solidi (A)* doi:10.1002/pssa.2210280110.

Kassing, R., and Kähler, E. (1974). The small signal behavior of SCLC-diodes with deep traps, *Solid State Communications* doi:10.1016/0038-1098(74)91169-7.

Kimerling, L. C. (1974). Influence of deep traps on the measurement of free-carrier distributions in semiconductors by junction capacitance techniques, *Journal of Applied Physics* 45(4), p. 1839, doi:10.1063/1.1663500, URL http://scitation.aip.org/content/aip/journal/jap/45/4/10.1063/1.1663500.

Lampert, M., and Mark, P. (1970). *Current Injection in Solids* (Academic Press).

Li, J. V., Nardes, A. M., Liang, Z., Shaheen, S. E., Gregg, B. A., and Levi, D. H. (2011). Simultaneous measurement of carrier density and mobility of organic semiconductors using capacitance techniques, *Organic Electronics: Physics, Materials, Applications* 12(11), pp. 1879–1885, doi:10.1016/j.orgel.2011.08.002, URL http://dx.doi.org/10.1016/j.orgel.2011.08.002.

Lungenschmied, C., Ehrenfreund, E., and Sariciftci, N. (2009). Negative capacitance and its photo-inhibition in organic bulk heterojunction devices, *Organic Electronics* 10(1), pp. 115–118, doi:10.1016/j.orgel.2008.10.011, URL http://linkinghub.elsevier.com/retrieve/pii/S1566119908002024.

Lüssem, B., Keum, C.-M., Kasemann, D., Naab, B., Bao, Z., and Leo, K. (2016). Doped organic transistors, *Chemical Reviews* doi:10.1021/acs.chemrev.6b00329, URL http://dx.doi.org/10.1021/acs.chemrev.6b00329.

Lüssem, B., Riede, M., and Leo, K. (2013). Doping of organic semiconductors, *Physica Status Solidi (A)* 210(1), pp. 9–43, doi:10.1002/pssa.201228310, URL http://dx.doi.org/10.1002/pssa.201228310.

Maddalena, F., de Falco, C., Caironi, M., and Natali, D. (2015). Assessing the width of Gaussian density of states in organic semiconductors, *Organic Electronics* 17, pp. 304–318, doi:10.1016/j.orgel.2014.12.001, URL http://linkinghub.elsevier.com/retrieve/pii/S1566119914005473.

Martens, H., Brom, H., and Blom, P. W. M. (1999). Frequency-dependent electrical response of holes in poly(p-phenylene vinylene), *Physical Review B Condensed Matter and Materials Physics* 60(12), pp. R8489–R8492, doi:10.1103/PhysRevB.60.R8489.

Montero, J. M., Bisquert, J., Garcia-Belmonte, G., Barea, E. M., and Bolink, H. J. (2009). Trap-limited mobility in space-charge limited current in organic layers, *Organic Electronics* 10(2), pp. 305–312, doi:10.1016/j.orgel.2008.11.017, URL http://linkinghub.elsevier.com/retrieve/pii/S1566119908002310.

Natali, D., and Caironi, M. (2012). Charge injection in solution-processed organic field-effect transistors: physics, models and characterization methods, *Advanced Materials* 24(11), pp. 1357–1387, doi:10.1002/adma.201104206, URL http://www.ncbi.nlm.nih.gov/pubmed/22354535.

Natali, D., and Caironi, M. (2016). Organic photodetectors, in *Photodetectors* (Woodhead Publishing), Chapter 7, ISBN 9781782424451, pp.195–254, doi:10.1016/B978-1-78242-445-1.00007-5.

Ogawa, S. (ed.) (2015). *Organic Electronics Materials and Devices* (Springer Japan, Tokyo), ISBN 978-4-431-55653-4, doi:10.1007/978-4-431-

55654-1, URL http://link.springer.com/10.1007/978-4-431-55654-1.

Okumoto, H., and Tsutsui, T. (2014). A source of negative capacitance in organic electronic devices observed by impedance spectroscopy: Self-heating effects, *Applied Physics Express* 7(6), doi:10.7567/APEX.7.061601.

Pace, G., Grimoldi, A., Sampietro, M., Natali, D., and Caironi, M. (2015). Printed photodetectors, *Semiconductor Science and Technology* 30(10), doi:10.1088/0268-1242/30/10/104006.

Ramachandhran, B., Huizing, H., and Coehoorn, R. (2006). Charge transport in metal/semiconductor/metal devices based on organic semiconductors with an exponential density of states, *Physical Review B* 73(23), p. 233306, doi:10.1103/PhysRevB.73.233306, URL http://link.aps.org/doi/10.1103/PhysRevB.73.233306.

Reineke, S., Thomschke, M., Lüssem, B., and Leo, K. (2013). White organic light-emitting diodes: Status and perspective, *Reviews of Modern Physics* 85(3), pp. 1245–1293, doi:10.1103/RevModPhys.85.1245, URL http://link.aps.org/doi/10.1103/RevModPhys.85.1245.

Schmeits, M. (2007). Electron and hole mobility determination in organic layers by analysis of admittance spectroscopy, *Journal of Applied Physics* 101(8), p. 084508, doi:10.1063/1.2719014, URL http://link.aip.org/link/JAPIAU/v101/i8/p084508/s1{\&}Agg=doi.

Sirringhaus, H. (2014). 25th Anniversary article: Organic field-effect transistors: The path beyond amorphous silicon, *Advanced Materials* 26(9), pp. 1319–1335, doi:10.1002/adma.201304346, URL http://doi.wiley.com/10.1002/adma.201304346.

Someya, T., Kaltenbrunner, M., and Yokota, T. (2015). Ultraflexible organic electronics, *MRS Bulletin* 40(12), pp. 1130–1137, doi:10.1557/mrs.2015.277, URL http://www.journals.cambridge.org/abstract{_}S0883769415002778.

Sung, M. J., Luzio, A., Park, W.-T., Kim, R., Gann, E., Maddalena, F., Pace, G., Xu, Y., Natali, D., de Falco, C., Dang, L., McNeill, C. R., Caironi, M., Noh, Y.-Y., and Kim, Y.-H. (2016). High-mobility naphthalene diimide and selenophene-vinylene-selenophene-based conjugated polymer: n-channel organic field-effect transistors and structure-property relationship, *Advanced Functional Materials* 26(27), pp. 4984–4997, doi:10.1002/adfm.201601144, URL http://doi.wiley.com/10.1002/adfm.201601144.

Takagi, K., Abe, S., Nagase, T., Kobayashi, T., and Naito, H. (2015). Characterization of transport properties of organic semiconductors using impedance spectroscopy, *Journal of Materials Science: Materials in Electronics* 26(7), pp. 4463–4474, doi:10.1007/s10854-015-3070-8, URL http://dx.doi.org/10.1007/s10854-015-3070-8.

Tang, Y., Peng, Y., Sun, L., Wei, Y., and Xu, S. (2015). Determining charge carrier mobility in Schottky contacted single-carrier organic devices by impedance spectroscopy, *EPL (Europhysics Letters)* 112(1), p. 17007, doi:10.1209/0295-5075/112/17007, URL http://stacks.iop.org/0295-5075/112/i=1/a=17007?key=crossref.d93dc16bc3b16 25af51db6cb7f0fc8e2.

Thejokalyani, N., and Dhoble, S. (2014). Novel approaches for energy efficient solid state lighting by RGB organic light emitting diodes: A review, *Renewable and Sustainable Energy Reviews* 32, pp. 448–467, doi:10.1016/j.rser.2014.01.013.

Tripathi, A. K., and Mohapatra, Y. (2010). Mobility with negative coefficient in PooleFrenkel field dependence in conjugated polymers: Role of injected hot electrons, *Organic Electronics* 11(11), pp. 1753–1758, doi:10.1016/j.orgel.2010.07.019, URL http://linkinghub.elsevier.com/retrieve/pii/S1566119910002491.

Walter, T., Herberholz, R., Müller, C., and Schock, H. W. (1996). Determination of defect distributions from admittance measurements and application to Cu(In,Ga)Se$_2$ based heterojunctions, *Journal of Applied Physics* 80(8), pp. 4411–4420, doi:10.1063/1.363401, URL http://scitation.aip.org.nthulib-oc.nthu.edu.tw/content/aip/journal/jap/80/8/10.1063/1.363401.

Zhang, B., Chen, Y., Neoh, K.-G., and Kang, E.-T. (2016). organic electronic memory devices, in *Electrical Memory Materials and Devices* (The Royal Society of Chemistry), ISBN 978-1-78262-116-4, Chapter 1, pp. 1–53, doi:10.1039/9781782622505-00001, URL http://dx.doi.org/10.1039/9781782622505-00001.

Chapter 10

Capacitance Spectroscopy for MOS Systems

Salvador Dueñas and Helena Castán

Department of Electronics,
University of Valladolid, Valladolid, ES47011, Spain

sduenas@ele.uva.es, helena@ele.uva.es

10.1 Introduction: A Historical Perspective

Capacitance studies of metal–oxide–semiconductor (MOS) capacitors have been extensively used since the early 60s of the past century to investigate the interface surface states, oxide charge and electron and ion phenomena in these structures. The simplicity of the MOS structures and the sensitivity of their capacitance characteristics have been advantageously used as a test vehicle for the study of the influence of the materials, fabrication technologies, and process parameters from the early days of the microelectronics up to now.

The original MOS capacitor was proposed by J. L. Moll in 1959 [1] as a voltage-control capacitance, known as varicap. In 1962, L. M. Terman [2] published a historical paper introducing this

Capacitance Spectroscopy of Semiconductors
Edited by Jian V. Li and Giorgio Ferrari
Copyright © 2018 Pan Stanford Publishing Pte. Ltd.
ISBN 978-981-4774-54-3 (Hardcover), 978-1-315-15013-0 (eBook)
www.panstanford.com

device as suitable to study the surface states at the semiconductor oxidized semiconductor surface. Probably, they did not imagine at that time that this device would be the core of the solid-state electronics revolution. From the beginning of the VLSI era, the Therman-Moll method has been profusely used to determine the interface trap time constant and concentration on MOS devices. This method consists of measuring the variations of gate capacitance as a function of frequency and bias voltage. Interface trap properties are obtained by comparing real and theoretical curves.

The first demonstration of using MOS capacitance voltage data to study electronic device grade oxides on silicon was presented by Grove, Deal, Snow, and Sah in 1965 [3]. The same authors presented the correct physics of the HFCV curve [4].

These techniques have been improved during five decades yielding to more sophisticated techniques, which will be presented in this chapter. Throughout the chapter, we will give detailed information about the theoretical basis, and examples of application of these techniques in a variety of MOS systems.

10.2 Capacitance Spectroscopy for MOS Systems

In this section, we summarize the capacitance spectroscopy techniques developed for the electrical characterization of MOS. These techniques provide detailed information of defects existing in the insulator bulk itself and interface states appearing at the insulator-semiconductor surface. These defects induce the apparition of several types of charges commonly named as fixed and mobile charges, border traps and interface states (Fig. 10.1).

Interface states appear at the insulator/semiconductor interface in the energy band gap of the semiconductor. These states are very close to the semiconductor in such a way that they can interact with free carriers existing at the semiconductor in very short times, from nanoseconds to milliseconds depending on their energy in the band gap. The interface states show a quasi-continuous distribution, D_{it}, in the band gap. The electrical nature of these states can be donor or acceptors with charge states positive or neutral for energies above the Fermi level, E_F, and neutral or negative under the Fermi level. Interface states trap free carriers in transistors channels, so degrading the carrier mobility by scattering.

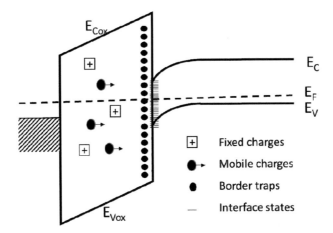

Figure 10.1 Energy band diagram and types of charges existing on a MOS capacitor.

Border traps are similar to interface states, but are located at inner positions inside the insulator, in such a way that they interact with the semiconductor bands as well, but with longer times.

Two kinds of traps and charges may occur at the insulator. Fixed charges due to traps originated from imperfections or contaminants, in the insulator atomic structure. These charges influence the charge, capacitance or voltage drop in the oxide depending on their location and energy depth. Finally, charges able to be displaced by an applied electric field are called mobile charges. These mobile charges are originated by ionic impurities such as Na^+, Li^+, K^+, and H^+. When these charges move, the charge centroid on the insulator shifts, therefore affecting the capacitance, flatband voltage of the MOS capacitors, or threshold voltage of MOS transistors.

These different types of charges can be measured and characterized by capacitance measurements of MOS capacitors as is discussed below. The capacitance of a MOS capacitor is defined as the change of charge due to a change of voltage, $C = dQ/dV$.

Capacitance measurement consist of applying a small-signal ac voltage and measuring the charge variation. The different types of charges existing on a MOS capacitor affect in different manner the term dQ in equation yielding to different capacitance spectroscopy techniques. These techniques can be classified in two main categories depending on whether they attend to time or frequency response of the MOS capacitance.

10.2.1 Interface States

Several methods exist to obtain defect densities at insulator/ semiconductor interface, such as deep level transient technique (DLTS), high and low (quasi static) frequency capacitance–voltage measurements and admittance spectroscopy. The main effect of interface states is to stretch-out the capacitance–voltage curves. The charge due to the interface states shifts the C–V curves, but the shift amount depends on the surface potential, φ_s, which changes with the applied gate voltage as is illustrated in Fig. 10.2. Traps are filled for energies under the Fermi level, but this change depends on the voltage bias. In Fig. 10.2 we can see that traps are emptied at inversion regime and gradually filled as the voltage goes to depletion and accumulation for n-type semiconductors.

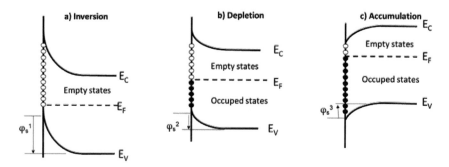

Figure 10.2 Band energy and interface state charge state on a MOS capacitor at inversion, depletion and accumulation regimes.

The total charge on the interface states induces a shift on the capacitor voltage given by

$$\Delta V_{G} = \frac{-Q_{it}(\varphi_{S})}{C_{OX}}.$$ (10.1)

The difference between the ideal and experimental $C–V$ curves of the MOS capacitors is a way to know the interface state distribution, D_{it}.

10.2.1.1 High-low frequency capacitance method

The frequency response of interface traps depends on their location at the band gap. As is well known, traps with energy close to the semiconductor bands have very low emission and capture time constants and can respond to high frequencies. In contrast, traps located in the middle of the gap have long response times, and only respond to very slow voltage variations. This fact is the basis of the high-low frequency capacitance method (HLCV) [5, 6], which provide the interface state density as a function of the energy in the semiconductor band gap.

A capacitance meter is usually employed to measure the high-frequency capacitance, C_{HF}. The quasi-static measurement of the low-frequency capacitance, C_{LF}, consists of recording the gate current while a ramp-voltage ($V_{G} = \alpha t$) is applied to the gate terminal. The ramp-voltage is chosen slow enough to allow all traps to respond to the voltage variation. The capacitance is obtained from the current I according to this expression:

$$I = \frac{dQ_{G}}{dt} = \left(\frac{dQ_{G}}{dV_{G}}\frac{dV_{G}}{dt}\right) = C_{LF} = \frac{dV_{G}}{dt} = \alpha C_{LF}$$ (10.2)

In the charge calculation, we include the term due to interface:

$$Q_{G} = -(Q_{S} + Q_{it})$$ (10.3)

And the capacitance will be now written as

$$C_{LF} = \frac{dQ_{G}}{dV_{G}} = -\frac{dQ_{S} + dQ_{it}}{dV_{OX} + d\varphi_{S}} = -\left(\frac{dV_{OX}}{dQ_{S} + dQ_{it}} + \frac{d\varphi_{S}}{dQ_{S} + dQ_{it}}\right)^{-1}$$

$$= (C_{OX}^{-1} + (C_{S} + C_{it})^{-1})^{-1}.$$ (10.4)

The interface trap density is given by

$$D_{it}(\varphi_S) = \frac{dQ_{it}}{q\,d\varphi_S} = \frac{C_{it}}{q} = \frac{1}{q}\left(\frac{C_{ox}C_G}{C_{ox}-C_G} - C_S\right). \tag{10.5}$$

At high frequency, interface traps do not respond, $C_{it} = \dfrac{dQ_{it}}{d\varphi_S} = 0$, and the capacitance at high frequency is given by

$$\frac{1}{C_{HF}} = \frac{1}{C_{ox}} + \frac{1}{C_S}, \tag{10.6}$$

which allow one to write Eq. (10.5) as

$$D_{it}(\varphi_S) = \frac{C_{ox}}{q}\left(\frac{C_{LF}}{C_{ox}-C_{LF}} - \frac{C_{HF}}{C_{ox}-C_{HF}}\right), \tag{10.7}$$

from which the interface state densities can be obtained for different values of the surface potential.

The surface potential, φ_S, can be obtained from

$$\frac{d\varphi_S}{dV_G} = 1 - \frac{C}{C_{ox}} \tag{10.8}$$

and keeping in mind that $\varphi_S = 0$ at the flatband voltage V_{FB}:

$$\varphi_S = \int_{V_{FB}}^{V_G}\left[1 - \frac{C}{C_{ox}}\right]dV_G \tag{10.9}$$

In Fig. 10.3, we can see an experimental curve obtained by M. Kuhn [6] for Au-Cr-SiO$_2$-Si capacitor.

A limitation of the quasi-static method is the very low level of currents at very low voltage sweep rate. A solution can be to employ large gate areas in order to increase the current. However, leakage currents on the insulator, which are as well proportional to the area, may have values even bigger than the interface states contributions. That is particularly important when the gate insulators are very leaky, as is the case of high-k oxides. In these cases, the quasi-static method is practically prohibited.

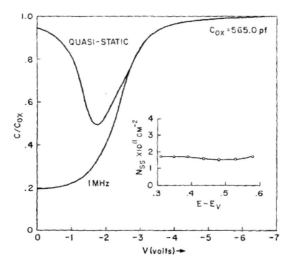

Figure 10.3 Combined high frequency (1 MHz) and quasistatic C–V curves for Au-Cr-SiO$_2$-Si capacitor (10 kΩ p-type Si, 500 A SiO$_2$) and (inset) the resulting surface state distribution as calculated from Eq. (10.7). In the energy range of 0.3–0.4 eV the measured surface state density is expected to be in error (smaller than the actual surface state density), because of surface-state dispersion effects in the 1 MHz curve (reprinted from [6]).

10.2.1.2 DLTS on MOS systems

In contrast to the steady-state C–V methods described so far, transient capacitance spectroscopy gives information by measuring how the non-steady-state high frequency capacitance changes with the time. Deep-level transient spectroscopy (DLTS) invented by Lang [7] is now widely used to detect traps of the so called "deep levels" in the band gap (see Chapter 4). Initially, the method utilized measurements of transient capacitance following the pulsed bias in a p-n junction or Schottky barrier diode to monitor changes in the charge state of defect centers. Schulz and Johnson [8] extended DLTS to study the charge emission from interface states in MOS structures. DLTS was initially developed to study deep levels in semiconductor bulk, but it can be used to characterize interface traps in MOS devices as well. The experimental setup is the same in both cases, but the data interpretation is more complex for interface traps, because interface traps forms a continuum in the band gap, whereas deep levels are discrete.

Historically, there are two ways to carry out the DLTS scanning: by varying the temperature keeping constant the scan filter or window rate, as was primarily proposed by D. V. Lang, or varying the scan filter while keeping constant the temperature. The advantages of both methods are combined in the single-shot DLTS (SS-DLTS). This technique consists of recording 1 MHz isothermal capacitance transients at different temperatures. Afterwards, the transients are processed by applying different window rates. The basis of SS-DLTS setup is described in Fig. 10.4. First, interface traps are filled by applying a positive voltage pulse which drives the MOS capacitor to accumulation. Then, the gate voltage returns to the limit between depletion and weak inversion, causing the traps emit carriers to the semiconductor and yielding capacitance transients, which are recorded for the DLTS processing. The entire capacitance transients are recorded and, consequently, the entire energy spectrum can be processed with only one temperature scan. The capacitance transients are processed as follows: two times t_1 and t_2 (the window rate) are selected. The difference in the capacitance value at these times is the DLTS correlation signal, which is given by [9, 10]

$$\Delta C = \frac{C(t_1)^3}{\varepsilon N_D} \frac{1}{C_{OX}} \int_{E_F^{t1}}^{E_F^{t2}} (\exp(-e_n t_1) - \exp(-e_n t_2)) D_{it} dE, \qquad (10.10)$$

where the emission rate, e_n, depends on the temperature and the level energy, E_T, according to the well-known Arrhenius law:

$$e_n = \sigma_n v_n N_c \exp\left(\frac{E_T - E_C}{kT}\right), \qquad (10.11)$$

where σ_n is the capture cross section, v_n is the electron thermal velocity and N_C is the effective state density at the silicon conduction band. In Eq. (10.10) only those states with emission rates close to the window rate have non-negligible contribution. Indeed, the term inside the integral of Eq. 10.10 (correlation function) has a maximum for interface traps having an emission rate of

$$e_n^{max} = \frac{\ln\left(\dfrac{t_2}{t_1}\right)}{t_2 - t_1} \qquad (10.12)$$

and are located at an energy of

$$E_T^{max} = kT \ln\left(\frac{\sigma_n v_n N_c (t_2 - t_1)}{\ln(t_2/t_1)}\right) \qquad (10.13)$$

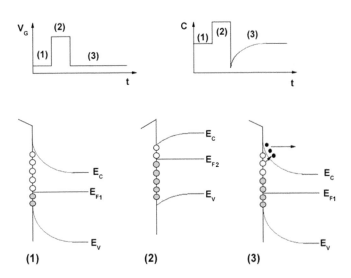

Figure 10.4 Capacitance transients described in an energy band diagram of an n-type MOS capacitor. (1) At the end of the previous transient interface traps are emptied above the Fermi level E_{F1}, (2) an accumulation gate voltage fills the traps up to the Fermi Level E_{F2} and (3) switching to inversion make traps between E_{F1} and E_{F2} to be emptied again.

In summary, the main contribution to $\Delta C(t)$ is due to traps with the energy given by Eq. (10.13) and decays very sharply when energy varies from the maximum. Therefore, Eq. (10.10) can be written as

$$\Delta C = \frac{C(t_1)^3}{\varepsilon_S N_D} \frac{kT}{C_{OX}} D_{it}(E_T^{max}) \ln\left(\frac{t_2}{t_1}\right), \qquad (10.14)$$

which allows to obtain the interface state density at E_T^{max}:

$$D_{it}(E_T^{max}) = -\frac{\varepsilon_s N_D}{kT \ln\left(t_2/t_1\right)} \frac{C_{ox}}{C(t_1)^3} \Delta C \qquad (10.15)$$

Equation (10.13) indicates that for a given window rate the energy is proportional to temperature. Therefore, low temperature transients provide D_{it} of states close to the majority carrier semiconductor band (conduction band for n-type or valence band for p-type). As temperature increases deeper states densities are obtained. Equation (10.15) indicates that D_{it} is proportional to $\Delta C/T$, that is, the sensitivity is lower for deeper states. Since SS-DLTS is a differential technique, its sensitivity is much higher than capacitance–voltage or conductance–voltage techniques. Typical sensitivities are in the range of 10^9 eV^{-1} cm^{-2}, which is lower than the state-of-the-art of thermal silicon oxide with silicon interface. One limitation of the DLTS is the difficulty of characterizing defects from the midgap to the minority carrier band. High temperatures are required to determine the state density for midgap energy levels. At so high temperatures, the generation/recombination processes occurring at the inversion channel induce the apparition of a false peak on the DLTS signals which disturbs the contribution of interface traps [11, 12].

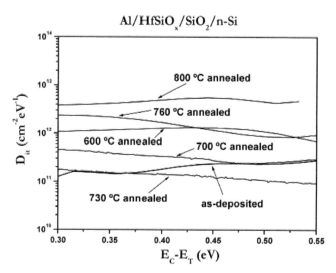

Figure 10.5 SS-DLTS Interface state profiles for Al/HfSi$_x$O$_y$/SiO$_2$/n-Si capacitors (reprinted from [13]).

Figure 10.5 is an example of SS-DLTS applied to the case of a hafnium silicate/silicon oxide on n-type silicon [13]. The silicate was deposited by atomic layer deposition (ALD). This study demonstrated that post deposition thermal annealing improves the quality of the interface up to certain temperatures. In contrast, non-uniformly distributed positive charge builds up in the dielectric stack, affecting the flatband voltage shift.

10.2.2 Border Traps and Disorder Induced Gap States

It is considered that one of the causes for the poor MOS characteristics with thermal oxide is the existence of near-interface oxide traps (NITs), which are sometimes expressed as border traps. These traps should be distinguished from interface state density located just at the interface. For example, it is reasonable to consider that the threshold voltage instability or hysteresis in the C–V curves of MOS capacitors and in the drain current–gate voltage (I_d–V_g) characteristics of MOSFETs is caused by the charge trapping of NITs. The main difference with interface states resides in the bigger value of the time constant when interacting with the semiconductor band. That is why border traps are usually named slow traps, whereas interface traps are called fast traps. However, it is reasonable to assume that both kind of defects have similar origin as was proposed in the disorder-induced gap states (DIGS) model of He et al. [14, 15]. When the insulator-semiconductor interface is formed the crystalline periodicity is broken in both materials, yielding to distortions like mechanical stress, loss of stoichiometry, point defects, etc. These defects induce a continuum of states distributed both in energy and in the spatial coordinate perpendicular to the interface (Fig. 10.6). In the He et al. model, defects are acceptors under a certain energy E_{H0}, and donors when they create energy levels above E_{H0}. Defects are not only localized at the interface, but they reach inner positions on the insulator (region B) and the semiconductor (region A). This model assumes several reasonable hypothesis. The extension of the region A is much lower than De Broglie's wavelength of free carriers. That means that the carrier emission and capture occur at the interface. This assumption has been confirmed by electron microscopy [16]. In contrast, the wave function at the insulator is assumed to decay exponentially, which is equivalent to assume a square barrier at the interface.

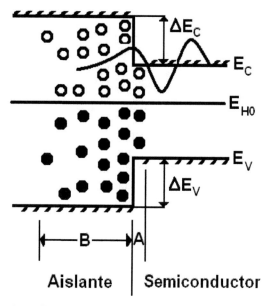

Figure 10.6 Disorder-Induced gap state distribution on an insulator/ semiconductor interface.

This model can explain the hysteresis usually observed in MOS devices. At stationary conditions, the energy states are emptied or filled depending on whether they are above or below the Fermi level (line $\overline{AB}$ in Fig. 10.7). In the case of a linear variation of the gate voltage which makes the Fermi level to increase with time, states over line $\overline{AB}$ will be gradually filled (either by electron capture or hole emission). The occupation factor $f(E, x, t)$ will follow the Shockley–Read–Hall statistics:

$$\frac{df(E,x,t)}{dt} = -[c_n(E,x)n_s(t) + e_p(E,x)][1 - f(E,x,t)], \qquad (10.16)$$

where $n_S(t)$ is the electron density at the interface. The electron capture, $c_n(E,x)$ and hole emission, $e_p(E,x)$, coefficients exponentially decay with the distance to the interface due to the tunneling barrier existing between the semiconductor and the trap [14]:

$$c_n(E,x) = \sigma_0^n v_{th}^n \exp\left(\frac{-2x\sqrt{2m_n^*\Delta E_c}}{\hbar}\right) \qquad (10.17)$$

$$e_p(E,x) = \sigma_0^p v_{th}^p \exp\left(\frac{-2x\sqrt{2m_p^*\Delta E_v}}{\hbar}\right) N_v \exp\left(-\frac{E-E_v}{kT}\right), \qquad (10.18)$$

where $\Delta E_{c,v}$ are the conduction and valence band energy barriers between the insulator and the semiconductor, $\sigma_0^{n,P}$, $v_{th}^{n,P}$, and $m_{n,P}^*$ are the electron and hole capture cross section, the thermal velocity and the effective mass of electron and holes at the semiconductor, and N_V is the density of states of the semiconductor valence band.

The solution of Eq. (10.18) is complex and can be written as

$$f(E,x,t) = 1 - \exp[-I_c(E,x,t)], \qquad \text{for } t > t_0, \qquad (10.19)$$

where t_0 is the moment when the Fermi level crosses the energy level E. The function $I_c(E,x,t)$ is complex and its expression is described in detail in reference [14]. Basically, $I_c(E,x,t)$ increases with time and decreases with the position x inside the insulator and can be well understood if we keep in mind that the process is tunneling assisted, that is, time consuming. At very long times $(t \to \infty)$, the function $I_c(E,x,t) \to \infty$ and the occupation factor is 1 (all traps are filled). Moreover, the probability of one trap to emit an electron or capture a hole exponentially diminishes with the distance, and, consequently, the filling time decreases with the distance $(I_c(E,x,t) \to \infty$ when $x \to \infty)$. In summary, the occupation factor profile at a given time will be as plotted in Fig. (10.7b).

In contrast, if we assume that levels are initially filled up to an energy E and the bias voltage is changed in such a way that the Fermi level diminishes, the profile of filled traps evolves in a very different manner (Fig. 10.7c). As the Fermi level decreases, the initially filled traps will empty by emitting electrons or capturing holes, and the occupation factor has to be written now as:

$$\frac{df(E,x,t)}{dt} = -[c_p(E,x)p_s(t) + e_n(E,x)]f(E,x,t), \qquad (10.20)$$

and similar to Eq. (10.18)

$$f(E, x, t) = 1 - \exp[-I_e(E, x, t)], \quad \text{for} \quad t > t_0, \qquad (10.21)$$

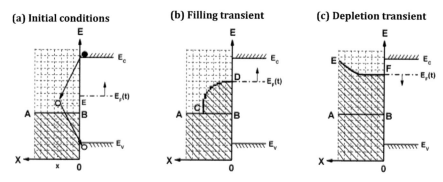

Figure 10.7 Dynamics of charge filling and depletion of DIGS states.

where t_0 is the moment when the Fermi level crosses the energy level E. The function $I_e(E, x, t)$ is similar to $I_C(E, x, t)$, increases with time and decreases with x. At very long times ($t \rightarrow \infty$), the function $I_e(E, x, t) \rightarrow \infty$ and the occupation factor is 0 (all traps are emptied). In contrast, it diminishes with the distance ($I_C(E, x, t) \rightarrow 0$ when $x \rightarrow \infty$), and the occupation factor will be 1 (traps remain filled for very long time) at locations very far away rom the interface. In summary, we can see in Fig. (10.7) that the occupation factor profile is clearly asymmetric depending on the direction of the gate voltage swing. When voltage ramps of finite durations are applied in both directions, it occurs that there is not enough time to empty all filled traps during the emission direction, or to fill all the empty traps during the capture ramp. The difference on the emitted and captured charge in both directions can hence explain the hysteresis mechanisms usually observed in C–V curves of MOS capacitors and in the drain current–gate voltage (I_d–V_g) characteristics of MOSFETs.

In the case of the gate voltage varying instantaneously, capacitance transients will occur where the contribution of all the traps has to be considered. In a case in which the Fermi level is varied between two values E_{F1} and E_{F2}, ($E_{F2} > E_{F1}$) the charge variation should be calculated as:

$$\Delta Q(t) = \int_{E_{F1}}^{E_{F2}} \int_{0}^{t_{ox}} qf(E,x,t) N_{DIGS}(E,x) d\varphi_S dx \qquad (10.22)$$

and, therefore, it will have many contributions with a very broad distribution of time constants. Consequently, to extract individual contribution of densities of DIGS traps at each depth and location is not possible in practice. This is the reason why an effective quantitative analysis method of these near interface traps has not been established. Recently, Fujino and Kita [17] have proposed an approximate method to extract the Near Interface Traps from capacitance transients on a SiO_2/SiC interface. They assume that the wide-spread response times, causes that the time-dependent capacitance change ($\Delta C(t)$) is better described using the "extended-Debye relaxation model" [18] instead of a conventional Debye relaxation model with a single time constant. In the extended-Debye relaxation model, $\Delta C(t)$, is expected to be expressed by the following equation if the gate bias changes from V_{trap} to V_{meas} at $t = 0$:

$$|C(t) - C(\infty)| = |(C(0) - C(\infty))| \exp[-(t_{eff})^{\beta}], \qquad (10.23)$$

where β is the stretched exponential factor ($0 < \beta < 1$) of the response time, t_{eff} is the effective response time of a group of traps with distributed response times. The equation is identical to a simple exponential function with a single time constant when β is unity, and decreases for broader time dependences, which expresses the integral of the contributions of all the relaxation processes with different response times distributed around t_{eff}. Equation (10.23) does not allow one to obtain the profile of the DIGS traps, but is a simple equation to describe the global response of the traps that dominantly contributes to the observed C–t characteristics.

A method to evaluate the profile of DIGS traps both in the distance from the interface and energy was developed by Dueñas et al. [19]. This method consists of deconvoluting the total contribution by carrying out the measurements at different frequencies. At a given frequency, traps with emission and/or capture time constants much lower than the ac signal frequency cannot follow the voltage variation, so reducing the capacitance transient to the contribution of the fastest traps.

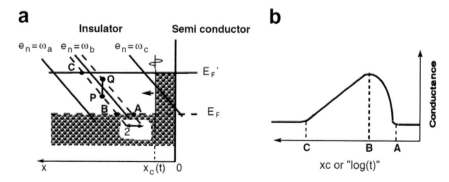

Figure 10.8 Schematic band diagram of an I–S interface illustrating the capture electrons by DIGS continuum states during a conductance transient. (b) General shape of the conductance transient.

One can imagine that making capacitance transient measurements in a broad range of frequencies the different contributions could be obtained. Nevertheless, the capacitance contribution of each component is so small that are no detectable in practice. In contrast, the real part of the MOS capacitor admittance (conductance) can be better used, because at each frequency the conductance transients have only contributions from traps with time constant equal to the signal frequency (see ref. [19]). Figure 10.8 is a schematic illustrating how conductance transients occur in a MOS structure on an n-type semiconductor substrate. At the beginning, a pulse is applied which drives the MOS capacitor from deep to weak inversion. E_F and E_F' are the locations of the Fermi level before and after the pulse. DIGS states with energies between E_F and E_F' trap electrons coming from the substrate conduction band. This process is tunneling assisted and, consequently, time consuming. Defects near the interface capture electrons before the ones located far away inside the dielectric. In Fig. 10.8, we define $x_C(t)$ as the distance covered by the electron front during the time t. The conductance is measured by applying ac signal of frequency ω_b. Only those defects with emission and capture rates close to ω_b contribute to the conductance (those located over equi-emission line $e_n = \omega_b$) and this contribution occurs just when the electron front reaches the region where these states are located [20]. Once the electron front reaches the point, the conductance signal is proportional to the DIGS states concentration profile. According the model introduced by He et al. model [15], the DIGS concentration decays with the distance. At the end, all the DIGS states susceptible to contribute to the conductance signal are filled and the conductance achieves its initial value.

The analytical model presented in reference [21] allows obtaining the DIGS state profile as a function of the spatial distance to the interface and the energy position by measuring conductance transients at different frequencies and temperatures. In the case of an n-MOS structure, the DIGS concentration is

$$N_{\text{DIGS}}(E(t), x_c(t)) = \frac{\Delta G_{SS}(t)}{0.4 q A \omega}, \tag{10.24}$$

where $E(t)$ is the energy of the DIGS states which a given time t during the transient contribute to the conductance variation. $x_C(t)$ is the distance covered by the front of tunneling electrons during the time t, and is given by, $x_C(t) = x_{on} \ln(\sigma_0 t_{th} n_s t)$, where $x_{on} = \frac{h}{2\sqrt{m_{eff} H_{eff}}}$ is the tunneling decay length, σ_0 is the carrier capture cross-section value for $x = 0$, v_{th} is the carrier thermal velocity in the semiconductor, and n_s is the free carrier density at the interface. Finally, m_{eff} is the electron effective mass at the dielectric and H_{eff} is the insulator semiconductor energy barrier for majority carriers, that is, the dielectric to semiconductor conduction band offset.

Figure 10.9 shows x_{on} for some high-k dielectrics electron effective mass and barrier height values have been obtained from references [22] and [23], respectively. One can see that x_{on} is higher for dielectrics in which H_{eff} and m_{eff} are low. In these cases, the tunneling front x_C is faster and, consequently, transients reach deeper locations in the dielectric. An important trend can be derived from this figure: as permittivity increases, tunneling decay length increases providing deeper DIGS profiles.

Finally, the energy positions of DIGS states in the band gap of the dielectric are obtained from the equi-emission line concept defined in Ref. [14] and considering that the measurement frequency is related to emission rate by $e_n = \omega/1.98$ [20]:

$$E' - E(x_c, t) = H_{eff} + kT \ln \frac{\sigma_0 v_{th} N_c}{\omega/1.98} - \frac{kT}{x_{on}} x_c(t) \tag{10.25}$$

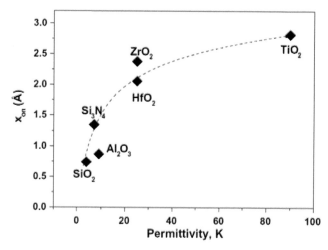

Figure 10.9 Tunneling decay length versus permittivity for several dielectrics.

When temperature decreases the emission rates of all interface states exponentially decrease, and the equi-emission lines shift approaching the interface. Thus, transients are modified in a similar way as when frequency is increased while keeping constant the temperature. DIGS three-dimensional profile or contour line maps can be obtained using Eqs. (10.24) and (10.25). As for the experimental sensitivity, temperature measurement involves an error of 0.1 K. Estimated errors of energy and defect concentration values on DIGS profiles are of about 10 meV and 5×10^{-9} eV^{-1} cm^{-2}, respectively. Estimated precision on DIGS depth is of about 2Å.

The experimental setup consists of a pulse generator to apply bias pulses, a lock-in analyzer to measure the conductance, and a digitizing oscilloscope to record conductance transients. Samples are cooled in darkness from room temperature to 77 K in a cryostat.

Figure 10.10 is an example of DIGS profiles obtained from conductance transients on MOS structures fabricated with ALD gadolinium oxide as dielectric.

Typically, DIGS states are localized in a shallow region within several tens of angstroms from the interface and at energies close to the conduction and valence bands of the semiconductor.

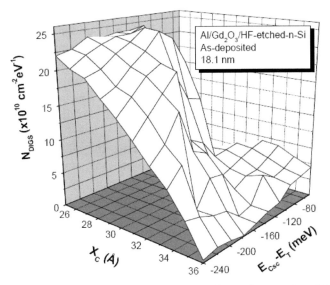

Figure 10.10 Example of DIGS profile: atomic layer deposited gadolinium oxide films.

10.2.3 Charge Traps in the Oxide: Slow Traps and Bulk Charges

Fixed and mobile charges cause serious performance degradation by shifting the threshold voltage, limiting transistor mobility, and reducing device lifetimes. Threshold voltage shifts are observed under positive bias, negative bias and hot-carrier stressing in high-k gate stacks. Charge trapping under positive bias stressing is known to be more severe compared to conventional SiO_2-based gate dielectrics [24]. It is believed to happen due to filling of pre-existing bulk traps. Charge trapping causes threshold voltage shifts and drive current degradation over device operation time. It also precludes accurate mobility (inversion charge) measurements due to a distortion of C–V curves. Negative bias temperature instability (NBTI)-induced threshold voltage shifts in high-k devices are also observed and are comparable to those observed in silicon-based oxide devices.

These traps cause flatband voltage transients in ultra-thin high-k dielectrics on silicon [25]. These transients can be experimentally obtained by measuring the gate voltage while keeping the capacitance constant at the initial flatband condition (C_{FB}). The experiments are carried out under no external stress conditions: zero electric field in the substrate, darkness conditions and no external charge injection. In these conditions, the only mechanism for defect trapping or detrapping is thermal activation, that is, phonons. We proved that the energy of soft-optical phonons in high-k dielectrics is obtained with this experimental approach.

The experimental setup to obtain the flatband voltage transients consists of a feedback system that varies the applied gate voltage accordingly to keep the capacitance at its flatband voltage value. The capacitance is measured with a capacitance meter and a programmable bias source provides the bias voltage to maintain the capacitance at the flatband value, C_{FB}.

The flatband voltage, V_{FB}, of a MOS capacitor is given by

$$V_{FB}(t) = \Phi_{MS} - \frac{Q_i}{C_{OX}} - \frac{1}{\varepsilon} \int_0^{t_{ox}} \rho_{ox}(x, t) x \, dx. \qquad (10.26)$$

When the charge density inside the insulator film, $\rho_{ox}(t)$, varied with time, t, or with the distance from the interface, x, the flatband voltage varies. In particular, trapping and detrapping on defects existing inside the dielectric will produce transient variations of the flatband voltage. According Eq. (10.26) voltage variations are opposite in sign to the charge ones. At flatband voltage condition, there are not electrons or holes directly injected from the gate or semiconductor, i.e., free charges move by hopping from trap to trap. Moreover, since no optical nor electrical external stimulus are applied, free charges must be originated from trapping or detrapping mechanisms of defects existing inside the dielectric and the energy needed to activate these mechanisms only can be provided as thermal energy, that is, phonons.

The experimental setup of this technique is identical to that used to capacitance–voltage technique. The only difference is that in order to obtain the flatband voltage transients we have implemented a feedback system that varies the applied gate voltage accordingly to keep the flatband capacitance value.

The experimental flatband voltage transients become faster when the dielectric thickness diminishes. Time dependences appear to be independent of the temperature. These two facts suggest that there are tunneling assisted process involved. The amplitude of the transients is thermally activated with energies in the range of soft-optical phonons usually reported for high-*k* dielectrics. We have proved that the flatband voltage transients are increasing or decreasing depending on the previous bias history (accumulation or inversion) and the hysteresis sign (clockwise or counter-clockwise) of the capacitance–voltage (*C–V*) characteristics of MOS structures.

To illustrate the technique, we have included in Fig. 10.11 some experimental results for the case of a sample of a 20 nm film of hafnium oxide deposited by ALD on silicon. The amplitude of the flatband voltage transients depends on temperature according to a law:

$$\Delta V_{FB}(T,t) \propto \exp-\left(\frac{\Phi_{ph}}{kT}\right), \tag{10.27}$$

where Φ_{ph} is the energy of the soft optical phonons of the dielectric.

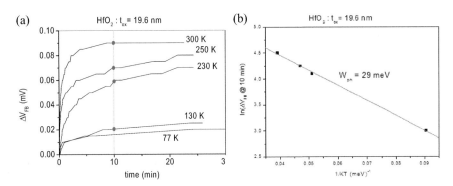

Figure 10.11 Example of DIGS profile: Atomic layer deposited gadolinium oxide films.

10.3 Examples of Capacitance Spectroscopy Application

In this section, we show some examples extracted from the specialized literature to display how the capacitance-spectroscopy techniques, which have been used from the early times of the MOS era, keep their relevance on the research of current and future generations of integrated circuits.

10.3.1 The Classics: SiO$_2$ on Silicon

Probably, one of the main reasons for the unpaired "MOS technology revolution" is the fact of silicon, an excellent semiconductor and one the most abundant elements on earth and its natural oxide (SiO$_2$) form an interface of high electronics quality and very thermodynamically stable. Many works can be found in the literature which demonstrate that capacitance spectroscopy techniques have been profusely used during more than five decades. Of particular interest are the pioneering works related in Section 10.1 [1–4].

Even today, with the change to high-k dielectric materials, unintentional silicon oxide and suboxides interlayer films often appear between the high-k films and the silicon surface. This motivates that the study of the silicon oxide and suboxides continues being a topic of great interest. Many efforts are devoted to understand the physical and chemical properties of these very thin SiO$_x$ films, and the oxidation process itself. This films not only interact with silicon substrate but also with the high-k dielectric, a new interface appear between the high-k and the IL films, inducing surface states at this interface, electron and hole energy barriers, tunneling mechanisms through the IL layer, and many other effects which have to be adequately characterized.

10.3.2 III–V Substrate MOS Capacitors

The scaling of Si-based metal-oxide-semiconductor field-effect transistors is approaching limitations that are determined by the properties of the materials that are currently employed in these devices. One potential solution that would allow further scaling is to replace Si with alternative semiconductor channel materials. In particular, III–V semiconductors have a lower electron effective mass than Si, allowing for high electron mobilities, and are thus of great interest for n-channel MOSFETs. To fabricate MOSFETs with III–V channels, suitable gate dielectrics must be developed, which must form a stable interface within the thermal budget of the transistor fabrication process, possess a high dielectric constant and sufficient band offsets with the semiconductor conduction band to allow for scaling and low gate leakage, and have a low interface trap state density D_{it}. While many high-k dielectric satisfy the two first requirements, achieving low enough interface trap densities has been proven to be extremely challenging.

Many examples of the application of capacitance spectroscopy on III-V semiconductor MOS devices can be found in the literature. As an illustrative example we can mention the work of J. Lin et al. [26] in which they study hysteresis on the capacitance voltage characteristics of Pd/HfO$_2$/InGaAs/InP MOS capacitors. They conclude that the traps responsible for the hysteresis are located near the HfO$_2$/InGaAs interface and not distributed throughout the oxide, indicating that the engineering of the interface transition region between the HfO$_2$ and the InGaAs is central to minimizing C–V hysteresis and thus device instabilities. In addition, the hysteresis amplitude increases with a power law dependence on stress time in accumulation and approaches to a plateau at sufficiently long stress times, consistent with trapping in pre-existing border traps which have a wide range of response time constants.

A detailed study of the different methods to quantify interface trap densities at dielectric/III–V semiconductors interface can be found in Ref. [27]. In this review, authors compare several techniques including conductance method, the high-low frequency CV, and Terman methods in the case of MOS capacitors with high-k/InGaAs interfaces.

10.3.3 High-k Dielectrics

The continuous miniaturization of complementary metal oxide-semiconductor (CMOS) technologies has led to unacceptable tunneling current leakage levels for conventional thermally grown SiO$_2$ gate dielectrics [28, 29]. During the last decades, a lot of efforts have been devoted to investigate alternative high-permittivity (high-k) dielectrics that could allow replacing SiO$_2$ and SiON$_x$ as gate insulators in MOS transistors [22]. The higher dielectric constant provides higher gate capacitances with moderated thickness layers; however, other requirements such as lower leakage currents, high breakdown fields, prevention of dopant diffusion, and good thermodynamic stability must also be fulfilled. Many high-k materials have been investigated during the last years. These dielectrics consist of single layers of metal oxides and silicates (e.g., HfO_2, ZrO_2, $HfSiO_x$, Gd_2O_3, Al_2O_3, TiO_2) directly deposited on n- or p-type silicon. A common practice is using combination of them in the form of multilayers, and gate stacks with silicon oxide or silicon nitride acting as interface layers which prevent from thermodynamic instabilities of directly deposited high-k films on silicon substrates.

Several problems must be fixed before these new gate dielectric materials could be used in fabrication. The first is the quality of the dielectric-semiconductor interface. Any high-k dielectric can provide an interface good as the SiO_2–Si system. Typically, high-dielectrics are not thermodynamically stable on silicon and unintentional silicon oxide interlayer films appear between the high-k dielectric and the silicon surface. Silicon oxide has a lower permittivity value than the high-k oxide so degrading the effective permittivity and the equivalent oxide thickness (EOT), which is defined as the thickness of a silicon oxide film having the same capacitance density as the high-k film deposited. The existence of the interlayer may induce artifacts on capacitance spectroscopy techniques such as C–V and DLTS, and the experimental results require careful interpretation.

An example can be found in [30]. In that work the influences of the silicon nitride blocking-layer thickness on the interface state densities (D_{it}) of HfO_2/SiN_x:H gate-stacks on n-type silicon were analyzed. The blocking layer consisted of 3 to 7 nm-thick silicon nitride films directly grown on the silicon substrates by electron-cyclotron-resonance assisted chemical-vapor-deposition (ECR-CVD). Afterwards, 12 nm-thick hafnium oxide films were deposited by high-pressure reactive sputtering (HPS). Interface state densities were determined by DLTS and by the high- and low-frequency capacitance–voltage (HLCV) method. The HLCV measurements provide interface trap densities in the 10^{11} cm^{-2}eV^{-1} range for all the samples. However, a significant increase of about two orders of magnitude was obtained by DLTS for the thinnest silicon nitride barrier layers. This discrepancy resulted to be an artifact due to the effect of traps located at the internal interface existing between the HfO$_2$ and SiN$_x$:H films. Because charge trapping and discharging are tunneling assisted, these traps are more easily charged or discharged as lower the distance from this interface to the substrate, that is, as thinner the SiN$_x$:H blocking layer. The trapping/detrapping mechanisms increase the amplitude of the capacitance transient and, in consequence, the DLTS signal has contributions not only from the insulator/substrate interface states but also from the HfO$_2$/SiNx:H interlayer traps, and the DLTS signal is overestimated according to the following experimentally obtained formula:

$$D_{it}^* = D_{it} + (\alpha - \beta\sqrt{E_c - E_T})F_{IL}, \qquad (10.28)$$

where D_{it}^* is the as-measured apparent interface state profile. D_{it} is the true trap interface state density profile that is the obtained at low electric filed values, F_{IL} is the electric field existing at the interface, and $(\alpha - \beta\sqrt{E_c - E_T})$ is a parameter associated to the electric field lowering of the energy barrier between the silicon conduction band and traps located at the inner layer interface. The true interface state density, D_{it}, is plotted at Fig. 10.20 as obtained for the lowest accumulation voltage values. These values do agree with those obtained when using HLCV technique. Moreover, this distribution shows a profile consisting of broad gaussian peaks, as is usually reported for silicon nitride films [31–33].

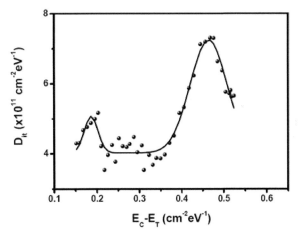

Figure 10.12 True interface state density profile as obtained at low electric fields (<1 MV·cm^{-1}).

The energy diagrams of the MOS structures under accumulation and inversion are displayed in Fig. 10.13. We also assume that defects exist at the HfO_2/SiN_x:H inner layer interface (IL). DLTS measurements consist of applying accumulation pulses to fill the interface states in the upper half of the semiconductor band gap followed by reverse pulses in which the interface states emit electrons to the conduction band yielding the capacitance transients that are conveniently recorded and processed to obtain the D_{it} distribution. If the SiN_x:H film is thin enough, tunneling between the semiconductor and the inner layer interface (IL) may occur. At accumulation, capturing electrons coming from the semiconductor band by direct tunneling fills IL states. Then, when the reverse pulse is applied, these defects emit the captured electrons to the semiconductor band. The emission process may occur in two different ways: IL states with energies above the silicon conduction band (light grey area) emit electrons by direct tunneling (A). On the other hand, for energies ranging from the Fermi level to the semiconductor conduction band (dark grey area) tunneling between the IL states and the interface states (B). These interface states can emit electrons to the conduction band in a similar way as occurs in conventional DLTS (C). Electrons emitted according to the (B) + (C) sequence increase the capacitance transient, obtaining an apparent increase in the measured interfacial state densities. Since all these mechanisms are tunneling assisted, as thinner the silicon nitride films as higher their probability.

Another cause of instability is due to charge trapping and detrapping inside the dielectric. Fixed and trapped charge cause serious performance degradation by shifting the threshold voltage, limiting transistor mobility and reducing device lifetimes. Threshold voltage shifts are observed under positive bias, negative bias and hot-carrier stressing in high-k gate stacks. Charge trapping under positive bias stressing is known to be more severe compared to conventional SiO_2-based gate dielectrics. It is believed to happen due to filling of pre-existing bulk traps. Charge trapping causes threshold voltage shifts and drive current degradation over device operation time. It also precludes accurate mobility (inversion charge) measurements due to a distortion of C–V curves. Negative bias temperature instability

(NBTI)-induced threshold voltage shifts in high-k devices are also observed and are comparable to those observed for silicon-based oxide devices.

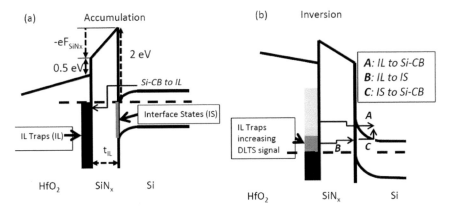

Figure 10.13 Energy band diagram of the HfO_2/SiN_x:H/n-Si MIS structures under accumulation (a) and inversion (b).

As mentioned in Section 10.2.2, border traps are defects located close to the interface that interact with the semiconductor bands and modify the capacitance device. The energy and position distributions of these traps are better determined by the conductance transient technique.

Traps located far away from the interface do not directly interact with the semiconductor, but when they move inside the dielectric, or change their occupation state, can modify the C–V curves, change the flatband voltage of MOS structure and induce electrostatic interaction with carriers in the inversion channel of MOS transistors. Flatband voltage transient technique, previously described, is suitable to characterize these type of defects.

As an example, we show here experimental results obtained in the case of HfO_2, which is one of the most promising high-k dielectrics due to its excellent thermal stability. Figure 10.14 shows three-dimensional DIGS plots for HfO_2 atomic layer deposited on n-Si and over p-Si using chloride as the metal precursor. DIGS states are located at energies close to the majority-carriers band edge of the semiconductor. On the other hand, no conductance transients were observed for samples thinner than 40 Å. Spatially distributed defect bands for films on both n and p types of silicon substrates. These defect bands could be due to oxygen vacancies: when the capacitor structure is terminated by the oxide-Si interface, the electric field existing in the dielectric film moves, for instance, oxygen vacancies (positively charged) to locations farther away to the interface in samples deposited on n-type silicon because of the difference in semiconductor band bending at the interface [34].

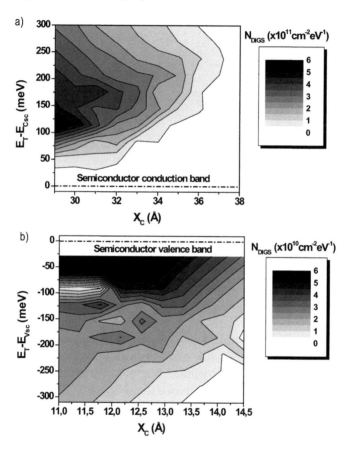

Figure 10.14 Contour plots of DIGS density obtained to 400°C-30 min. annealed $Al/HfO_2/p$-Si (oxide grown at 450°C) and $Al/HfSi_xO_y/n$-Si (silicate grown at 400°C).

An example of flatband voltage transient technique is displayed in Figs. 10.15–10.17 to show the information that can be extracted from the transients as well as the parameters affecting to their amplitude, shape, and time constant. The main parameters affecting the transients are the experimental temperature, the dielectric film thickness, the dielectric material itself and, finally, the bias voltage, and the setup time just before the flatband voltage condition is established in the sample.

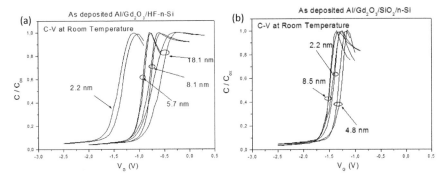

Figure 10.15 Normalized C-V curves of Al/Gd$_2$O$_3$/HF-etched-Si (a) and Al/Gd$_2$O$_3$/SiO$_2$/Si (b) with different Gd$_2$O$_3$ thicknesses, measured at room temperature.

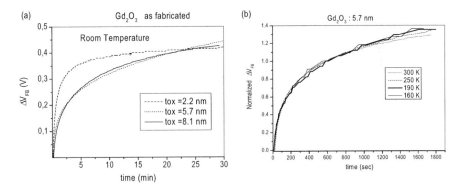

Figure 10.16 Flatband voltage transients at different Gd$_2$O$_3$ thickness, t_{ox}, (a) and temperature (b) of Gd$_2$O$_3$-based MIS structures.

Figure 10.15 shows capacitance–voltage curves obtained at room temperature for as-deposited Al/Gd$_2$O$_3$/ Si(a) and Al/Gd$_2$O$_3$/ SiO$_2$/Si (b) MOS structures with different Gd$_2$O$_3$ thickness. V_{FB} is negative in all cases indicating the existence of positive charge in the dielectric. In Fig. 10.15a, we see that V_{FB} moves to less negative values with thickness indicating that the charge centroid is closer to the interface for thicker films. That means that traps are preferentially created in the very first dielectric layers. Moreover, in Fig. 10.15b we see that when a SiO_2 film is present, V_{FB} shows more negative values and a weaker thickness dependence than when Gd_2O_3 films are directly deposited on HF-etched silicon. That must be due to the existence of non-mobile charge trapped at the interface between the high-k and SiO_2 films. V_{FB} transients for different thicknesses (Fig. 10.16a) reveal time constants increasing with thickness below 5.7 nm. That indicates the existence of charge displacement mechanisms: the thinner the films the lower the distances to be covered for the mobile charges to reach the gate and/or the insulator-semiconductor interface. In Fig. 10.16b, the transients have been normalized at their value at 600 sec. It is clear that the time constant is independent of the temperature, indicating that tunneling mechanisms are involved in the conduction. As for temperature dependency of V_{FB} transients, we have recorded transients at several temperatures (Fig. 10.17a) and we observed that their magnitude follows an Arrhenius plot (Fig. 10.17b) with activation energy in the range of the soft-optical phonon energies (W_{PH}) usually reported for high-k dielectrics [35]. From our fits for different samples we have obtained that for Gd_2O_3 these energies are of about 55±10 meV. These values were obtained for both annealed and as-deposited samples and for Gd_2O_3 film thicknesses from 2 to 20 nm.

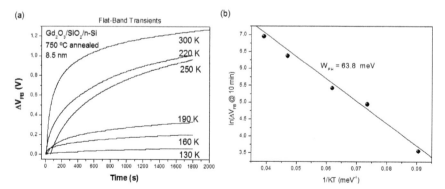

Figure 10.17 Flatband voltage transients at different temperatures (a) and Arrhenius plot of the transient amplitude at 10 min (b) for an $Al/Gd_2O_3/SiO_2/Si$ sample.

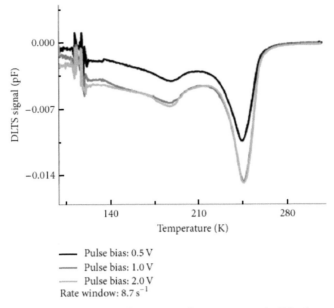

Figure 10.18 DLTS spectra of MOS sample containing Si NCs dependent on fitting pulse bias: as the pulse bias is rising, the transient signal rises too until saturation (Ref. [37]).

From all these observations, one can conclude that the flatband voltage transients under conditions without external stress are originated by phonon-assisted tunneling between localized states: Phonons produce the ionization of traps existing in the band gap of the insulator. Electrons and/or holes generated in this way move by hopping from trap to trap until they reach a defect location and neutralize the charge state of this defect. It is important to point out that the electrons (or holes) do not enter the conduction (or valence) band of the dielectric and the conduction takes place within the band gap.

10.3.4 Localized States: Silicon Nanocrystals Studied by DLTS

Conventional FLASH memories suffer from problems such as short retention, endurance, and SILC (stress induced leakage current), due to charge loss in the floating gate. Memory-cell structures employing discrete traps as the charge storage media have received much attention as promising candidates to replace conventional nonvolatile memories for future high capacity and low power consuming memory devices. Nanocrystal memory devices employing distributed nanodots as storage elements have shown great potential in device applications [36]. A floating gate composed of individual nanocrystals reduces the problem of charge loss encountered in conventional floating-gate devices and allows further scaling down of tunnel oxides thereby lowering working bias and faster write/erase speeds. Silicon nanocrystals create localized states on the insulator which must be conveniently characterized in order to qualify the fabricated devices. In a recent work, Lv and Zhao [37] have demonstrated the convenience of using DLTS to observe thermal emission from silicon nanocrystals around 300 K for certain rate windows. DLTS measurements prove that nanocrystals create an energy level with an activation energy of 0.6 eV and a capture cross section of 10^{-13} cm^2. Authors concluded that these data are consistent with the properties of Si NCs measured by optical methods, such as photoluminescence Auger saturation [38]. These results could be a clue that the trapping mechanism in MOS systems containing Si NCs is related to the quantum levels of the Si quantum dots at around 300 K. This trapping mechanism also is suitable for other types of quantum dots embedded in the dielectric matrix as a system. DLTS supplied information other than obtained as high frequency C–V measurement on other electrical characterizations. Figure 10.18 shows experimental DLTS curves at several bias voltage. The fact of the DLTS peak being independent on the pulse bias demonstrates that the level detected is indeed a localized level.

10.3.5 Radiation Effects

Advanced microelectronic devices are extensively used in radiation environments in applications involving radiology equipment, space navigation and communications, high-energy physics experiment, and so on. The continuous exposure to the ionizing radiation may produce failure effects, especially in recent CMOS technologies, which introduce high-k materials as gate dielectrics in transistors. Therefore, it is of main interest to understand the stability and long-term reliability of these materials before their effective incorporation into commercial ICs. It is known that ionizing radiation generates electron-hole pairs in the gate dielectric of CMOS devices [39, 40]. Some of the radiation-induced charge recombines and does not affect the device behavior; however, a significant amount of the radiation-induced charge can become trapped at micro-structural defects in the dielectric, causing leakage current to flow in the OFF state condition. This will result in an increase in the static power supply current of integrated circuits and may cause their failure. In addition to dielectric charge build-up, ionizing radiation can also induce defects in the insulator-semiconductor interface, thus changing the interface trapping properties of devices [41]. In fact, interface trapped charge causes threshold voltage shifts and mobility degradation in transistors. All these effects can decrease the drive of transistors and degrade the timing parameters of ICs.

To illustrate the radiation effects, we show here an example [42] in which 10 nm-thick films of high-k dielectrics films (Al_2O_3, HfO_2, and nanolaminates of both materials) were grown by ALD on type n- and type p-silicon. MOS capacitors where submitted to 2 MeV electron irradiations of three different fluencies (10^{14}, 10^{15} and 10^{16} e cm^{-2}), with total ionizing doses of about 2.5, 25 and 250 Mrad (Si), respectively, on an electron accelerator at Takasaki-JAEA in Japan. Irradiation was carried out at room temperature on unbiased samples. Table 10.1 summarizes the samples used in this experiment.

Table 10.1 MIS capacitors fabricated for electrical characterization

			Irradiated samples		
			Fluence (e/cm^2)		
		Fresh			
Substrate	Gate dielectric	samples	10^{14}	10^{15}	10^{16}
p-Si (4–50 Ωcm)	Al_2O_3	1 fresh	1d	1e	1f
	HfO_2	2 fresh	2d	2e	2f
	Nanolaminate	3 fresh	3d	3e	3f
p-Si (0.1–1.4 Ωcm)	Al_2O_3	4 fresh	4d	4e	4f
	HfO_2	5 fresh	5d	5e	5f
	Nanolaminate	6 fresh	6d	6e	6f

Figure 10.19 reveals that as the dose increases, more positive charge is accumulated in the bulk dielectric (curves moves to more negative values). Moreover, *C–V* curves show counterclockwise hysteresis with similar amplitude for all the samples, regardless of the irradiation dose. This fact indicates that the concentrations of slow states inside the oxide and near the oxide/semiconductor interface remain constant after electron irradiation, in good agreement with the results previously reported by A. Y. Kang et al. [39].

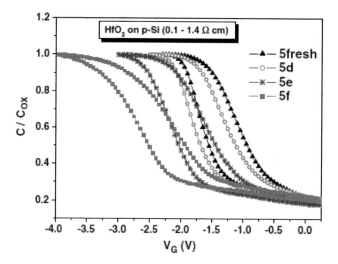

Figure 10.19 Normalized 1 MHz *C–V* curves measured at room temperature of HfO_2 on p-Si (0.1–1.4 Ω cm) submitted to several 2 MeV electron irradiation doses.

As for the interface state density, Fig. 10.20 reveals that electron irradiation do not degrade the oxide/semiconductor interface for low irradiation doses. However, for high doses, the interface trap density increases.

With regard to traps near the interface, one could think that slow (or DIGS) states near high-*k*/Si interface could be formed in the same way as interface states increases. However, conductance transients (Fig. 10.21) show that the DIGS density remained almost constant after irradiation. In contrast, flatband voltage presents a transient behavior, the transient being slightly larger when the irradiation dose increases (Fig. 10.22). This instability could be due to mobile charge remaining inside the insulator after electron irradiation and could be also responsible of the hysteresis increase in C–V curves. In HfO$_2$-based samples, the hysteresis in C–V curves remained constant in a large value, and flatband voltage transient does not depend on irradiation fluency.

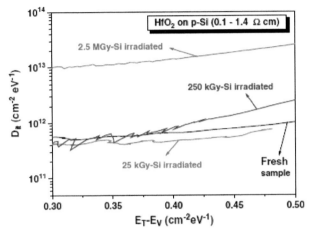

Figure 10.20 DLTS interface profiles of HfO$_2$ on p-Si (0.1–1.4 Ω cm) submitted to several 2 MeV electron irradiation doses.

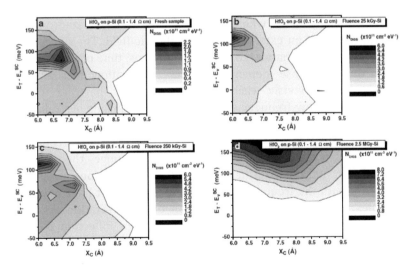

Figure 10.21 DIGS profiles of HfO$_2$ on p-Si (0.1–1.4 Ω cm) submitted to several 2 MeV electron irradiation doses.

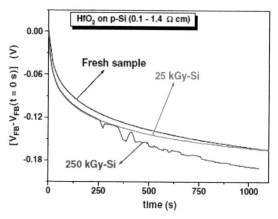

Figure 10.22 Flatband voltage transients for of HfO_2 on p-Si (0.1–1.4 Ωcm) submitted to several 2 MeV electron irradiation doses.

10.3.6 Radio-Frequency Complex Capacitance Spectroscopy

A deep knowledge of dielectric properties could provide a wider insight of dielectric nature. In this sense, RF impedance spectroscopy measurements allow one to detect the dipolar relaxation of the dielectrics. A dielectric material is a non-conducting substance whose bound charges are polarized under the influence of an externally applied electric field. The dielectric relaxation can be characterized in two domains: time or frequency. The time response depends on the polarization mechanism involved. Analogously, each relaxation process have its own frequency. At low frequencies, relaxation mechanisms are related to interface polarization. Charge dipoles appearing at material imperfections (surfaces, grain boundaries, and interfaces) contain dipoles. Externally applied electric fields change the dipole orientation and polarize the material. This polarization results to increase with the frequency. Here, the material must have natural dipoles which can rotate freely. When dipole relaxation occurs, the real part of the permittivity, ε', shows an inflection point, where the imaginary part, ε'', is maximum. In a capacitor, ε' and ε'' are proportional to the capacitance and conductance signals, respectively. Hence, measuring the complex impedance of the device at high frequencies is a direct method to study the permittivity relaxation [43].

This study requires the use of an RF Impedance Analyzer, allowing to realize measurements up to the GHz range. This kind of equipment has been recently introduced in the market and nowadays is possible to make up this characterization. Capacitance and conductance of MOS devices are obtained by scanning the frequency of the RF signal while keeping the gate voltage at a given value. A whole RF characterization is obtained by varying the voltage from accumulation to inversion regime. The influence of the gate voltage on the RF characteristics is obtained in this way. Figure 10.23 plots RF admittance curves of a W/HfO$_2$/Si MOS structure. The most noticeable point is the fact that the frequencies of the inflection point of the capacitance signal and the maximum of the conductance signal depend on the bias voltage: more positive voltages yield to higher relaxation frequencies. In this case, MOS capacitors are in the inversion regime for positive bias and in accumulation for negative ones. The main conclusion is that the inversion layer at the interface channel affects to the dipole relaxation in such a way that it occurs at higher frequencies. In accumulation, the voltage drop in the oxide is equal to the applied gate voltage, whereas in depletion or inversion regime, part of the applied voltage drops in the semiconductor layer close to the interface. Hence, higher electric field exists on the accumulation regime and dipole orientation is more effective in this regime, and dipoles could not respond to so high frequencies as in the inversion regime. Figure 10.24 shows this effect from a three-dimensional point of view.

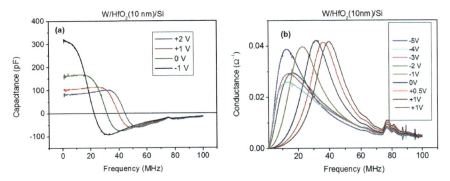

Figure 10.23 Frequency variations of capacitance (a) and conductance (b) for a W/HfO$_2$ (10 nm)/Si MIS capacitor at different voltage values.

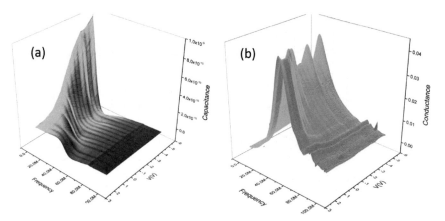

Figure 10.24 Three-dimensional plots showing frequency and voltage variations of capacitance (a) and conductance (b) for a W/HfO$_2$ (20 nm)/Si MIS capacitor.

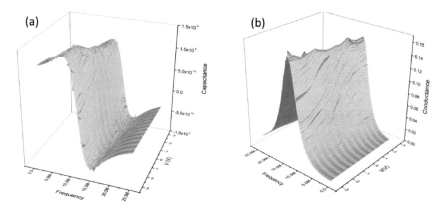

Figure 10.25 Three-dimensional plots showing frequency and voltage variations of capacitance (a) and conductance (b) for a Ni/HfO$_2$ (20 nm)/Si MOS capacitor.

In order to check the influence of top electrode material on this effect, the same measurements were carried out on similar samples with nickel instead of tungsten as top electrode (Fig. 10.25). In this case, relaxation occurs at lower frequencies (15 MHz) and no influence of voltage bias on the dipole relaxation frequency values was observed. This can be due to some Fermi level pinning effect in the nickel samples. Also, it can be related with the fact that nickel ions diffuse inside the insulator. These charged ions create local electric fields that interact with insulator dipoles in such a way that relaxation frequency occurs at lower frequencies. Local electric field dominates over the external applied field, and the resonance frequency results independent of the externally applied voltage.

10.4 Low Dimension Limits of Capacitance Spectroscopy

In sub 4 nm oxide layers, C–V measurements provide the same information, but the interpretation of the data requires considerable caution. The "classical model" reaches its limit of validity and quantum mechanical corrections must be included. The Maxwell–Boltzmann approximation does not describe the charge density in the accumulation and inversion layers, and must be substituted by the Fermi–Dirac statistics. Moreover, the very strong band-bending appearing in the interface at the inversion regime and the electrostatic potential in the semiconductor, produce the apparition of a very narrow potential well at the interface barrier. The correct analytical treatment requires solving the complex coupled effective mass Schrodinger and Poisson equations self-consistently. Modern device technology uses highly doped channel and ultra-thin gate oxides in MOSFETs. In consequence, a high electric field exists in the direction vertical to the silicon/silicon oxide interface. Although the high electric field in the vertical direction can keep the charges in the channel under gate control against the influence of drain potential, it confines the movement of carriers in a narrow potential well. From quantum theory, the energy of the channel carriers can only take discrete values and not a continuous energy distribution as described by classical device physics. Moreover, quantization also causes a redistribution of carrier density close to the semiconductor-insulator interface as compared to that of the classical prediction. Thus, it is critical to model accurately this quantization effect in the MOS devices and understand the relationship between the charge density and the gate bias. From the quantization effect analysis, it becomes clear that the distribution of inversion charge is quite different from that predicted by classical analysis. For example, in the classical model, inversion charge changes with potential exponentially, resulting in a maximum concentration at the channel surface [44]. In quantum mechanics, by considering the potential barrier of the gate oxide, the wave function diminishes at the interface, leading to zero charge density. As a result, the density peak of inversion charges shifts away from the interface. Meanwhile, inversion charges are distributed on the split sub bands instead of having a continuous distribution. In the quantum analysis, inversion charge can be described as a two-dimensional gas at discrete energy levels.

The appropriate charge profile must be considered for an accurate gate capacitance model. Through a gate capacitance model that incorporates the energy quantization effect, a better understanding of the behavior of MOSFETs in different operational regions can be achieved. Device parameters justified quantum mechanically can be further studied based on the gate capacitance model. In the classical model, the inversion layer capacitance is intentionally ignored, because it assumes that inversion charges concentrate at the surface of the channel, which causes the inversion layer capacitance to be much larger than C_{ox}. However, in sub-90 nm devices, the inversion charge density peak is located inside the channel, and C_{inv} needs to be specified taking energy quantization into consideration. This results in a finite inversion layer capacitance. In cases when the oxide layer is thick and the inversion layer capacitance is much larger than the oxide layer capacitance, the error in total gate capacitance by ignoring the inversion layer capacitance is negligible. However, oxide thickness is greatly reduced in sub-90 nm MOSFETs, where the oxide layer capacitance becomes comparable to the inversion layer capacitance. In this situation, the gate capacitance will be noticeably reduced by taking into account the quantum-induced inversion layer capacitance.

In summary, the semiconductor bands discretization, the non-zero channel inversion thickness, and the distribution of electrons in the reverse zone have to be taken into account for an adequate interpretation of capacitance spectroscopy in low dimension devices.

10.5 Summary

This chapter is an overview of the application of capacitance spectroscopy techniques to study the MOS systems. These techniques have been widely used since the beginning of the "MOS era" to quantify traps and defects in the different regions of these structures. Surface states at the insulator-semiconductor interface are measured by capacitance measurements at high and low frequency, and deep-level transient spectroscopy. Slow traps near the interface can be obtained by deconvoluting the conductance and capacitance transients as a function of time and frequency. Traps located in the dielectric bulk are quantified in energy and location by recording voltage transients while keeping constant the capacitance at flatband conditions. Several examples of recent works applying these techniques probe that these techniques continue being useful today. Finally, the limitation of capacitance measurements when the device dimensions reach quantum limits is discussed.

References

1. J. Moll, Variable capacitance with large capacitance change, *IRE 1959 Western Convention Show*, vol. 3, pp. 32–36, 1959.

2. Terman, An Investigation of surface states at a silicon/silicon oxide interface employing metal-oxide silicon diodes, *Solid-State Electronics*, vol. 5, pp. 285–299, 1962.

3. A. Grove, B. Deal, E. Snow, and C. Sah, Investigation of thermally oxidized silicon surfaces using metal-oxide-silicon diodes, *Solid-State Electronics*, vol. 8, no. 2, pp. 145–163, 1965.

4. A. Grove, E. Snow, B. Deal, and C. Sah, Simple physical model for space-charge capacitance of metal-oxide semiconductor structures, *J. Appl. Phys.*, vol. 36, no. 5, pp. 2458–2460, 1964.

5. R. Castagné and A. Vapaille, Description of the SiO_2/Si interface properties by means of very low frequency MOS capacitance measurements, *Surf. Sci.*, vol. 28, pp. 157–193, 1971.

6. M. Kuhn, A quasistatic technique for MOS C-V and surface state measurements, *Solid-State Electronics*, vol. 13, pp. 873–885, 1970.

7. D. Lang, Deep level transient spectroscopy: A new method to characterize traps in semiconductors, *J. Appl. Phys.*, vol. 45, p. 3023, 1974.

8. M. Schulz and N. Johnson, Evidence for multiphonon emission from interface states in MOS structures, *Solid State Commun.*, vol. 25, no. 7, pp. 481–484, 1978.

9. K. Yamasaki, M. Yoshida, and T. Sugano, Deep level transient spectroscopy of bulk traps and interface states in Si MOS diodes, *Japanese J. Appl. Phys.*, vol. 18, pp. 113–122, 1979.

10. N. Johnson, Measurement of semiconductor-insulator interface states by constant capacitance deep-level transient spectroscopy, *J. Vacuum Sci. Technol.*, vol. 21, pp. 303–314, 1982.

11. K. Wang, MOS interface-state measurements using transient capacitance spectroscopy, *IEEE Trans. Electron Devices*, vol. 27, pp. 2231–2239, 1980.

12. D. Vuillaume and J. Bourgoin, Characterization of SiO_2–Si interface states: Comparison between transient capacitance and conductance techniques, *J. Appl. Phys.*, vol. 58, pp. 2077–2079, 1985.

13. S. Dueñas, H. Castán, H. García, L. Bailón, K. Kukli, M. Ritala, M. Leskelä, M. Rooth, O. Wilhelmsson, and A. Harsta, Experimental investigation of the electrical properties of atomic layer deposited hafnium-rich films on n-type silicon, *J. Appl. Phys.*, vol. 100, pp. 094107-1– 094107-9, 2006.

14. L. He, H. Hasegawa, T. Sawada, and H. Ohno, A self-consistent computer simulation of compound semiconductor metal-insulator-semiconductor C-V curves based on the disorder-induced gap-state model, *J. Appl.*, vol. 63, pp. 2120–2130, 1988.

15. L. He, H. Hasegawa, T. Sawada, and A. Ohno, A computer analysis of effects of annealing on InP insulator semiconductor interface properties using MIS C-V curves, *Japanese J. Appl. Phys.*, vol. 27, pp. 512–521, 1988.

16. H. Hasegawa, L. He, H. Ohno, T. Sawada, T. Haga, Y. Abe, and H. Takahasi, Electronic and microstructural properties of disorder-induced gap states at compound semiconductor-insulator interfaces, *J. Vacuum Sci. Technol. B*, vol. 5, pp. 1097–1107, 1987.

17. Y. Fujino and K. Kita, Estimation of near-interface oxide trap density at SiO_2/SiC metal-oxide-semiconductor interfaces by transient capacitance measurements at various temperatures, *J. Appl. Phys.*, vol. 120, pp. 085710 (1–8), 2016.

18. A. Jonscher, Dielectric relaxation in solids, *Journal of Physics D*, vol. 32, pp. R57–R70, 1999.

19. S. Dueñas, R. Peláez, H. Castán, R. Pinacho, L. Quintanilla, J. Barbolla, I. Mártil, and G. González-Díaz, Experimental observation of conductance transients in Al/SiNx:H/SiAl/SiNx:H/Si metal-insulator-semiconductor structures, *Appl. Phy. Lett.*, vol. 71, pp. 826–828, 1997.

20. J. Barbolla, S. Dueñas, and L. Bailón, Admittance spectroscopy in junctions, *Solid-State Electronics*, vol. 35, no. 3, pp. 285–297, 1992.

21. H. Castán, S. Dueñas, J. Barbolla, E. Redondo, N. Blanco, I. Mártil, and G. González-Díaz, Interface quality study of ECR-deposited and rapid thermal annealed silicon nitride Al/SiNx:H/InP and Al/SiNx:H/In$_{0.53}$Ga$_{0.47}$As structures by DLTS and conductance transient techniques, *Microelectron. Reliability*, vol. 40, p. 845, 2000.

22. G. Wilk, R. Walace, and J. Anthony, High-k gate dielectrics: Current status and materials properties considerations, *J. Appl. Phys.*, vol. 89, pp. 5243–5275, 2001.

23. G. Lucovsky, J. Hong, C. Fulton, Y. Zou, R. Nemanich, H. Ade, D. Scholm, and J. Freeouf, Spectroscopic studies of metal high-*k* dielectrics: transition metal oxides and silicates, and complex rare earth/transition metal oxides, *Phys. Status Solidi B*, vol. 241, p. 2221, 2004.

24. S. Zafar, A. Kumar, E. Gusev, and E. Cartier, Threshold voltage instabilities in high-/spl kappa/gate dielectric stacks, *IEEE Tans. Dev. Mater. Rel.*, vol. 5, no. 1, p. 45, 2005.

25. S. Dueñas, H. Castán, H. García, L. Bailón, K. Kukli, T. Hatanpää, M. Ritala, and M. Leskela, Experimental observations of temperature-dependent flat band voltage transients on high-*k* dielectrics, *Microelectron. Reliab.*, vol. 47, no. Dueñas S, Castán H, García H, Bailón L, Kukli K, Hatanpää T, Ritala M, and Leskela M, pp. 653–656, 2007.

26. J. Lin, S. Monaghan, K. Cherkaoui, I. Povey, E. O'Connor, B. Sheehan, and P. Hurley, A study of capacitance–voltage hysteresis in the HfO$_2$/InGaAs metal-oxide-semiconductor system, *Microelectronic Engineering*, vol. 147, p. 273–276, 2015.

27. R. Engel-Herbert, Y. Hwang, and S. Stemmer, Comparison of methods to quantify interface trap densities at dielectric/III-V semiconductor interfaces, *J. Appl. Phys.*, vol. 108, pp. 124101 (1–15), 2010.

28. *The International Technology Roadmap for Semiconductors*, edition 2013, http://public.itrs.net.

29. H. Wong and H. Iwai, On the scaling issues and high-j replacement of ultrathin gate dielectrics for nanoscale MOS transistors, *Microelectronics Eng.*, vol. 83, p. 1867'1904, 2006.

30. S. Dueñas, H. Castán, H. García, A. Gómez, L. Bailón, M. Toledano-Luque, I. Mártil, and G. González-Díaz, Effect of interlayer trapping and detrapping on the determination of interface state densities on high-k dielectric stacks, *J. Appl. Phys.*, vol. 107, p. 114104, 2010.

31. A. Aberle, S. Glunz, and W. Warta, Impact of illumination level and oxide parameters on Shockley–Read–Hall recombination at the Si-SiO$_2$ interface, *J. Appl. Phys.*, vol. 71, no. 9, pp. 4422–4431, 1992.

32. R. Hezel, K. Blumenstock, and R. Schiirner, Interface states and fixed charges in MNOS structures with APCVD and plasma silicon nitride, *J. Electrochem. Soc.*, vol. 131, no. 7, pp. 1679–1683, 1984.

33. J. Schmidt, F. Schuurmans, W. Sinke, A. Aberle, and S. Glunz, Observation of multiple defect states at silicon-silicon nitride interfaces fabricated by low-frequency plasma-enhanced chemical vapor deposition, *Appl. Phys. Lett.*, vol. 71, no. 2, p. 252'254, 1997.

34. S. Dueñas, H. Castán, H. García, J. Barbolla, K. Kukli, J. Aarik, and A. Aidla, The electrical-interface quality of as-grown atomic-layer-deposited disordered HfO$_2$ on p-and n-type silicon, *Semiconductor Sci. Technol.*, vol. 19, p. 1141, 2004.

35. M. Fischetti, D. Neumayer, and E. Cartier, Effective electron mobility in Si inversion layers in metal–oxide–semiconductor systems with a high-k insulator: The role of remote phonon scattering, *J. Appl. Phys.*, vol. 90, no. 9, p. 4587, 2001.

36. S. Tiwari, F. Rona, K. Chan, L. Shi, and H. Hanafi, A silicon nanocrystal based memory, *Appl. Phys. Lett.*, vol. 68, p. 1377, 1996.

37. T. Lv and L. Zhao, The Si nanocrystal trap center studied by Deep Level Transient Spectroscopy (DLTS=, *J. Nanomater.*, vol. 2014, pp. 748487 (1–6), 2014.

38. D. Kovalev, J. Diener, H. Heckler, G. Polisski, N. Künzer, and F. Koch, Optical absorption cross sections of Si nanocrystals, *Phys. Rev. B*, vol. 61, no. 7, p. 4485, 2000.

39. A. Kang, P. Lenahan, and J. Conley, The radiation response of the high dielectric-constant hafnium oxide/silicon system, *IEEE Trans. Nucl. Sci.*, vol. 49, p. 2636, 2002.

40. J. Felix, D. Fletwood, R. Schrimpf, J. Hong, G. Lucovsky, J. Schwank, and M. Shaneyfelt, Total-dose radiation response of hafnium-silicate capacitors, *IEEE Trans. Nucl. Sci.*, vol. 2002, p. 3191, 2002.

41. J. Rafí, F. Campabadal, H. Ohyama, K. Takakura, I. Tsunoda, M. Zabala, O. Beldarraina, M. González, H. García, H. Castán, A. Gómez, and S. Dueñas, 2 MeV electron irradiation effects on the electrical characteristics of metal–oxide–silicon capacitors with atomic layer deposited Al_2O_3, HfO_2 and nanolaminated dielectrics, *Solild-State Electronics,* vol. 79, p. 65, 2013.

42. H. Castán, L. Fuentes, H. García, S. Dueñas, L. Bailón, F. Campabadal, J. Rafí, M. González, K. Takakura, I. Tsunoda, and M. Yoneoka, Hole trap distribution on 2 MeV electron irradiated high-k dielectrics, *J. Vac. Sci. Technol. B,* vol. 33, p. 032201, 2015.

43. P. Debye, *Polar Molecules*, Chemical Catlago Company, New York (USA), 1929.

44. K. Yang, K. Ya-Chin, and H. Chemming, eds. Quantum effect in oxide thickness determination from capacitance measurement, *1999 Symposium on VLSI Technology. Digest of Technical Papers.,* pp. 77–78, 1999.

Chapter 11

Capacitance Spectroscopy in Single-Charge Devices

Alessandro Crippa,[a] **Marco Lorenzo Valerio Tagliaferri,**[b] **and Enrico Prati**[c]

[a]*Université Grenoble Alpes and CEA INAC-PHELIQS,*
F-38000 Grenoble, France
[b]*QuTech, TU Delft, 2600 GA Delft, The Netherlands*
[c]*Istituto di Fotonica e Nanotecnologie, Consiglio Nazionale delle Ricerche,*
Piazza Leonardo da Vinci 32, I-20133 Milano, Italy

enrico.prati@cnr.it

11.1 Introduction

Since the 1980s, the control of individual excess charge carriers has been achieved in miniaturized devices where the charge is confined simultaneously along the three dimensions. The nanostructures suitable for such a control can be fabricated by using several different materials and by employing different confinement approaches. They are universally known as quantum dots because, by virtue of their nanometric size, their energy spectrum is quantized (discrete) according to the rules of quantum

Capacitance Spectroscopy of Semiconductors
Edited by Jian V. Li and Giorgio Ferrari
Copyright © 2018 Pan Stanford Publishing Pte. Ltd.
ISBN 978-981-4774-54-3 (Hardcover), 978-1-315-15013-0 (eBook)
www.panstanford.com

mechanics. Such objects are also called artificial atoms and they are described as three dimensional quantum wells.

There are two main reasons justifying the interest for confinement of charge carriers along the three axes. The first is associated to scaling of electron devices according to Moore's law, as transistors of size below the 110 nm node progressively behave according to effects determined by gate confinement. At 5 nm, a non-monotonic conductance as a function of the gate voltage reflects the discrete spectrum of the quantum well even at room temperature. On the other hand, storing individual electrons and holes in semiconductor nanostructures is interesting for the potential use of such quantum objects for Beyond CMOS applications concerning quantum memories, quantum bit, single photon emission and detection and so on.

Additional features are associated to specific materials used as substrate to host excess charge. Depending on their biasing regime imposed by surrounding control electrodes, quantum dots can host either electrons in superconductors (Haviland et al., 1994), metals (Fulton and Dolan, 1987), and semiconductors or holes in semiconductors, respectively (Kouwenhoven et al., 2001). The quantum dots discussed in this chapter are not mere nanoparticles with small radius deposited on a surface: they are confinement centers embedded in functional devices, which include at least three electrode terminals. Generally, two electrodes (sometimes called leads) with a metal-like density of states, acting as source and drain, allow a fA-µA range current to flow across the quantum dot. A metal gate grants the control of the excess charge in the quantum dot via electric coupling across a dielectric material. Here is where the capacitance sets in: the control potentials applied to the three terminals are delivered, in the case of the leads, through two tunnel junctions between the leads and the quantum dot, and via a dielectric, with no carrier exchange, in the case of the gates (Devoret and Glattli, 1998). The tunnel junctions consist of a parallel of a capacitance of few attofarad (accounting for the nanometric gap between dot and leads) and a MΩ resistance modeling the tunnel events through the capacitor of before. Therefore, the fine-tuning of the potential makes possible to control the charge state of the quantum dot via these attofarad scale capacitances.

Charge-dependent quantum transport across the dot allows to quantify relevant quantities, e.g., the conversion factor between gate voltage and energy scale (lever arm), intended for a quantitative spectroscopy of the dot itself (Kouwenhoven et al., 2001).

The relevant length scale of the devices used in this chapter is set by the operating temperature. To observe single-charge effects by exploiting the discrete nature of the spectrum, the charging energy E_{ch} and the energy separation between two states Δ have to be larger than few kT, where k is the Boltzmann constant and T is the temperature (Beenakker, 1991). The charging energy is the amount of energy needed to add a charge carrier to a quantum dot, when in Coulomb Blockade regime. As discussed in detail in Section 11.3, the charging energy is proportional to the total capacitance of the dot C_Σ. Since $E_{ch} = e^2/C_\Sigma$ and approximating the dot to a disc capacitor of diameter d, it can be easily obtained that $E_{ch} = e^2/4\varepsilon_0\varepsilon_r d$ with ε_0 the vacuum and ε_r the relative permittivity respectively. It is thus possible to link the dot size to E_{ch}. From the quantum description of an electron confined in a one-dimensional potential well of length L the energy level spacing Δ is proportional to L^{-2} (Ashcroft and Mermin, 1976).

Using the same approximation of a disc capacitor, in the case of silicon, a dot with a diameter of 20 nm has a level spacing of about 1 meV. Consequently, most of the devices discussed in the following are operated at cryogenic temperature (below 10 K), where a dot size in the range of tens-hundreds of nanometers is sufficient to satisfy the conditions mentioned above. As extreme case, the discrete spectrum of silicon quantum dots of 2–5 nm size has been revealed at room temperature (Shin et al., 2010).

The next section covers a number of examples of how to fabricate quantum dots from different materials. Section 11.3 covers the quantitative analysis of the stability diagram of a quantum dot and presents how the constant interaction model enables to extract capacitances and lever arm factors of the dot. Section 11.4 discusses the concept of quantum capacitance in zero dimensional objects.

11.2 Single-Charge Quantum Dots in Semiconductor Devices

There are several methods and materials to fabricate zero dimensional objects. The vertical confinement may be achieved by engineering the relative alignment of conduction band minima of different materials (as AlGaAs/GaAs heterostructures), or by a physical constraint given by a two dimensional monolayer of a semiconductor material (such as graphene), so that a uniform two-dimensional sheet of free carriers is obtained. Once such a layer is confined, the shape of the quantum dot can be tailored by creating lateral energy barriers so they become isolated by contacts and by free charges of surrounding materials. Alternatively, the lateral confinement can be assessed even by physically imposing constraints obtained by either the etching or the ion beam lithography of the substrate. As a consequence, dots are referred to as electrostactically or physically defined, whether belonging to the first or second category. Finally, one can rely on individual donors and acceptors by exploiting their isolated states in the bandgap of the hosting semiconductor, so that ionization is prevented at cryogenic temperatures. There are differences and similarities between an interface trap defect, an isolated donor atom and a quantum dot nanostructure. In common, they share their capability to host an individual electron when their ground state is set below the Fermi energy of a reservoir of electrons in their proximity, by tuning its energy value with a gate. On the other hand, a defect and a donor in bulk silicon (provided the temperature is sufficiently low so that it does not ionize) have a limited available occupation, which is generally 1 or 2 electrons. If the donor is close to a quantum dot state thanks to proximity to interface, it may hybridize so it becomes able to store more than two electrons via the quantum dot. A quantum dot may store any number of electrons, until the gate voltage induces the breakdown of the oxide. From the technological point of view, there is scarce control of position of intentionally induced defects, while donors and more effectively quantum dots can be assessed more straightforward. We should also mention that, at room temperature, and partially at cryogenic temperature, the capture of an electron by an interface defect involves multiphonon processes, associated to a rearrangement of the crystal lattice in its surrounding, which is detrimental for those applications relying on coherent states such as quantum bits for quantum computing. In the following, different materials and methods are presented to fabricate electron and hole quantum dots.

11.2.1 Single-Electron Quantum Dots in Heterostructures

Few-electron effects and Coulomb blockade features have been obtained originally in electrostatically defined GaAs lateral quantum dots. Here the word lateral is used in contrast to vertical quantum dots, as the transport occurs along the plane where a two-dimensional electron gas (2DEG) is formed and for the employment of lateral gate geometry. Such quantum dots are fabricated from heterostructures of GaAs and AlGaAs grown by molecular beam epitaxy and by doping one AlGaAs layer with silicon to introduce free electrons (Kouwenhoven et al., 1981). The resultant band engineering operation allows for the accumulation of electrons at the GaAs/AlGaAs interface as a thin (few nm) sheet. The free electrons have typically a low density (1–5×10^{15} m^{-2}). The large screening length allows to locally deplete the 2DEG by applying an electric field from metal gate electrodes biased at negative voltages (Fig. 11.1). The islands of electrons are isolated from the 2DEG by a suitable design of the gate structure. A similar method has been later employed in Si/SiGe heterostructures. As an example, double quantum dots for individual charge control have been formed by depleting charge from a two-dimensional electron gas within a Si/SiGe undoped heterostructure with lithographically patterned electrostatic gates (Simmons et al., 2007; Maune et al., 2012).

11.2.2 Single-Charge Quantum Dots in CMOS Devices

Classical three-terminal devices such as the well-known field-effect transistor (FET) have been widely studied since the 1960s. The quantum counterpart of FET, the single-electron transistor (SET), is actually a more recent discovery (Fulton and Dolan, 1987). A SET consists of a nanometric transistor where few electrons can be spatially confined, behaving as a quantum dot, in a three-terminal device. A FinFEt is shown, as an example, in Fig. 11.2a. The polarization of the contacts allows a current to flow through the charge island itself (see Fig. 11.2b,c). A single-hole transistor (SHT) represents its complementary device. Silicon quantum dots can be induced in CMOS devices realized by industrial processes. Complete elimination of disorder at the silicon/oxide interface is very challenging and the atomic-like shell structure of the orbitals of a silicon dot is further complicated by the sub-orbitals related to valley physics (De Michielis et al., 2012), which are strongly affected by confinement. In two-dimensional transistors, the vertical confinement occurs at the interface between silicon and the oxide. A two layered gate stack has also been employed (Lim et al., 2011). The lower layer, separated from the channel by means of an Al_2O_3, depletes the charges and it contributes to form barriers in the silicon. The upper layer controls the number of energy levels below the Fermi energy of the conduction electrons in the contacts, and drives the total number of charges N in the quantum dot. A CMOS fully compatible approach in pre-industrial technology (Prati et al., 2012) has been proposed by confining electrons in a well defined by lateral doping modulation, where the barriers are consequence of undoped silicon below spacers which are wrapping the gate. The lateral confinement is achieved by the nanowire shape of the channel covered by a three-side gate. This technique has the advantages of the compactness, with regard to previous works on Si, including additional upper gate. Multiple quantum dots have been obtained by a hybrid approach combining electron beam lithography of silicon nanowire and 10 nm-wide top metal gates (Tagliaferri et al., 2016). By exploiting the bulk contact as a back-gate in addition to the top gate in industrial CMOS transistors below 100 nm channel size, it is possible to separately control by two distinct electric fields the proximity of the charge island to the silicon/oxide interface and therefore to tune its coupling with the contacts (Turchetti et al., 2016). Single-electron effects at room temperature have been observed by using deep-trench and pattern-dependent oxidation on FinFETs, suitable to form a single-electron transistor with a Coulomb island size of less than 5 nm (Shin et al., 2010).

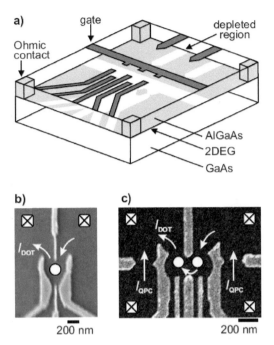

Figure 11.1 (a) Schematic view of a lateral quantum dot device defined by metal electrodes deposited at the surface of a GaAs/AlGaAs heterostructure. Metal gate electrodes (dark gray) are used to depleted carriers by applying a negative voltage (white) in the 2DEG (light gray). The light gray island near the farthest right gate represents the quantum dot. The contacts (light gray columns) are used for bonding wires to electrically access the 2DEG. (b) Scanning electron micrographs of a few-electron single-dot device. (c) Scanning electron micrographs of a few-electron double-dot device. The two gates on the sides are used to create two quantum point contacts, serving as electrometers by the current I_{QPC}. Reprinted with permission from R. Hanson, L. P. Kouwenhoven, J. R. Petta, S. Tarucha, and L. M. K. Vandersypen, *Rev. Mod. Phys.* 79, 1217–1265, 2007. Copyright 2007 by the American Physical Society.

11.2.3 Quantum Confinement Based on Single Dopants

In silicon devices such as commercial transistors the doping is generally treated under the hypothesis of a uniform approximation, which means that discretization of atoms randomly distributed in the lattice is neglected.

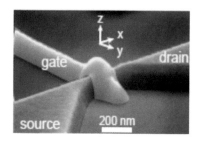

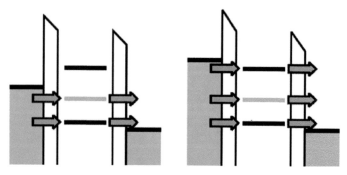

Figure 11.2 (a) Scanning electron micrograph of a nanodevice (in particular, a FinFET) employed as single-electron transistor. The quantum dot is located in the narrow region between source and drain and it is electrostatically tuned by the gate voltage. Reprinted by permission from Macmillan Publishers Ltd: G. P. Lansbergen, R. Rahman, C. J. Wellard, I. Woo, J. Caro, N. Collaert, S. Biesemans, G. Klimeck, L. C. L. Hollenberg, and S. Rogge, *Nat. Phys.* 4, 656–661, copyright 2008. (b, c) The quantum dot is isolated from source and drain by high potential barriers. Source-drain conduction is mediated by the discrete levels of the dot. A net current flow is measured when one or more levels (two for panel b), three for panel c)) of the dot allow a single charge to tunnel through the barriers and, via the dot, reach one of the reservoirs. Reprinted with permission from R. Hanson, L. P. Kouwenhoven, J. R. Petta, S. Tarucha, and L. M. K. Vandersypen, *Rev. Mod. Phys.* 79, 1217–1265, 2007. Copyright 2007 by the American Physical Society.

On the contrary, the small size of silicon nanowire FETs allows to correlate the discrete nature of doping to macroscopic parameters like threshold voltage variability, subthreshold conductance by means of low-temperature transport spectroscopy via isolated dopants. When the device consists of a silicon wire of around $10 \times 10 \times 50$ nm^3 even at a doping density of 10^{18-19} cm^{-3}, the number of dopants expected in such a nanometric box is 5–50 (Leti et al., 2011). An isolated donor atom in bulk silicon such as phosphorous (P) and arsenic (As)—and similarly an acceptor like boron (B)—is formally equivalent to an hydrogen atom in vacuum with a rescaled Bohr radius of $r_{\text{dopant}} = \dfrac{\varepsilon}{m^*} r_{\text{Bohr}}$ (Prati and Morello, 2013). The ground state energy is around 40–50 meV below the conduction band (above the valence band for B) and therefore ionized at room temperature. Silicon has indirect bandgap and therefore the ground state is six-fold degenerate because of the valley degeneracy, in addition to the two-fold spin degeneracy. When the atom lies in the proximity of the interface between the silicon wire and the oxide, the degeneracy of the ground state is lifted thus creating a richer variety of effects (Prati, 2011; Crippa et al., 2015). At cryogenic temperature, the presence of an impurity atom whose bound state wavefunction partially overlaps the wavefunction of the conduction carriers of both the source and the drain (therefore sitting around the center of the channel) provides two current channels determined by the ground state GS and an excited state ES. Up to two electrons can be bound to a donor atom in bulk silicon by employing an external electric field imposed by a control metal gate. The binding energy of the second electron is of the order of a few meV to the conduction band and no additional electrons can be trapped because of the Coulomb repulsion (Thomas et al., 1981). An array of N atoms may determine N-ples of states generating two impurity Hubbard bands (Prati et al., 2012, 2016). Experiments with donors in the channel of a transistor coupled to top and back gate and to two split top gates have been carried out so individual charge control of donors was possible. More complex topologies where a single donor is coupled from the side of the channel and it is controlled by a side gate have been explored (Mazzeo et al., 2012). The channel of the transistor serves as a single-electron electrometer capacitively coupled with the drain, the source, the top gate and the side gate by C_d, C_s, C_g, and C_{side}, respectively, while the donor only with the top and the side gates $C_{g(D)}$ and $C_{\text{side}(D)}$.

11.2.4 Quantum Dots Based on Graphene

Graphene is a two-dimensional material based on a hexagon lattice of carbon atoms, which naturally provides charge carriers confined in a plane. Graphene has been employed to create both single-electron and single-hole quantum dots by adjusting the biasing of control gates. Such devices have been fabricated based on single-layer graphene flakes, for instance from mechanical exfoliation of bulk graphite. According to the fabrication process of (Stampfer et al., 2008), the flakes are deposited on a highly doped silicon substrate with some thick silicon oxide layer of some tens of nanometers on top. In order to pattern the isolated graphene flake an electron beam (e-beam) lithography is employed in combination with Ar/O_2 reactive ion etching. Next, the electrodes are formed by an additional e-beam and lift-off step for patterning Ti/Au (2 nm/50 nm). In the example in Fig. 11.3, source (S) and drain (D) contacts are connect via 50 nm-wide constrictions to the graphene island. Such constrictions are separated by about 750 nm. Such graphene quantum dots are experimentally characterized at a temperature of a few Kelvin.

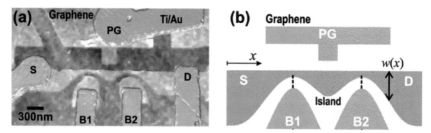

Figure 11.3 (a) Scanning force microscope image of a graphene single electron/single-hole transistor device. The graphene layer and the metal electrodes are highlighted. The minimum feature size is ≈50 nm. (b) Schematic illustration of the tunable graphene quantum dot device. Reprinted (adapted or reprinted in part) with permission from C. Stampfer, E. Schurtenberger, F. Molitor, J. Guttinger, T. Ihn, and K. Ensslin, *Nano Lett.* 8, 2378–2383. Copyright 2008 American Chemical Society.

11.3 Capacitance Characterization of Quantum Dots

As explained in the previous section, quantum dots can be defined by different approaches. As a consequence, devices with different architectures have been investigated. Among them, the simplest is a three-terminal device, where two terminals give and collect the carriers tunneling through the dot and the third one controls the electrostatics where the dot is formed. The energy spectrum of the dot becomes accessible at proper energy scales, like temperature and bias voltage values, by transport measurements: the discretization of charge in the quantum dot, placed between the emitter and collector terminals, causes a strongly non-linear $I_{SD} - V_G$ characteristic, displaying peaks of current spaced by regions of no current. The phenomenon underlying this kind of trace is known as Coulomb blockade (Beenakker, 1991).

   This section explains how bias spectroscopy can be used as a tool to determine the capacitance of a quantum dot. Both working principles of quantum dot functional devices and the constant interaction model based on classical electrostatics are quantitatively discussed to link the experimental measurements with the capacitive network of the dot with the surroundings. Next, the concepts of multi-terminal devices and double quantum dots are introduced. For simplicity, the discussion is limited to the case of electrons in semiconductor quantum dots (QDs) and no assumption on the confinement nature (electrostatic or physical) is made; however, what follows can be easily adapted to all the single-charge devices mentioned in Section 11.1.

11.3.1   Quantum Dots in Three-Terminal Devices

An elementary sketch of a three-terminal device, as described above, is shown in Fig. 11.4: a single quantum dot is capacitively coupled (via C_G) to the gate electrode, whereas the charge exchange between the dot and the electrodes takes place via two tunnel junctions (the island is said tunnel coupled to the source and drain electrodes—sometimes called reservoirs). The source (drain) tunnel barriers are modeled as the parallel of a capacitor $C_{S(D)}$ and a resistor $R_{S(D)}$. Such tunnel junctions are non-conducting layers (e.g., insulators or depleted regions) sandwiched between the dot and each reservoir. When these barriers are opaque enough, i.e., a relatively high potential separates the dot from the leads, a sequence of tunneling events of single electrons through the dot originates a measurable source-drain current dependent on the states available inside the dot. The transparency of the tunnel junctions strongly depends on the dot position inside the channel, as the probability of a tunnel event has an exponential dependence on the barrier width (Ashcroft and Mermin, 1976).

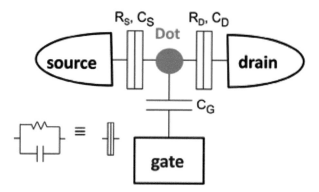

Figure 11.4 Schematic circuit of a QD in a three-terminal device, the capacitance network consists of the source, drain and gate capacitances C_S, C_D, and C_G. The tunnel coupling of the dot to source (drain) is modeled as the parallel of the resistance $R_{S(D)}$ and the capacitance $C_{S(D)}$.

The many-electron problem describing electrons in solids is here reduced to a single particle (electron) moving in a potential landscape describing the solid (Ashcroft and Mermin, 1976). As a consequence, a quantum dot owns a discrete energy spectrum formed by a finite number of single particle levels, with a density of states described by Dirac delta functions centered at the energy levels. In absence of an external magnetic field, the energy levels of the quantum dot are spin-degenerate: because of Pauli exclusion principle, only two electrons can have the same energy, provided having opposite spin.

In such an equivalent circuit model, the resistors just allow the charge to tunnel from the reservoirs to the dot, while the interaction between the dot and the environment is described only by means of the capacitances, as described in detail in next section.

Differently from the dot, source and drain can be regarded as metal-like regions where electrons are distributed according to the Fermi–Dirac distribution in a continuum of single particle levels:

$$f_{FD}(E) = \frac{1}{e^{[E-\mu_{S(D)}]/kT} + 1},$$ (11.1)

where E is the electron energy, $\mu_{S(D)}$ the source (drain) chemical potential. The distribution $f_{FD}(E)$ gives the probability that an energy level E is occupied. It returns a probability 1 for $E \ll \mu_{S(D)}$ and 0 for $E \gg \mu_{S(D)}$; the smoothness of such probability transition is set by the temperature T of the Fermi sea in source and drain contacts. It is worth noting that the identity $f_{FD}(\mu_{S(D)}) = 1/2$ defines the chemical potential $\mu_{S(D)}$ as the Fermi energy, i.e., the occupied energy level with the highest energy at $T = 0$. For a current to flow in the SET channel, non-equilibrium thermodynamics requires a difference between μ_D and μ_S. The misalignment is $-|e|V_{SD} = \mu_S - \mu_D$, where $-|e|$ is the electron charge, V_{SD} is defined bias voltage and it is controlled by applying a net voltage across source and drain electrodes.

Two different energy scales characterize the spectrum of a quantum dot: the energy level spacing $\Delta = \varepsilon_n - \varepsilon_{n-1}$ between (two-fold spin degenerate) single particle levels, and the charging energy $E_{ch} = \frac{e^2}{2C_\Sigma}$ needed to add an additional electron to the dot, with $C_\Sigma = C_G + C_D + C_S$ the total capacitance of the dot. Δ reflects the discretization of the orbitals of the quantum dot; E_{ch} counts for the charge Coulomb repulsion and comes from classic electrostatics. Usually in semiconductor quantum dots Δ ranges from tens of μeV to few meV and E_{ch} from few to tens of meV. Therefore, the discretization of charge and energy is observed only for $kT < \Delta$, E_{ch}, i.e., at temperature below 10 K. Whenever the condition $\Delta \simeq kT \ll E_{ch}$ is satisfied, the dot is defined metallic, whereas when $kT < \Delta \ll E_{ch}$, it is regarded as a properly semiconductor quantum dot. In both cases, since $kT < E_{ch}$, thermal fluctuations are not able to populate the dot. As anticipated, the tunnel barriers have to be opaque enough to make negligible the uncertainty on the energy scale on the charge measurement timescale. Quantitatively, since the tunnel barriers are modeled as the parallel of a capacitance and a resistance, the characteristic time of a tunneling event is set by the RC constant of the tunnel junction. Taking into account Heisenberg uncertainty principle $E \cdot t \geq \hbar/2$, where $\hbar$ is the reduced Planck constant, and replacing E with E_{ch} and t with the RC constant, the electron number is well defined when $R_{S(D)} > h/e^2 \simeq 25$ kΩ. In this regime the dot is in weak coupling with the reservoirs. In such conditions, the number N is fixed until the gate electrode provides an amount of energy equal to the charging energy to win the Coulomb repulsion between electrons.

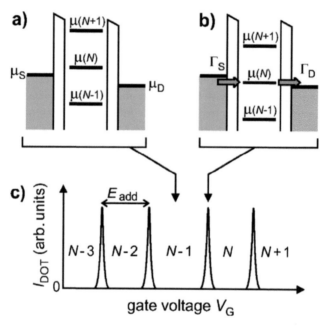

Figure 11.5 (a) Potential landscape of a quantum dot when the current is blocked and $N - 1$ electrons are trapped in the quantum dot. (b) As soon as the gate voltage aligns $\mu(N)$ to the Fermi level of the contacts, a current can flows and the charge number varies between $N - 1$ and N, giving rise (c) to a current peak in the $I_{SD} - V_G$ characteristic. Reprinted with permission from R. Hanson, L. P. Kouwenhoven, J. R. Petta, S. Tarucha, and L. M. K. Vandersypen, *Rev. Mod. Phys.* **79**, 1217–1265, 2007. Copyright 2007 by the American Physical Society.

Basically, these two conditions define the Coulomb Blockade regime in a metallic dot, as depicted in Fig. 11.5 for the case of $N - 1$ electrons trapped in the dot. Since the number of the confined charge is undefined it is labeled as N hereafter. N electrons remain stuck inside the dot, which ultimately is described by an electrochemical potential $\mu(N)$. As soon as the gate provides an amount of energy equal to the charging energy, an extra electron is admitted. All of a sudden the number of electrons in the dot rapidly oscillates between N and $N + 1$ since $\mu(N) = \mu(N + 1)$. At the end of a cycle $N \leftrightarrow N + 1 \leftrightarrow N$ an electron is transferred from source to drain: some current has flown through the dot. The charge flow is sequential as a single electron moves per time; the typical rate of this from source to drain motion is between MHz and THz (current e.g., from fA to nA). The $I_{SD} - V_G$ characteristic is a succession of current peaks, located at the gate voltages for which $\mu(N) = \mu(N + 1)$, spaced by E_{ch}. An $I_{SD} - V_{SD}$ curve reflects the charge discretization as well: each time that a new electrochemical potential enters the bias window V_{SD}, an electron can be added to the dot and the current I_{SD} is increased abruptly and the characteristic has a peculiar staircase behavior.

11.3.2 Constant Interaction Model

Once explained the working principles of a single-charge device, it is possible to develop a simple theory, known as constant interaction model (Kouwenhoven et al., 2001), to describe quantitatively charge transport phenomena and to extrapolate the capacitances of quantum dots from $I_{SD} - V_G$ and $I_{SD} - V_{SD}$ characteristics.

The constant interaction model is based on two assumptions, in addition to those already made to describe a quantum dot embedded in a three-terminal device. First, both the Coulomb interaction between the electrons in the dot and that between them and those in the leads are described by means of only the total capacitance C_Σ. This capacitance is assumed independent of N. Second, the single particle energy level spectrum of the dot is independent from the Coulomb interaction between electrons, thus from N.

Given N electrons in the dot, the potential on the dot is

$$V_{dot}(N) = -|e|\frac{N}{C_\Sigma} + V_{ext},\qquad(11.2)$$

where the second term accounts for background charges and the effective charge induced on the dot by the three terminals, so $V_{ext} = (-Q_{bg} + C_G V_G + C_D V_D + C_S V_S)/C_\Sigma$. It is to be noted that even if the charge induced by the three electrodes can be tuned continuously, the actual charge on the dot changes only by integer value. The electrostatic energy is the work done to add the charges on the dot or $U(N) = \int_0^{-Ne} V_{dot}(Q')dQ'$. From Eq. (11.2) and accounting for the discrete energy spectrum of the dot, its total energy in the ground state is

$$U(N) = \sum_{i=1}^{N} \varepsilon_i + \frac{(|e|N)^2}{2C_\Sigma} - |e|NV_{ext},\tag{11.3}$$

where the first term is the sum over the single particle levels of the dot. The electrochemical potential of the charge island with N electrons is defined as

$$\mu(N) = U(N) - U(N-1) = \varepsilon_N + \left(N - \frac{1}{2}\right)\frac{e^2}{C_\Sigma} - |e|V_{ext}.\tag{11.4}$$

From Eq. (11.4), $\mu(N)$ is effectively an electrochemical potential: the first term is the chemical potential of the dot and the other describe the electrostatic potential. The energy required to add an extra electron to the dot is the addition energy:

$$E_{add} = \mu(N+1) - \mu(N) = \Delta(N) + \frac{e^2}{C_\Sigma}\tag{11.5}$$

and it can be observed that Eq. (11.5) reduces, in metallic dots, to the double of the charging energy. On the other hand, in semiconductor QD E_{add} depends on the parity of N because of the two-fold spin degeneracy of the single particle levels: when an odd electron is added to the dot, it occupies a new orbital so the addition energy has both terms of Eq. (11.5); an even electron is added to an half occupied level and the term $\Delta(N)$ is dropped.

Two regimes can be identified with respect to the applied bias V_{SD}. At small bias $V_{SD} \ll \Delta$, E_{ch}, the dot contains N charges and the current is Coulomb blocked for $\mu(N) < \mu_S$, $\mu_D < \mu(N+1)$. By sweeping the gate voltage of an amount ΔV_G, $\mu(N+1)$ can be aligned to μ_S and a net current can flow (see Fig. 11.5., where the case of $N-1$ electrons in the dot is shown): the charge oscillates between the configurations with N and $N+1$ electrons, so that $\mu(N, V_G) = \mu(N + 1, V_G + \Delta V_G)$. Recalling Eq. (11.4), the spacing ΔV_G between two subsequent current peaks is

$$-|e|\alpha\Delta V_G = \Delta(N) + \frac{e^2}{C_\Sigma}, \qquad (11.6)$$

where α is the lever arm factor for the conversion between gate voltage and energy ($\alpha = C_G/C_\Sigma$). At first order, the reservoirs lever arm factors are assumed to be unitary, in a sense that V_{SD} is directly converted in energy. The position of current peaks can be obtained from $U(N + 1) = U(N)$ yielding $-|e|\alpha V_G = \varepsilon_N + \left(N - \frac{1}{2}\right)\frac{e^2}{C_\Sigma}$. The peak position is thus independent from the bias voltage, which affects only the peak width.

Since the dot total capacitance contains also the leads capacitances, the low bias regime is not enough to extrapolate all the capacitances of the network. The so far missing relations can be obtained by setting the bias as an independent variable. A two-axis current map can be obtained by sweeping the gate voltage and stepping the bias as in Fig. 11.6c. Such a stability diagram allows to completely characterize a quantum dot in a three-terminal device. As shown in Fig. 11.6d, the fingerprint of Coulomb blockade is the presence of rhombus of blocked current, the so-called Coulomb diamonds. The width (i.e., the spacing between two peaks) in voltage of a diamond is converted in energy through the lever arm factor and it corresponds to the diamond height, defined as the voltage distance of the diamond vertex from zero bias. Conversely, the edges of the diamonds represent two opposite situations. At the edge with positive slope the electrochemical potential is aligned to the drain chemical potential, at the other edge to the source one. In particular for $V_D = 0$ and $V_{SD} = V_S$, the slopes are $m_- = -C_G/(C_\Sigma - C_S)$ and $m_+ = C_G/C_S$ and the lever arm results in $\alpha = (m_+|m_-|)/(m_+ + |m_-|)$. As a consequence, the dot capacitances can be obtained just measuring the height of a Coulomb diamond and the slopes of its edges. From the height the charging energy and thus the total capacitance C_Σ are easily obtained. The capacitances C_S, C_D, and C_G are then extracted solving the three-equation linear system composed by the definitions of the charging energy and slopes.

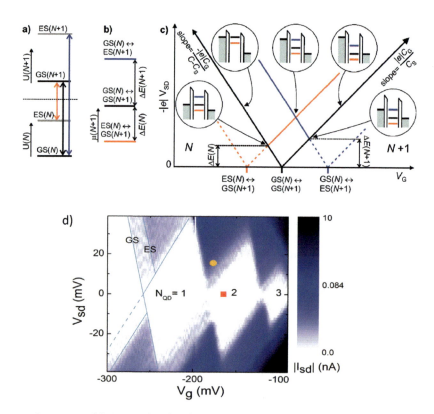

Figure 11.6 (a) Energy levels of the dot with N and $N + 1$ electrons. The arrows represent the possible transitions and the associated energy is shown in (b) with the same color code. (c) Schematic diagram of a stability diagram $V_G - V_{SD}$, where the transitions in (b) happen along the current lines depicted with the same color. Schematic diagrams show the dot electrochemical potential positions. Here C_Σ is indicated as C. Reprinted with permission from R. Hanson, L. P. Kouwenhoven, J. R. Petta, S. Tarucha, and L. M. K. Vandersypen, *Rev. Mod. Phys.* 79, 1217–1265, 2007. Copyright 2007 by the American Physical Society. (d) Example of a stability diagram experimentally measured. Reprinted with permission from A. Crippa, M. L. V. Tagliaferri, D. Rotta, M. De Michielis, G. Mazzeo, M. Fanciulli, R. Wacquez, M. Vinet, and E. Prati, *Phys. Rev. B* 92, 035424–1–11, 2015. Copyright 2015 by the American Physical Society.

When V_{SD} becomes bigger than Δ, transport via excited states is allowed. In this case, the dot oscillates between the ground state of one charge configuration and an excited state of the other. Thus, if, for example, the system oscillates between

ES(N + 1) and GS(N), a line parallel to the GS(N + 1)–GS(N) transition and separated by Δ appears, ending on the region with N + 1 electrons trapped in the dot (see Fig. 11.6c).

11.3.3   Multi-Terminal Devices

So far only the simplest architecture to realize a single-charge device has been investigated. For a more complete discussion, also multi-terminal devices will be briefly explained. In a multi-terminal device with one quantum dot and M gates, the total capacitance is just the sum of M + 2 terms or $C_\Sigma = C_S + C_D + \sum_{i=1}^{M} C_i$, where C_i is the capacitance between the dot and the i-th gate. The discussion in Sections 11.3.1 and 11.3.2 is still valid for each gate: when sweeping one gate voltage the term V_{ext} accounts also for the charge induced by all the other gates, which are kept fixed.

However, a single stability V_{SD} – V_G diagram is insufficient to measure all the capacitances of the dot and additional V_{Gi} – V_{Gj} maps have to be measured. Let us consider the simplest case of a four terminal device, for example a MOS transistor whose substrate can be employed as a back gate (Prati et al., 2011). Hence, here $C_\Sigma = C_S + C_D + C_G + C_B$, but the slopes of the Coulomb diamonds are still $-C_G/(C_\Sigma - C_S)$ and C_G/C_S and the missing relation is obtained from a V_G – V_B map. Since the electrostatics of the channel is controlled by both gates, Coulomb oscillations appear as current lines whose slope depends on the coupling to the two gates. Two extreme cases of horizontal and vertical lines may be detected when the charge transitions are fully controlled by only one of the two gates. Generally, the dot is coupled to both gates and the slope of the current strip is, at first order, the ratio of the two dot-gate capacitances. In a more general picture, being V_x and V_y the gate voltages on the two axes of a stability diagram, the lines slope is $-C_x/C_y$. Such relation can be obtained recalling that the addition energy is the same while sweeping different gates: thus the energy spacing between two successive lines is the same along the two directions (i.e., $-|e|\alpha_x \Delta V_x = -|e|\alpha_y \Delta V_y$) and exploiting the definition of line slope $\Delta V_y/\Delta V_x$. An example of such a map is reported in Fig. 11.7a.

11.3.4 Double-Quantum-Dot Devices

Multi-terminal devices are also used to control multiple dots, whose presence emerges in stability diagrams measured with respect to a couple of gates. For example, in Fig. 11.7a, two different patterns of lines are recognizable by the different slopes, related to two confinement centers in the channel, as shown in the schematic in Fig. 11.7b. The two sets of current lines define the condition of resonant transport when the electrochemical potential of each dot is separately aligned to source and drain chemical potentials. In this case, the capacitive coupling between the islands is negligible and the two dots are independent. When an interdot (or mutual) capacitance C_m is present the model developed has to be refined. Generally speaking, double-quantum-dot devices can be obtained in different configurations. As for classical circuits, two dots are in series if a current can flow sequentially through them, such that the two dots are tunnel coupled. By contrast, two dots are in parallel when either of them is independently connected to the leads. The intermediate case reads as follows: both the dots are tunnel coupled to both reservoirs and to each other.

In four-terminal double-dot devices, each gate controls one quantum dot via a capacitive coupling; the cross-capacitance describes, more realistically, the coupling of each gate with the other dot, but here it is neglected to keep the model simple. Let us concentrate, for the sake of simplicity, to the case of two quantum dots in series, whose capacitance network is depicted in Fig. 11.8a.[1]

At small bias (e.g., when $V_{SD} \approx 0$ and $\mu_S \approx \mu_D$), four different configurations can be obtained depending on the alignment of the chemical potentials of the dots (μ_1 and μ_2) and the reservoirs (μ_R for the sake of clarity in the following). When $N_{1(2)}$ electrons are confined in the first (second) dot and there are no available levels the charge is fixed in both dot and the condition $\mu_1(N_1)$, $\mu_2(N_2) < \mu_R < \mu_1(N_1 + 1)$, $\mu_2(N_2 + 1)$ is verified. When only the chemical potential of one dot is aligned to the reservoir while the other dot is in a blocked configuration (i.e., $\mu_{1(2)} = \mu_R$ and $\mu_{2(1)}(N_2) < \mu_R < \mu_{2(1)}(N_2 + 1)$), the current is still blocked, since

---

[1] A more detailed discussion on double dot in series can be found in van der Wiel et al. (2003).

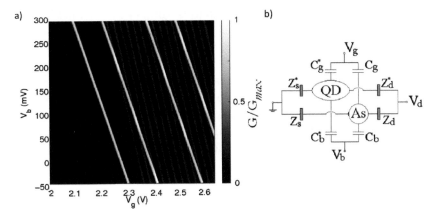

Figure 11.7 (a) Stability diagram showing the current line typical of a four-terminal devices. In the background, a series of fainter current lines is visible and is attributed to a second dot. Since in both cases the lines are straight, there is no capacitive coupling between the two dots. Thus, the two different patterns can be regarded as independent conductive channels. (b) Schematic circuit of a the two-dot four-terminal device with no interaction between the two confinement centers. In this example QD and As represent a quantum dot and an arsenic atom respectively. Reproduced from E. Prati, M. Belli, S. Cocco, G. Petretto, and M. Fanciulli, *Appl. Phys. Lett.* 98, 053109–1–3, 2011, with the permission of AIP Publishing.

the current from source to drain requires sequential hopping from the first to the second dot (see Figs. 11.8a and 11.5). However, because of higher-order processes or direct tunnel coupling with both leads, a current can be measured giving rise to current lines in a $V_{G1} - V_{G2}$ stability diagram, with G1(2) the gate controlling the first (second) dot, as reported in Fig. 11.8b. The lines slope depends on which dot is conducting, as shown in Fig. 11.8d. Since such lines are related to the (dis)charging of only one of the two dots, they are called single-dot lines hereafter. Near the crossing points, the mutual capacitance plays an important role. The single-dot current lines are shifted at every crossing point with the lines related to the second dot, which show a similar displacement (Tagliaferri et al., 2016) and defining the honeycomb pattern typical of coupled double dots shown in Fig. 11.8b. The displacement of a single-dot line is caused by the change in the double-dot electrostatics due to the addition of an extra electron on the other dot. The line connecting the two parts of the same single-dot line defines the condition of alignment between the electrochemical potentials of the two dots and is often referred to as interdot transition line. From a stability map $V_{G1} - V_{G2}$, it is possible to extrapolate some of the capacitances of the network in Fig. 11.8a. The spacing between the lines that identify the charging of the first(second) dot is $\Delta V_{G1(2)} = |e|/C_{1(2)}$, where $C_{1(2)} = C_{S(D)} + C_{G1(2)} + C_m$ is the total capacitance of the first (second) dot. The shift of the lines related to the first (second) dot are, instead $\Delta V_{G1(2)}{}^m = \Delta V_{G1(2)} C_m / C_{2(1)}$.

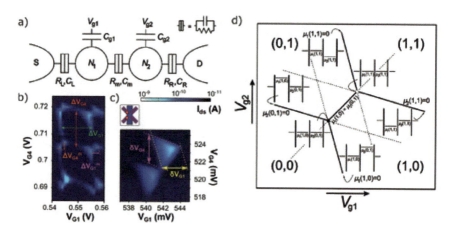

Figure 11.8 (a) Schematic circuit of two quantum dots in series. (b) Example of the honeycomb pattern typical of a double dot at small bias. Here the second gate is labeled G4 instead of G2, because of the device studied in (Betz et al., 2016) was a five-gate nanotransistor. (c) Bias triangles that are observed at the triple points when some bias is applied. (d) Energy diagram showing the alignment of the different chemical potential along the different edges of a honeycomb. (a) and (d) reprinted with permission from W. G. van der Wiel, S. De Franceschi, T. Fujisawa, S. Tarucha, and L. P. Kouwenhoven, *Rev. Mod. Phys.* 43, 1–22, 2003. Copyright 2003 by the American Physical Society; (b) and (c) reproduced from A. Betz, M. L. V. Tagliaferri, M. Vinet, M. Brostrom, M. Sanquer, A. J. Ferguson, and M. F. Gonzalez-Zalba, *Appl. Phys.* 108, 203108–1–3, 2016, with the permission of AIP Publishing.

A third relation can be obtained by applying a finite bias. Near the triple points, i.e., the points where the condition $\mu_S = \mu_1 (N_1) = \mu_2(N_2) = \mu_D$ is satisfied, a current can flow whenever both $\mu_1(N_1)$ and $\mu(N_2)$ are in the bias window. Such regions are called bias triangles; the reason is clearly visible in Fig. 11.8c. The height of these triangles is $|e|V_{SD} = |e|\delta V_{G1(2)}C_{G1(2)}/C_{1(2)}$. Thus from a stability diagram at finite bias is possible to extract the capacitances $C_{1(2)}$, C_m, and $C_{G1(2)}$. The remaining capacitances of the network shown in Fig. 11.8a can be determined by switching the device in single-dot configuration (Betz et al., 2016), and then by measuring C_S and C_D as already discussed in Section 11.3.2.

The knowledge of the capacitances of multi-terminal and multiple dot systems is paramount to have a complete picture of such nanoscaled functional devices, thereby allowing for a quantitative description. Information on the charging energy and level spacing, as well as dot position and coupling to each terminal, can be obtained only from a capacitance characterization.

11.4 Quantum Capacitance

The explanation in the previous section has followed a semiclassical approach: starting from electrostatics principles (Eq. (11.2)), it has been shown that stability diagrams provide a mean to quantify the capacitance among quantum dots and surrounding electrodes. However, to achieve a complete modeling of a single-charge device the so-called quantum capacitance has to be additionally considered. In the following, the general definition of quantum capacitance is briefly introduced by illustrating how it originates from the quantum treatment of the carrier concentration in any low dimensional system. Special attention is dedicated to the analysis of zero-dimensional objects in single-charge devices, like Cooper-Pair transistors and single-electron transistors; finally, the typical experimental setup to measure the quantum capacitance in semiconductor devices is outlined.

11.4.1 Quantum Capacitance for a Two-Dimensional Electron Gas in MOS Devices

The seminal concept of what nowadays is called quantum capacitance has been introduced in 1984–1985 in the context of small superconductor–insulator–superconductor junctions

(Josephson junctions) (Likharev and Zorin, 1985; Averin et al., 1985). The linearity of the current response as a function of voltage applied across the junction was not observed in small junctions; here current fluctuations were successfully expressed by inductive and capacitive terms, the latter highly sensitive to the charge flow from one superconducting section to the other one. This voltage-dependent effective capacitance turned out to be related to the second derivative, or curvature, of the energy bands involved in the carriers transport.

A few years later, Luryi (1988) developed an equivalent circuit model for devices operating with two-dimensional (2D) metallic planes, like the electron gas in a quantum well or the inversion layer of a MOSFET. He identified a contribution in the capacitance seen by the top gate contact of a MOS-like structure as due to the density of states (DOS) of the two-dimensional metal sandwiched in the semiconductor. In such example, the different DOS of the bulk metal providing the carriers and the 2D metal leads to a capacitance lower than the parallel-plate limit. He called this new capacitor of the equivalent circuit of his system "quantum capacitance."

Since then, this name has been exhaustively used for any capacitance as a function of carrier density in a low dimensional system. The ultimate reason for such interest is that the accurate knowledge of the influence of a gate voltage on the electron density in narrow semiconductor channels is of major relevance to scale down the size of the devices, as it will be discussed later.

Extensive studies regarding the quantum capacitance concept can be found in the modeling of nanoscale transistors where 2D systems are formed, like in commercial FET channel under strong-inversion conditions (John et al., 2004; Granzner et al., 2010) or in graphene sheets (Dröscher et al., 2010) or when 1D geometries are analyzed, as nanowire FETs (Razavieh et al., 2013) and carbon nanotubes devices (Ilani et al., 2006). Finally, the interaction between charge states in 0D objects, like Cooper-pair boxes (Duty et al., 2005) or semiconductor quantum dots (Gonzalez-Zalba et al., 2015), ultimately relies on quantum capacitance.

To understand what the quantum capacitance is, let's consider a standard n-MOSFET (see Fig. 11.9a). No specific assumptions are required about the device size for the moment. Source and drain contacts are supposed 3D metallic reservoirs of electrons characterized by a chemical potential μ. The top gate voltage V_G tunes the electric field in the semiconductor channel. By means of V_G, the conduction band edge E_C can be shifted below the chemical potential of the leads, so that an electron sheet is tightly confined in the z direction by the gate field and spread along x and y directions of the channel. Such inversion layer is referred to as a two-dimensional electron gas (2DEG). However, a limited number of states per energy unit is available for a certain applied V_G. This concept is expressed by a low density of states (DOS) of the semiconductor material constituting the channel. Assuming a simple parabolic dispersion near the bottom of the conduction band, the 2D DOS per unit surface takes the expression

$$g(E) = \frac{m^*}{\pi\hbar^2}\Theta(E - E_c), \tag{11.7}$$

where m^* is the effective mass of the electrons determined by the semiconductor band structure. The density of charge (i.e., number of electrons per unit area) is

$$n_s = \int g(E) f_{FD}(E) dE, \tag{11.8}$$

where $g(E)$ and $f_{FD}(E)$ have been introduced in Eqs. (11.7) and (11.1), respectively.

The step function Θ in $g(E)$ makes inaccessible the energy states below the bottom of conduction band E_C of a semiconductor, where $n_s = 0$. However, by tuning the gate potential, E_C can be downshifted below the chemical potential μ to start populating the channel.

Figure 11.9b shows the move of E_C due to the change of the channel potential V_c: the electronic states energetically available inside the channel are $\mu - (E_C - |e|V_c)$ for sufficiently large V_c. According to Eq. (11.8), the density of charge n_s in the semiconductor is obtained by integrating the filled states with the 2D DOS.

As a last step, the effective voltage inside the channel V_c remains to be evaluated. Two limiting cases can be envisioned: if no carriers are already present inside, the channel behaves like an insulator, so its potential V_c is as the gate potential V_G. On the contrary, if many electrons are confined, the channel is a conductor with the same potential as the contacts.

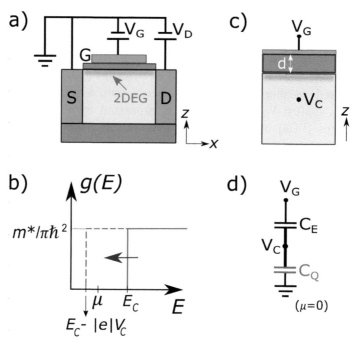

Figure 11.9 (a) Sketch of the cross section of a MOSFET showing the gate-induced formation of the 2DEG at the semiconductor-oxide interface. (b) Density of states of the 2DEG $g(E)$ as a function of the energy inside the channel. The gate polarization $-|e|V_c$ shifts the onset of the conduction band E_C below the chemical potential μ of the leads: the states of the semiconductor are then energetically allowed, and the 2DEG can be formed. (c) The oxide forms a parallel-plate capacitor between the 2DEG and the gate electrode. (d) Equivalent circuit for the capacitive network inside a MOS channel when the inversion layer is formed.

At intermediate regime, the 2DEG partially screens V_G inside the channel: $V_c = V_G - |e|n_s/C_E$. Here $C_E = \varepsilon S/d$ is the electrostatic capacitance of the capacitor having as negative plate the 2DEG of area S and as positive plate the gate contact, d is the thickness and ε the dielectric constant of the dielectric, as displayed by Fig. 11.9c.

The rate of variation of n_s with respect of V_G is

$$\frac{d(|e|n_s)}{dV_G} = \frac{d(|e|n_s)}{dV_c}\frac{dV_c}{dV_G}. \tag{11.9}$$

From Eq. (11.8) the term named quantum capacitance (which here is capacitance per unit of area) can be evaluated:

$$\frac{d(|e|n_s)}{dV_c} = e^2\frac{m^*}{\pi\hbar^2} \equiv C_{Q,2D} \tag{11.10}$$

Finally, by inserting this result in Eq. (11.9), one obtains

$$\frac{d(|e|n_s)}{dV_G} = \frac{C_{Q,2D}C_E}{C_{Q,2D}+C_E}. \tag{11.11}$$

The quantum capacitance $C_{Q,2D}$ for 2D electron systems (like inversion layers in FETs or states in quantum wells) appears as an effective capacitance in series with the electrostatic term C_E (see Fig. 11.9d). Its character is "quantum" in the sense that it originates from the wave-like behavior of the electrons filling the semiconductor's DOS: only specific energy states are allowed because of the Pauli exclusion principle to which electrons are subjected. The more electrons are present in the 2DEG, the more energy is required to keep filling it. This concept is emphasized by Eq. (11.10): a low DOS corresponds to a low carrier population in the channel and, thus, a small C_Q, and vice versa.

In the situation just described, the fact that C_E and C_Q are in series has important implications. For large oxide thickness d, the control of the channel potential is given by C_E as $C_E \ll C_Q$; the opposite limit $C_E \gg C_Q$ can be reached for narrow channels, where the DOS available to allow carriers motion is low. This brings to the surprising conclusion that in ultrascaled devices, C_Q controls the channel potential, and not the physical C_E (at least for V_G exceeding the threshold voltage considerably).

C_Q derives from charge dynamics and not from geometrical parameters of the circuit (like, for example, the capacitances defined in Section 11.3). A last remark for any electronic system is that the quantum capacitance is differential: the electron concentration must exhibit a dependence on a tunable voltage (like V_G), so that a non-zero derivative with respect to such voltage comes out.

11.4.2 The Zero-Dimensional Case: Derivation and Measurements

In the following, geometrical and quantum capacitances are assembled in a circuit model accounting for single-charge exchanges within zero-dimensional charge islands. Historically, such theory has been developed in parallel to the discovery and first implementations of RF-SETs. An RF-SET is essentially a SET such those described in Section 11.3 with a radio-frequency (RF) resonant circuit connected to source, drain or gate contact. As explained at the end of this section, RF-SETs represent a powerful sensor for capacitive fluctuations induced by single-charge motions: the phase shifts of the back-reflected RF signal are sensitive to the quantum capacitance associated to a single-charge transition.

A quantum mechanical system with two possible charge states is now being considered. For brevity, one is named "empty" and the other "filled", according to the presence of a single charge inside the 0D object. Such two-level system is extensively treated in many books on quantum mechanics and scientific papers (refer to references Duty et al. (2005), Shevchenko et al. (2015), and Sillanpää et al. (2006) for further details).

The two levels $E_{\pm}$ are chosen symmetric with respect a convenient origin of the energy axes $\varepsilon_0 = 0$, which represents the charge degeneracy point, i.e., where the system has not to pay any energy cost to accept the charge inside. Hereafter, ε_0 is called detuning parameter and counts the energy difference between having and not having the charge into the island. Ideally the two levels are supposed to have equal energy at zero detuning, but it is forbidden by quantum mechanics: a gap opens over there (the levels are said to anticross). Since a tunnel coupling t exists between such states, and the island can get charged at proper energy conditions, the energies of the two-level system are

$$E_{\pm} = \pm\frac{1}{2}\sqrt{\varepsilon_0^2 + (2t)^2}. \tag{11.12}$$

E_- denotes the lower energy (ground) and E_+ the upper energy (excited) state, and $2t$ turns out to be the energy gap when the levels are not detuned ($\varepsilon_0 = 0$). For the ground and excited state as a function of detuning, see Fig. 11.10a, continuous and dashed lines, respectively.

To allow the charging and discharging of the island, let's now introduce in the model two metal-like reservoirs and consider a device similar to the one of Fig. 11.4. From electrostatics, the quantum expectation value for the charge added Q_g modifies the potential of the island itself, so that

$$\langle Q_g \rangle = \frac{C_G (C_S + C_D)}{C_\Sigma} V_G + e \langle N \rangle \frac{C_G}{C_\Sigma} \tag{11.13}$$

where N is the electron number. The brackets denote a quantum expectation value due to the stochastic nature of the tunneling events to and from the island. $\langle N \rangle$ is displayed in Fig. 11.10b. The resulting effective capacitance is

$$C_{\text{eff}} = \frac{C_G (C_S + C_D)}{C_\Sigma} + \frac{e C_G}{C_\Sigma} \frac{\partial \langle N \rangle}{\partial V_G} = C_E + C_Q. \tag{11.14}$$

The first term represents a capacitance set by the electrostatic couplings with the environment, as in Section 11.3; the second term is the quantum capacitance and counts for the dynamics, i.e., has to be considered when the average number of charges is modified by V_G. Here C_Q and C_E have the dimension of a capacitance instead that of a capacitance per unit of area.

C_Q contributes to C_{eff} where some current flows through the device (region of the yellow circle in Fig. 11.6d); contrarily C_E is approximately constant in the entire $V_{SD} - V_G$ plane of the stability diagrams, so it takes the same value when the device is either conductive or in Coulomb blockade (respectively, yellow circle and red square in the same figure).

In the basis defined by {empty, filled} charge states, the charge expectation values is either 0 or 1; therefore, $\langle N \rangle = 0 \cdot P_0 + 1 \cdot P_1 = P_1$, where $P_0 (P_1)$ is the probability of having 0 (1) electron in the island. It is now useful to explicit the dependence of $\langle N \rangle$ on the energy detuning ε_0, as in Shevchenko et al. (2015):

$$\langle N \rangle = \frac{1}{2}\left(1 + \frac{\varepsilon_0}{\sqrt{\varepsilon_0^2 + (2t)^2}}(P_+ - P_-)\right),\qquad (11.15)$$

where $P_\pm$ is the probability of occupation of the $E_\pm$ the state. Then V_G can be written in the form of the energy detuning ε_0, through the relation $-\alpha \cdot \partial V_G = \partial \varepsilon_0$, so that

$$C_Q = \alpha \frac{\partial \langle N \rangle}{\partial V_G} = -\alpha^2 \frac{\partial \langle N \rangle}{\partial \varepsilon_0}$$

$$= \frac{\alpha^2}{2}\left(\frac{dZ}{d\varepsilon_0}\frac{\varepsilon_0}{\sqrt{\varepsilon_0^2 + (2t)^2}} + Z\frac{(2t)^2}{(\varepsilon_0^2 + (2t)^2)^{3/2}}\right)\qquad (11.16)$$

$Z \equiv P_- - P_+$ is the difference between the occupation probabilities of the ground and excited state of the two-level system of Eq. (11.12).

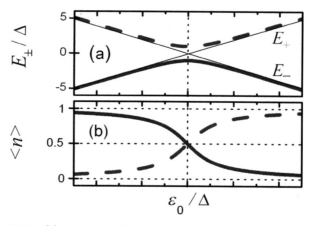

Figure 11.10 (a) ε_0 gate-voltage dependence of the energy levels $E_\pm$; the axes have no units as are normalized by the energy factor Δ equivalent to tunnel coupling t. (b) Average electron number $\langle n \rangle$ ($\langle N \rangle$ in the main text). Reprinted with permission from S. N. Shevchenko, D. G. Rubanov, and F. Nori, *Phys. Rev. B* **91**, 165422-1–9, 2015. Copyright 2015 by the American Physical Society.

The first term of Eq. (11.16) counts for transitions whose probability depends on the detuning, whereas the second one is mainly related to band curvature around the anticrossing point at $\varepsilon_0 = 0$. It is worth noting that such definition of C_Q is the

0D counterpart of Eq. (11.10), with the remarkable difference that now C_Q is in parallel with the geometric term C_E. Depending on the sweep velocity of the detuning parameter, adiabatic and non-adiabatic transitions can take place: the former follows the energy path given by E_-, so that the system is kept always in the energy ground state; the latter populates the excited state E_+ by a rapid change in the detuning parameter.

In the case of adiabatic transitions Z decreases from 1 at $\varepsilon_0 < 0$, it reaches a positive minimum at $\varepsilon_0 = 0$ and then it increases for $\varepsilon_0 > 0$. It implies that the product between ε_0 and $dZ/d\varepsilon_0$ of Eq. (11.16) is always positive. However, in the purely adiabatic limit $dZ/d\varepsilon_0 \approx 0$ since the excited state is never populated. Thus, only just the contribution of the band curvature survives:

$$C_Q \approx \frac{\alpha^2}{2} Z \frac{(2t)^2}{(\varepsilon_0^2 + (2t)^2)^{3/2}}. \tag{11.17}$$

For non-adiabatic transitions, let's first consider the case where there is no energy gap (i.e., $2t = 0$) and the two energy levels cross at ε_0. Such configuration models those transitions between a discrete state and a continuum of states: at the vicinity of the crossing point there is always an empty state of the continuum available. In the literature, such inelastic transitions are responsible of the Sisyphus dissipation when the driving frequency is comparable to the tunnel rate (Gonzalez-Zalba et al., 2015). In this form, the capacitive contribution is called tunnel capacitance and from Eq. (11.16) the expression reads

$$C_Q \approx C_t = \frac{\alpha^2}{2} \frac{\varepsilon_0}{|\varepsilon_0|} \frac{dZ}{d\varepsilon_0}. \tag{11.18}$$

The last case to consider is when the system is driven non-adiabatically in presence of an energy gap. Such a case might imply rapid pulses on ε_0 or microwave irradiation at frequency f such that $2t/hf \lesssim 1$. In such a case, the whole expression of Eq. (11.16) has to be taken into account. The competition of the two terms leads to a sign-changing behavior near resonance of the observable C_Q.

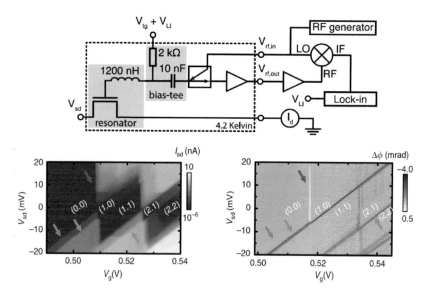

Figure 11.11 (a) Reflectometry setup for dispersive readout on a SET by homodyne detection, reproduced from J. Verduijn, M. Vinet, and S. Rogge, *Appl. Phys. Lett.* 104, 102107-1–4, 2014, with the permission of AIP Publishing. (b) Coulomb diamonds measured by source-drain current I_{SD} as a function of gate voltage V_G and bias V_{SD}. The numbers indicate electron occupancies in two quantum dots, respectively. (c) Same diagram as (b) but extracted by the reflectometry setup. The dispersive response $\Delta\Phi$ of the resonator is visible on the charge transition lines, i.e., where $\partial\langle N\rangle/\partial V_G$ is maximized. Reprinted by permission from Macmillan Publishers Ltd: M. F. Gonzalez-Zalba, S. Barraud, A. J. Ferguson, and A. C. Betz, *Nat. Comm.* 6, copyright 2015.

Experimental measurements of the quantum capacitance in semiconductor single-charge devices are usually accomplished in RF-SETs by dispersive readout. SET and LC circuit are mounted close each other and connected at ultra-low temperature in the chamber of a cryostat, as shown in Fig. 11.11a. The LC resonant circuit is formed by a chip inductor L (typically of the order of few hundreds of nH) with a capacitance C_p, usually in the pF range. Such capacitance is related to the parasitics of the inductor in combination with stray capacitances of the chip hosting the sample. Sometimes, C_p is partially tuned by varactor diodes to improve the 50 Ω matching of the load (the SET). Anyway, the purpose of the tank circuit is to provide a resonant network sensitive to small changes in the load capacitance. Typically, a voltage RF tone of few µV is generated at room temperature and transmitted to the tank circuit at the LC resonant frequency $f_r = 1/2\pi\sqrt{LC_p}$. The dissipative and dispersive responses of the resonator are related to changes in the SET admittance (Cottet et al., 2011); in particular, the phase response depends on the effective change in the load capacitance due to the quantum capacitive contribution in correspondence of single-charge transitions. The signal back reflected by the resonator is amplified at low and room temperature, then it is fed into a quadrature demodulator which yields the quadrature and in-phase components (Gonzalez-Zalba et al., 2015). Variations of the phase $\Delta\Phi$ between the applied and the reflected signal give finally access to the quantum capacitance through

$$\Delta\Phi \simeq -\pi Q C_Q/C_p, \tag{11.19}$$

where Q is the quality factor of the resonator.

Figure 11.11b shows the comparison between probing the Coulomb diamonds by transport (via I_{SD}) and capacitance spectroscopy (by dispersive signal $\Delta\Phi$). This technique, known as radio frequency reflectometry, is currently being employed for capacitive sensing in frontier devices for quantum information and computation.

11.5   Conclusions

In this chapter, we have presented techniques for capacitance spectroscopy on electronic devices in presence of quantum mechanical effects due to tight charge confinement. The discussion has spanned single-charge devices, a huge family containing devices of different architectures and materials where quantum dots can be formed, as well as commercial ultrascaled transistors where the quantum capacitance becomes crucial for their operability. The constant interaction model provides a simple description of the mechanisms by which external voltage potentials can control the charge state of a quantum dot via the capacitances that connect the dot to the electrodes of the single-electron transistor.

The model is usually applied in transport spectroscopy to determine the capacitive network of the dot by means of the stability diagrams with the characteristic pattern of Coulomb diamonds.

In addition to these geometric capacitances, the quantum capacitance due to single-charge dynamics can be probed by means of radio frequency reflectometry. Combined with transport spectroscopy, this technique represents the state of the art for single-charge detection in frontier devices exploitable for drastically innovative applications like quantum computation.

References

Ashcroft, N. W., and Mermin, N. D. (1990). *Solid State Physics* (Brooks Cole).

Averin, D. V., Zorin, A. B., and Likharev, K. K. (1985). Bloch oscillations in small Josephson junctions, *Sov. Phys.-JETP*, 61(2), (American Institute of Physics), pp. 407–413.

Betz, A. C., Tagliaferri, M. L. V., Vinet, M., Broström, M., Sanquer, M., Ferguson, A. J., and Gonzalez-Zalba, M. F. (2016). Reconfigurable quadruple quantum dots in a silicon nanowire transistor, *App. Phys. Lett.*, 108, pp. 203108-1–3.

Beenakker, C. W. J. (1991). Theory of Coulomb-blockade oscillations in the conductance of a quantum dot, *Phys. Rev. B* 44, pp. 1646–1656.

Cottet, A., Mora, C., and Kontos, T. (2011). Mesoscopic admittance of a double quantum dot, *Phys. Rev. B*, vol. 83(12), (American Physical Society), p. 121311.

Crippa, A., Tagliaferri, M. L. V., Rotta, D., De Michielis, M., Mazzeo, G., Fanciulli, M., Wacquez, R., Vinet, M., and Prati, E. (2015). Valley blockade and multielectron spin-valley Kondo effect in silicon, *Phys. Rev. B*, 92, pp. 035424-1–11.

De Michielis, M., Prati, E., Fanciulli, M., Fiori, G., and Iannaccone, G. (2012). Geometrical effects on valley-orbital filling patterns in silicon quantum dots for robust qubit implementation, *Appl. Phys. Expr.*, 5(12), p. 124001.

Devoret, M. H., and Glattli, C. (1998). Single-electron transistors, *Phys. World,* 11, pp. 29–33.

Dröscher, S., Roulleau, P., Molitor, F., Studerus, P., Stampfer, C., Ensslin, K., and Ihn, T. (2010). Quantum capacitance and density of states of graphene, *Appl. Phys. Lett.*, 96(15), (AIP Publishing) p. 152104.

Duty, T., Johansson, G., Bladh, K., Gunnarsson, D., Wilson, C., and Delsing, P. (2005). Observation of quantum capacitance in the Cooper-pair transistor, *Phys. Rev. Lett.*, 95(20) (American Physical Society), p. 206807.

Fulton, T. A., and Dolan, G. J. (1987). Observation of single-electron charging effects in small tunnel junctions, *Phys. Rev. Lett.*, 59, pp. 109–112.

Gonzalez-Zalba, M. F., Barraud, S., Ferguson, A. J., and Betz, A. C. (2015). Probing the limits of gate-based charge sensing, *Nat. Commun.*, 6 (Macmillan Publishers Limited), doi:10.1038/ncomms7084.

Granzner, R., Thiele, S., Schippel, C., and Schwierz, F. (2010). Quantum effects on the gate capacitance of trigate SOI MOSFETs, *IEEE Trans. Electron Devices*, 57(12), pp. 3231–3238.

Hanson, R., Kouwenhoven, L. P., Petta, J. R., Tarucha, S., and Vandersypen, L. M. K. (2007). Spins in few-electron quantum dots, *Rev. Mod. Phys.*, 79, pp. 1217–1265.

Haviland, D. B., Harada, Y., Delsing, P., Chen, C. D., and Claeson, T. (1994). Observation of the resonant tunneling of Cooper pairs, *Phys. Rev. Lett.*, 73, pp. 1541–1544.

Ilani, S., Donev, L. A., Kindermann, M., and McEuen, P. L. (2006). Measurement of the quantum capacitance of interacting electrons in carbon nanotubes, *Nat. Phys.*, 2(10), (Macmillan Publishers Limited) pp. 687–691.

John, D. L., Castro, L. C., and Pulfrey, D. L. (2004). Quantum capacitance in nanoscale device modeling, *J. Appl. Phys.*, 96(9), (AIP Publishing) pp. 5180–5184.

Kouwenhoven, L. P., Austing, D. G., and Tarucha, S. (2001). Few-electron quantum dots, *Rep. Prog. Phys.*, 64, pp. 701–736.

Kouwenhoven, L. P., Austing, D. G., Tarucha, S. (2010). Electron transport through double quantum dots, *Rep. Prog. Phys.*, 64, pp. 701–736.

Lansbergen, G. P., Rahman, R., Wellard, C. J., Woo, I., Caro, J., Collaert, N., Biesemans, S., Klimeck, G., Hollenberg, L. C. L., and Rogge, S. (2008). Gate-induced quantum-confinement transition of a single dopant atom in a silicon FinFET, *Nat. Phys.*, 4, pp. 656–661.

Leti, G., Prati, E., Belli, M., Petretto, G., Fanciulli, M., Wacquez, R., Vinet, M., and Sanquer, M. (2011). Switching quantum transport in a three donors silicon Fin-field effect transistor, *Appl. Phys. Lett.*, 99, p. 242102.

Likharev, K. K., and Zorin, A. B. (1985). Theory of the Bloch-wave oscillations in small Josephson junctions, *J. Low Temp. Phys.*, 59 (Plenum Publishing Corporation), pp. 347–382.

Lim, W. H., Yang, C. H., Zwanenburg, F. A., and Dzurak, A. S. (2011). Spin filling of valleyorbit states in a silicon quantum dot, *Nanotechnology*, 22(33), p. 335704.

Luryi, S. (1988). Quantum capacitance devices, *Appl. Phys. Lett.*, 52(6), (American Institute of Physics), pp. 501–503.

Maune, B. M., Borselli, M. G., Huang, B., Ladd, T. D., Deelman, P. W., Holabird, K. S., Kiselev, A. A., Alvarado-Rodriguez, I., Ross, R. S., Schmitz, A. E., Sokolich, M., Watson, C. A., Gyure, M. F., and Hunter, A. T. (2012). Coherent singlet-triplet oscillations in a silicon-based double quantum dot, *Nature*, 481, p. 344.

Mazzeo, G., Prati, E., Belli, M., Leti, G., Cocco, S., Fanciulli, M., Guagliardo, F., and Ferrari, G. (2012). Charge dynamics of a single donor coupled to a few-electron quantum dot in silicon, *Appl. Phys. Lett.*, 100(21), p. 213107.

Prati, E. (2011). Valley blockade quantum switching in silicon nanostructures, *J. Nanosci. Nanotechnol.*, 11, pp. 8522–8526.

Prati, E., Belli, M., Cocco, S., Petretto, G., and Fanciulli, M. (2011). Adiabatic charge control in a single donor atom transistor, *App. Phys. Lett.*, 98, pp. 053109-1–3.

Prati, E., De Michielis, M., Belli, M., Cocco, S., Fanciulli, M., Kotekar-Patil, D., and Tettamanzi, G. C. (2012). Few electron limit of n-type metal oxide semiconductor single electron transistors, *Nanotechnology*, 23(21), p. 215204.

Prati, E., Hori, M., Guagliardo, F., Ferrari, G., and Shinada, T. (2012). Anderson-Mott transition in arrays of a few dopant atoms in a silicon transistor, *Nat. Nano*, 7(7), pp. 443–447.

Prati, E., Kumagai, K., Hori, M., and Shinada, T. (2016). Band transport across a chain of dopant sites in silicon over micron distances and high temperatures, *Scient. Rep.*, 6, p. 19704.

Prati, E., Morello, A. (2013). Quantum information in silicon devices based on individual dopants. *Single-Atom Nanoelectronics*, pp. 5–39 (Pan Stanford Publishing).

Razavieh, A., Janes, D. B., and Appenzeller, J. (2013). Transconductance linearity analysis of 1-D, nanowire FETs in the quantum capacitance limit, *IEEE Trans. Electron Devices*, 60(6), pp. 2071–2076.

Shevchenko, S. N., Rubanov, D. G., and Nori, F. (2015). Delayed-response quantum back action in nanoelectromechanical systems, *Phys. Rev. B*, 91(16), (American Physical Society) p. 165422.

Shin, S. J., Jung, C. S., Park, B. J., Yoon, T. K., Lee, J. J., Kim, S. J., Choi, J. B., Takahashi, Y., Hasko, D. G. (2010). Si-based ultra-small multiswitching single-electron transistor operating at room-temperature, *Appl. Phys. Lett.*, 97, p. 103101.

Sillanpää, M., Lehtinen, T., Paila, A., Makhlin, Y., and Hakonen, P. (2006). Continuous-time monitoring of Landau-Zener interference in a Cooper-pair box, *Phys. Rev. Lett.*, 96(18), (American Physical Society), p. 187002.

Simmons, C. B., Thalakulam, M., Shaji, N., Klein, L. J., Qin, H., Blick, R. H., and Eriksson, M. A. (2007). Single-electron quantum dot in Si/SiGe with integrated charge sensing, *Appl. Phys. Lett.*, 91(21), p. 3103.

Stampfer, C., Schurtenberger, E., Molitor, F., Guttinger, J., Ihn, T., and Ensslin, K. (2008). Tunable graphene single electron transistor, *Nano Letters*, 8(8), pp. 2378–2383.

Tagliaferri, M. L. V., Crippa, A., De Michielis, M., Mazzeo, G., Fanciulli, M., and Prati, E. (2016). A compact T-shaped nanodevice for charge sensing of a tunable double quantum dot in scalable silicon technology, *Phys. Lett. A*, 380, pp. 1205–1209.

Thomas, G. A., Capizzi, M., DeRosa, F., Bhatt, R. N., and Rice, M. T. (1981). Optical study of interacting donors in semiconductors, *Phys. Rev. B*, 23(10), pp. 5472–5494.

Turchetti, M., Homulle, H., Sebastiano, F., Ferrari, G., Charbon, E., and Prati, E. (2016). Tunable single hole regime of a silicon field effect transistor in standard CMOS technology, *Appl. Phys. Exp.*, 9(1), p. 014001.

van der Wiel, W. G., De Franceschi, S., Fujisawa, T., Tarucha, S., and Kouwenhoven, L. P. (2003). Electron transport through double quantum dots, *Rev. Mod. Phys.*, 43, pp. 1–22.

Verduijn, J., Vinet, M., and Rogge, S. (2014). Radio-frequency dispersive detection of donor atoms in a field-effect transistor, *Appl. Phys. Lett.*, 104(10) (AIP Publishing), p. 102107.

SECTION IV: EMERGING TECHNOLOGIES

Chapter 12

Scanning Capacitance Microscopy

Jian V. Li[a] and Chun-Sheng Jiang[b]

[a]*Department of Aeronautics and Astronautics,*
National Cheng Kung University, Tainan, Taiwan
[b]*National Renewable Energy Laboratory,*
Golden, Colorado 80401, USA

jianvli@mail.ncku.edu.tw, chun.sheng.jiang@nrel.gov

This chapter reviews the measurement principle, physics, and applications of scanning capacitance microscopy (SCM). We present the scientific motivations behind the development of the SCM technique. After discussing the technical challenges of nanoscale capacitance measurement, we describe the engineering solutions incorporated in present-day SCMs. We review the applications of SCM to various materials and devices. Finally, the limitations of SCM and corresponding progresses are discussed.

12.1 Introduction

The motivation for the scanning capacitance microscopy (SCM) lies in a persistent need to better understand the inhomogeneous distribution of electrical properties at increasingly small scales, especially for the technologically important integrated

Capacitance Spectroscopy of Semiconductors
Edited by Jian V. Li and Giorgio Ferrari
Copyright © 2018 Pan Stanford Publishing Pte. Ltd.
ISBN 978-981-4774-54-3 (Hardcover), 978-1-315-15013-0 (eBook)
www.panstanford.com

circuits. Conventional techniques such as secondary ion mass spectrometry (SIMS) and capacitance–voltage (CV) profiling are routinely used to characterize the one-dimensional doping profiles of semiconductors at the macroscopic scale. Prior to the development and maturation of SCM, however, there had been no suitable characterization tools for two-dimensional mapping of doping in semiconductors with sub 100 nm spatial resolution.

Scanning probe microscopy (SPM) is an ideal platform for nano-characterization of electrical properties because of the adequate spatial resolution stemming from a localized contact between the tip and the sample that also provides a conductive path. However, SPM modes that predated SCM are largely limited to the carrier transport in the DC regime, which are not useful for independently extracting the carrier density from their dynamic response. Scanning tunneling microscopy (STM), for example, offers information on electronic structures but its working principle decides that its application is mainly in the atomic scale. Thus the STM technique is not suitable for characterization of dynamic carriers with a scalable spatial resolution of from ten to hundreds nm as demanded by the modern microelectronics. SPM-based scanning Kelvin probe force microscopy (SKPFM) [1] is able to extract electrical potential but offers no information to charge response.

The capacitance–voltage profiling technique, as described in Chapters 4 and 10, inspects the charge response to a varying electric potential. The CV method can be applied to junctions of semiconductor–semiconductor (PN), metal–semiconductor, and metal–insulator–semiconductor (MIS) types to extract the one-dimensional depth profile of the free carrier density, depletion region width, and built-in voltage. This is the physics basis of the SCM technique.

The CV principle and nanoscale contacting capability of an SPM led to the development CV measurement on MIS structure by an SPM, which is the essence of present-day SCM instrumentation.

12.2 Instrumentation

This section describes the considerations and construction of the SCM that obtains 2D mapping of doping with nanoscale spatial resolution. The focus is on how to realize the conventionally macroscopic-scale CV technique on an SPM platform to enable extraction of localized electrical properties with nanoscale spatial resolution. The following discussion is aimed at understanding of the technical issues and development of their solutions in an SCM.

12.2.1 The Challenge of Small Capacitance

We conduct a quantitative survey of landscape of all relevant capacitances present in an SCM experiment. The first and foremost notable quantity is the exceedingly small capacitance between the SPM tip and the sample, C_S. The following is an estimate of C_S using Si as an example. The depletion width of the semiconductor contacted by the SPM tip is ~100 nm if we take a doping of 10^{16} cm^{-3} and a dielectric constant of 12 for the calculation. Assuming no 3D fringe effects for the time being, the associated depletion capacitance is ~10^{-7} F/cm^2. That is, C_S is 0.01 fF (10^{-17} F) for a 100 nm by 100 nm tip-sample interaction area. For CV, the variation of capacitance can be ~50% of C_S. Therefore, the capacitance measurement circuit for a CV experiment with 8-bit digital resolution needs to resolve 0.02 aF at a few fixed frequencies. This is pushing the limit of modern-day electronic circuits.

It is worth noting that the coupling capacitor C_C between the cantilever body and the sample may reach 10–100 fF. This capacitor C_C is much higher that the tip-sample capacitance C_S and poses a serious challenge to reliably measurement, if not properly treated.

12.2.2 Instrumentation Development

The objective is to develop an apparatus for two-dimensional imaging spectroscopy of the electronic properties in modern electronic devices such a MOS transistor or a pn junction solar cell using SPM. With this instrument, each pixel of an X–Y image in the real space contains the local carrier density. The measurement electronics couple the X–Y plane spatial scanning with AC capacitance measurement scanning over voltage to achieve CV imaging. These functionalities significantly extend the boundaries of SPM instrumentation for characterizing doping in materials. Enabling this instrument are several novel concepts and components.

For the ultra-high-sensitivity capacitance measurement required for the CV measurements, the RCA capacitance detection scheme is commonly employed as the basis of SCM. The RCA capacitance detection scheme incorporates the tip-sample capacitance as part of the capacitance branch of a parallel LC circuit (Fig. 12.1). Because of the resonant nature of the LC circuit, the tip-sample capacitance sensitively tunes the resonant characteristics, or frequency response, of the LC circuit. The LC circuit is inserted between a UHF oscillator, whose driving frequency (e.g., 915 MHz) is chosen near the resonant frequency of the LC circuit, and a rectifying detector circuit. In a sense, the LC circuit functions as a filter network whose transfer function determines the coupling efficiency of energy between the oscillator (source) and the detector (measurement). The output voltage of the rectifying detector circuit therefore evaluates the shift of the frequency response of the LC circuit and, hence, measures the *change* of the tip-sample capacitance (Fig. 12.2), assuming everything else in the LC circuit is unchanged.

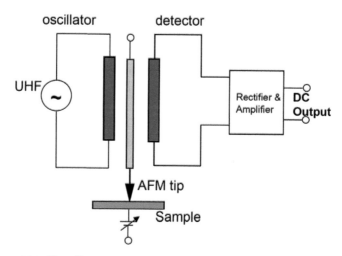

Figure 12.1 The illustrative schematic of an RCA capacitance detection scheme used for SCM.

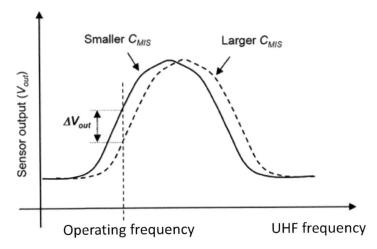

Figure 12.2 A qualitative illustration of the frequency response of the RCA LC circuit shifted by a change in the tip-sample.

The RCA scheme can achieve the required capacitance sensitivity: <0.01 aF (10^{-20} F) out of a total of ~10 aF (10^{-17} F) for steady-state CV measurement. According to estimates made earlier, this is sufficient for achieving 10 nm spatial resolution. Moreover, the coupling capacitance is rejected.

Figure 12.3 shows a diagram of the SCM experiment applied to the n-type semiconductor. A variable bias voltage is applied between the tip and the sample to modulate the capacitance between them. The SPM tip and sample interaction is characterized as a metal–insulator–semiconductor junction. The insulator layer on top of the semiconductor may be unintentionally formed. More often than not, though, it is prepared intentionally by a certain growth method and may be specially treated by a certain post-growth procedure to enhance the quality and reliability of the measurement. The variable bias voltage contains two parts: DC and AC. The DC bias voltage determines the operating points and regions of the metal–insulator–semiconductor capacitor. The AC bias voltage is used to modulate and demodulate the charge response with a lock-in amplifier thereby allowing detection of the capacitance with better signal-to-noise ratio.

In an SCM measurement, the SCM tip, surface oxide, and sample surface form a metal–oxide–semiconductor structure.

A DC bias is applied between the probe and the sample, Vs, and an additional AC bias allows the SCM signal dC/dV to be measured using a lock-in amplifier (Fig. 12.4). The carrier density is reversely proportional to the SCM signal. In the case of a higher carrier concentration, the high-frequency capacitance depletes at a higher saturation capacitance value with applying a reverse Vs, thus resulting a smaller slope of the $C-V$ curve and smaller dC/dV value, comparing with a lower carrier concentration.

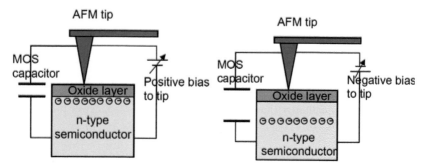

Figure 12.3 The illustrative diagram of an experiment setup of SCM applied to a n-type semiconductor with (a) a positive and (b) a negative tip bias voltage.

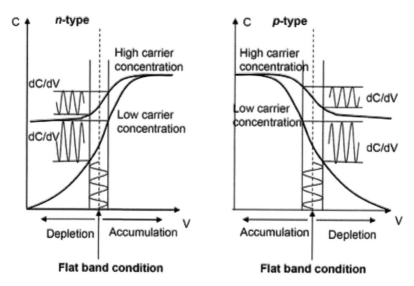

Figure 12.4 The capacitance–voltage response of an MIS capacitor consisting of a piece of (a) n-type and (b) p-type semiconductor.

12.3   Applications

12.3.1   Doping in Si Integrated Circuits

There are many examples in this regard. The information yielded by SCM for this application needs to be quantitative. This requirement necessitates the use of calibration and modeling methodologies, which has been extensively reviewed by Williams [2] and Kopanski [3].

12.3.2   Doping in Other Semiconductors

Besides silicon, SCM has been used to determine the doping of other semiconductors. An incomplete list of these semiconductors includes InP, InGaAs, InGaN, SiC, and CdTe. Here we will use CdTe as example. As shown in Fig. 12.5, an effective insulation layer can be developed on top of MBE-grown single-crystal CdTe material, leading to the successful measurement of 2D doping profile in that material.

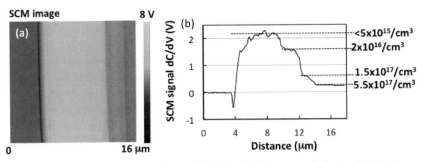

Figure 12.5 (a) The 2D profile and (b) the 1D linescan of doping in MBE-grown single-crystal CdTe semiconductor.

12.3.3   Failure Analysis

Multiple mechanisms can cause the doping profiles to deviate from the original design therefore lead to failure of integrated circuits. Prior to the advent of SCM, 2D doping–related failures were detected by selective (wet) etching and were therefore difficult to identify. The superior spatial resolution and sensitivity of SCM enables a dry measurement of 2D doping profiles.

Together with its companion modes of AFM operations, SCM has become a major technique for failure analysis of dopant-related causes. The information yielded by SCM for this application is often qualitative instead of quantitative.

12.4   Limitations

SCM has been very successful in determining doping in Si-based MIS structures. However, it is constrained by a few issues. While some of these limitations are either methodological or technological, and some are fundamental.

12.4.1   Calibration and Modeling

First, the SCM measurement is an indirect measurement of doping requiring an auxiliary calibration procedure or modeling. Sometimes, calibration is difficult because a lack of independent techniques to determine the 2D doping profiles at very small scales and in complex device structures.

12.4.2   RF Frequency

Second, the SCM instrument operates at a fixed frequency of near 1 GHz. This frequency is suitable for silicon but may be too high for other semiconductor materials. For example, the dielectric relaxation frequency of polycrystalline CdTe (doping 1e14 cm^{-3}, epsilon 10.2) may be as low as 10 MHz. For semiconductors with even lower carrier density or carrier mobility, e.g., most organic semiconductors, this problem is even more severe. This means at 1 GHz, the majority carriers are not responding to thus invalidating the working principle of SCM.

12.4.3   Dielectric Requirement

SCM requires a high-quality insulator layer to form the MIS structure, which is critical for the probe/sample capacitance measurement to be unaffected by the conductance through the probe/sample. Meanwhile, the insulating layer has to be adequately thin so that the MIS capacitance is dominated by the semiconductor but not by the insulating layer. The search for a high-quality and thin insulating layer, which is by itself a research project for certain semiconductors, may become the technical barrier preventing or delaying SCM from being applied to materials other than silicon.

12.4.4   Non-Absolution Capacitance Measurement

Above all, SCM does not typically yield the absolute value of the tip-sample capacitance. Instead, it measures the change or derivative of this capacitance due to the resonant nature of the capacitance-detecting scheme. This limitation excludes the usage of the traditional Mott–Schottky method for the extraction of carrier density. The CV technique is also used to extract the fixed charge and density of states at the interface of a metal–insulator–semiconductor structure. This function is conspicuously missing in the SCM mode of operation. The reason is that the SCM resonant circuit measures only the derivative of capacitance (i.e., dC/dV) but not the capacitance itself, which is insufficient for recovering D_{it} and Q_t without additional information.

12.5   Recent Progress

Among the limitations reviewed in the previous section, some are either technological or methodological. They have inspired continuous research effort aiming at overcoming these limitations. This section reviews a few examples of such progress.

12.5.1   Scanning Capacitance Spectroscopy

Scanning capacitance spectroscopy (SCS), a measurement of SCM dC/dV as a function of Vs at designated locations in the SCM image, provides a richer characterization of the sample response. In unipolar regions, the sign of dC/dV does not depend on Vs and is positive for p-type material and negative for n-type. The dC/dV–Vs spectra show a single peak at the flat-band voltage. When the probe is positioned in the junction area, the spectra consist of a peak and a valley with opposite signs, like a mixture of the two unipolar spectra, corresponding to U-shaped capacitance–voltage curves. These low-frequency-like C–Vs curves are due to mixture of carrier flows from both sides of the junction, and the dominant carrier depends on Vs. The tip screening length is greatly enhanced in the depletion region, and an overlapping response from the p and n regions occurs. A qualitative model was proposed for a heavy-asymmetrical junction, where the light doping is more dominant over the spectra than the heavy doping and the *C*–Vs spectra show the characteristic U-shape from the electrical junction to metallurgical junction.

12.5.2 Thin-Film PV Insulator

SCM has not been widely applied to materials other than silicon [4], mainly because of the challenge in creating a high-quality insulating layer, which is critical for the probe/sample capacitance measurement to be unaffected by the conductance through the probe/sample. The growth of a high-quality insulating layer between the metal probe and the semiconductor sample, if it is not silicon and has not been sufficiently explored with the SMC technique, may require significant amount of research. There have been development of sample preparation procedures to make a high-quality thin insulating layer on the cross-sectional surfaces of $Cu(In,Ga)Se_2$ (see Fig. 12.6) and $Cu(Zn,Sn)Se_4$ semiconductors. This process involves thermal oxidation of the sample surface to grow native oxide, which results in a high-quality thin layer (~10 nm) of oxide that is adequately insulating [5].

12.5.3 Low-Frequency Operation

As mentioned earlier, the RCA capacitance detection scheme is based on a resonant circuit thus operating at a high frequency of near 1 GHz. There is therefore an interest for measuring the small capacitance due to the tip-sample interaction in low-to-medium frequency (kHz to MHz). Attempts have been made to use low-frequency SCM setup via lock-in amplification. Alternative, sensitive capacitance measurement is feasible by employing trans-impedance amplifier-based electronics. An important benefit of the trans-impedance amplifier-based capacitance measurement scheme is that the capability of measuring the absolute capacitance, instead of the change of capacitance as required by the resonant circuit detection scheme. Details of the low-to-medium frequency capacitance measurement electronics and results are described in Chapter 14.

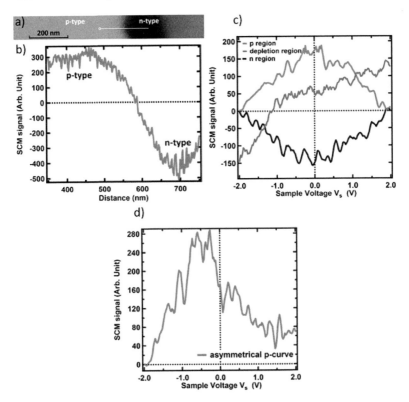

Figure 12.6 (a) An SCM image taken on a cross section of a CIGS/CdS device; (b) a line profile of SCM along the white line in (a); (c) a typical p-type SCS spectrum from CIGS, an n-type spectrum from the n-type area and transitional spectrum in the depletion region; and (d) asymmetrical p-type spectrum taken in a p-type area with a low-quality insulating layer.

12.6 Summary

The SCM technique is based on very sensitive measurement of the change of tip-sample capacitance, which is essentially a metal–insulator–semiconductor capacitor. With proper calibration and modeling, high-resolution 2D doping profiles can be accurately quantified. The SCM instrument has also been very useful in the integrated circuit industry for failure analysis. Like any other technique, SCM faces a number of limitations, some of which are technological and methodological and have inspired continuous improvement of this technique.

References

1. Nonnenmacher, M., O'Boyle, M. P., and Wickramasinghe, H. K. (1991) Kelvin probe force microscopy, *Allp. Phys. Lett.*, 58, p. 2921.

2. C. S. Jiang, Microscopic electrical characterization of inorganic semiconductor-based solar cell materials and devices using AFM-based techniques, in *Scanning Probe Microscopy in Nanoscience and Nanotechnology*, vol. 2, book series: Nanoscience and Technology, (2011).

3. Williams, C. C. (1999) Two-dimensional dopant profiling by scanning capacitance microscopy, *Annu. Rev. Mater. Sci.*, 29, pp. 471–504.

4. Kopanski, J. J. (2007) Scanning capacitance microscopy for electrical characterization of semiconductors and dielectrics, in *Scanning Probe Microscopy* vol 1, eds., Willardson, R., and Weber, E., Chapter 4 (Springer, New York) pp. 88–112.

5. Jiang, C.-S., Heath, J. T., Moutinho, H. R., and Al-Jassim M. M. (2011) Scanning capacitance spectroscopy on n⁺-p asymmetrical junctions in multicrystalline Si solar cells, *J. Appl. Lett.*, 29, p. 014514.

6. Xiao, C., Jiang, C.-S., Moutinho, H. R., Levi, D., Yan, Y., Gorman, B., and Al-Jassim M. M. (2016) Locating the electrical junctions in $Cu(In,Ga)Se_2$ and $Cu_2ZnSnSe_4$ solar cells by scanning capacitance spectroscopy, *Prog. Photovolt. Res. Appl.* DOI: 10.1002/pip.2805.

Chapter 13

Probing Dielectric Constant at the Nanoscale with Scanning Probe Microscopy

Laura Fumagalli[a] and Gabriel Gomila[b,c]

[a]*School of Physics and Astronomy, The University of Manchester, Oxford Road, Manchester, M13 9PL, UK*
[b]*Institut de Bioenginyeria de Catalunya (IBEC), C/Balidiri i Reixac 15-21, Barcelona, 08028, Spain*
[c]*Departament d'Electrònica, Universitat de Barcelona, C/Martí i Franquès 1, Barcelona, 08028, Spain*

laura.fumagalli@manchester.ac.uk, ggomila@ibecbarcelona.eu

13.1 Introduction

Dielectric polarization, represented by the dielectric constant (or permittivity), ε_r, is an intrinsic property of matter that plays a fundamental role in many fields of research, from materials science and technology to chemistry and biology. It represents the intrinsic response of a material to an applied electric field, which depends on materials composition, structure and phase. In electronic devices, it is inherently linked to charge storage and

Capacitance Spectroscopy of Semiconductors
Edited by Jian V. Li and Giorgio Ferrari
Copyright © 2018 Pan Stanford Publishing Pte. Ltd.
ISBN 978-981-4774-54-3 (Hardcover), 978-1-315-15013-0 (eBook)
www.panstanford.com

transport. In biology, it modulates crucial electrostatic interactions between biomolecules such as DNA and proteins, and influences their shape and structure. Yet, quantifying local dielectric properties using scanning probe microscopy (SPM) [1, 2] has remained a long-standing challenge because the local signal to detect is extremely weak and strongly affected by non-local contributions as well as geometric/size artefacts.

In the last years, measurement of dielectric properties using scanning probes, which we will refer to here as scanning dielectric microscopy, has been demonstrated at both low [3–12] and high frequencies [13–17]. Here we will review our work at low frequencies (< 100 MHz) in which we showed that the dielectric constant of different samples—from thick [6] insulating films and bacteria [18, 19] to molecularly thin layers [5, 20] and macromolecules [21]—can be precisely measured with a scanning probe using two approaches, ac current-sensing [4, 5, 22] and electrostatic force-sensing [6, 7, 23–27]. In particular, by combining low-noise capacitive detection with quantitative numerical analysis, we showed that the dielectric constants of 10 nm-radius nanoparticles can be quantitatively obtained from ultraweak polarization forces [23]—a resolution currently unparalleled. We also showed that the dielectric constants can be used as fingerprints to recognize nano-objects of identical shape but different chemical composition [23], which would be impossible from topography. We determined the dielectric constants of important biological samples, from single bacteria [18, 19] to viruses [21, 23] and ultrathin biolayers [5, 20]. Importantly, we experimentally resolved the dielectric constant of DNA [21], which had remained unknown owing to the lack of tools able to measure it. We also demonstrated that dielectric constant quantification can be extended to liquid environment at 1–100 MHz [28–30], which will enable the study of electrochemical systems and biomolecules in their natural environment.

This review is divided as follows. After introducing the main challenges of low-frequency measurements in Section 13.2, we will present the theoretical framework (Section 13.3) and calibration procedure of the probe geometry (Section 13.4) that we implemented to interpret the data. The main experimental steps that we follow to measure the local dielectric constant free from artefacts are summarized in Section 13.5. In Section 13.6, we will review our results based on the current-sensing approach, while we will focus on our results using the electrostatic force approach in Section 13.7. Finally, we will explain the extension to liquid environment in Section 13.8, followed by the conclusions.

13.2 Why Probing Dielectric Constant Is So Difficult?

With the invention of SPM in the 1980s, many scanning probes microscopes have been developed to access electrical properties of solids and liquids at the nanoscale simultaneously to structural properties [1, 2]. The first and simplest one is the scanning tunnelling microscope (STM), probing the tunnelling current through an atomically sharp asperity of a metallic wire with atomic spatial resolution. However, STM has a number of limitations, e.g. the electrical image is coupled to the surface topography and limited to conductive samples or to insulating samples as thin as few atomic layers. To overcome these limitations, several electrical techniques based on the AFM have been introduced in the late 1980s, probing local electrical properties with a small sharp tip mounted on a cantilever. They are advantageous because the electrical information is decoupled from topography. Furthermore, they can be applied to virtually any type of samples (metallic and insulating) by combining different sensing technologies (electrical, mechanical and optical). Therefore, electrical AFM techniques such as conductive-AFM, scanning capacitance microscopy (SCM), electrostatic force microscopy (EFM) and Kelvin Probe Force microscopy (KPFM), which probe impedance (capacitance), electrostatic force and contact potential, respectively, have rapidly gained popularity [2].

Yet, measuring the dielectric constant using a scanning probe had proven challenging for long. In the 2000s, despite electrical-AFM techniques were routinely used for characterization of solid and soft (biological) materials, probing dielectric properties remained a formidable challenge as it requires to quantitatively extract the dielectric constant from capacitance at low frequencies (or the dielectric function using optical approaches at higher frequencies—not treated here). First, one needs to quantify the local capacitance probed by the very tip apex which is in the order of 1–10 aF. This is an extremely small signal, in the range of resolution offered by best benchtop instrumentation. Second, as depicted in Fig. 13.1, the local capacitance is buried in major contributions of the cantilever in the order of 1 pF— more than five orders of magnitude larger. Therefore, to probe local dielectric properties, one needs first to implement new instrumental solutions to achieve a sensitivity better than that offered by commercial equipment, and then to apply new measurement approaches able to resolve local capacitance changes over stray capacitance contributions and quantify them. Furthermore, once local capacitance changes are detected, the impact of geometry of the tip/sample needs to be perfectly taken into account to isolate the material polarization contribution from geometrical effects. Finally, if measurements need to be carried out in electrolytic environment, as in the case of biological applications, things are more difficult owing to the ionic contribution of the liquid solution and the complexity of the solid–electrolyte interface, which have to be properly treated.

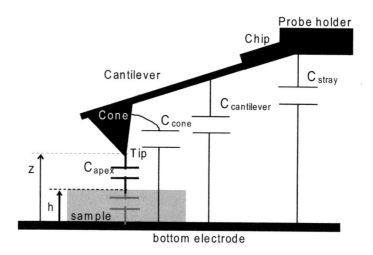

Figure 13.1 Simplified schematics of the AFM probe on an insulating layer, showing the main capacitance contributions (tip apex, cone, cantilever, and probe holder) in scanning dielectric microscopy at low frequencies (< 100 MHz).

To be able to extract the local dielectric constant, we first developed a fully customized instrumentation mounted on a

AFM, based on a wide-bandwidth low-noise current amplifier and lock-in detection, able to resolve and precisely quantify the local attoFarad capacitance probed by the AFM tip apex [32], as further detailed in Section 13.6. We note that traditional SCM [33], based on different instrumental solutions at higher frequencies, had not allowed that. To quantify the dielectric constant from local capacitance data, we introduced a probe calibration procedure and the theoretical model which allow, first, to precisely quantify the tip radius [4, 22]—the main geometrical parameters that represents a major obstacle to dielectric quantification—and the local dielectric constant of the sample. By using this approach based on low-noise *current sensing* and achieving capacitance resolution of ~150 zF/√Hz for 1 V ac applied signal in the frequency range 10 kHz–1 MHz, we obtained dielectric constant images of thin insulating film, from tens of nanometre thicknesses [4] down to molecular monolayers [5], showing spatial resolution around 50–100 nm [5], well below the micrometer limit of standard dielectric measurements that are based on using microelectrodes instead of using a scanning probe. In order to further increase both the electrical and lateral resolution of the technique and measure smaller objects such as nanoparticles and macromolecules, we adopted the *electrostatic force-sensing mode*, as detailed in Section 13.7. By applying measurement procedures similar to those applied in current-sensing mode (probe calibration, constant-height imaging mode) and measuring small forces down to picoNewton level—corresponding to few zeptoFarads in capacitance—we showed that the dielectric constants of objects as small as 20 nm in diameter could be precisely obtained [23].

13.3   Theoretical Modelling

Both sensing approaches considered here, current- and force-sensing, share the same theoretical framework. In particular, here we treat the case when an ac voltage $v_{ac}\cos(\omega t)$ of amplitude, v_{ac}, and angular frequency, ω, is applied between a conductive AFM probe and a conductive substrate on which the sample is placed. We note that force-sensing can also be performed by applying a dc voltage [7] but this is less accurate. This is because it achieves lower electrical resolution than in ac mode. Furthermore, in addition to dielectric properties, it is sensitive to surface potential changes that should be carefully taken into account, and to thermal drift and instabilities, therefore it is more difficult to implement. In ac regime instead, we only need to detect either the ac current amplitude, i_{ac}, flowing through the probe at ω (current-sensing mode) or, alternatively, the mechanical oscillation of the cantilever, $A_{2\omega}$, at the second harmonics, 2ω, of the excitation (**force-sensing mode**), from which we directly obtain the capacitance or the capacitance first derivative, respectively, and through that the dielectric properties of the sample. In particular, for a purely insulating system, one has, in the case of **current-sensing,**

$$i_{ac}(\omega) = \omega C(\omega) v_{ac} \tag{13.1}$$

and, in the case of **force-sensing,**

$$A_{2\omega} = \frac{1}{4k} \frac{\partial C(\omega)}{\partial z} v_{ac}^2, \tag{13.2}$$

respectively, where k is the spring constant of the cantilever. Here, C is the total capacitance between the AFM probe, including the probe chip and probe holder which can hardly be perfectly shielded, and the sample substrate. In general, this can be divided into different contributions

$$C = C_{apex} + C_{cone} + C_{cant} + C_{stray}, \tag{13.3}$$

where C_{apex}, C_{cone}, C_{cant} and C_{stray} are the tip apex, tip cone, cantilever and stray contributions to the total capacitance, respectively, as depicted in Fig. 13.1. In the case of force-sensing, the stray contributions of the chip and probe holder are not detected, therefore the total capacitance reduces to

$$C = C_{apex} + C_{cone} + C_{cant}. \tag{13.4}$$

To extract the local dielectric constant, in principle, one should quantify the capacitance (or the capacitance first derivative) and isolate the local capacitance probed by the very tip apex, C_{apex}, from the other contributions. However, this is a virtually impossible task, as the local capacitance is orders of magnitude smaller than the other contributions, which in turn depend on multiple geometric parameters associated to the detailed probe/sample geometry and arrangement—probe inclination, conductive areas of the probe/sample, edge effects, etc.—which are difficult if not impossible to quantify. Absolute values are therefore disregarded in the data analysis. On the other hand, dielectric quantification can be achieved by simply considering the spatial variations of capacitance, ΔC, or capacitance first derivative, $\Delta C'$, when the tip is moving either vertically towards the sample (approaching) or laterally over the sample (scanning). Equation (13.3) will be then written in terms of spatial variations as

$$\Delta C = \Delta C_{apex} + \Delta C_{cone} + \Delta C_{cant} + \Delta C_{stray}, \qquad (13.5)$$

where ΔC is the difference in capacitance at the position (x,y,z) with respect to a reference position, $\Delta C = C(x,y,z) - C(x_0,y_0,z_0)$. Similarly, in the case of force-sensing, Eq. (13.4) is rewritten as

$$\Delta C' = \Delta C'_{apex} + \Delta C'_{cone} + \Delta C'_{cant}, \qquad (13.6)$$

where for brevity we use C' to indicate the first derivative of capacitance.

To proceed using Eqs. (13.5) and (13.6) and obtain the nanoscale dielectric properties of the sample, the accurate theoretical model of the probe-sample capacitance (or capacitance derivative) is needed. However, while this is normally available or easy to derive for macroscale measurements that can use planar electrodes, typically requiring simple parallel-plate analytical expressions, in the case of measurements on the nanoscale with a scanning probe, finding the correct theoretical model that describes the electrostatics of the probe-sample system is not trivial. The electrostatic interaction between the sharp conducting tip attached at the end of a micro cantilever and the planar conductive substrate on which the sample is typically supported can be rather complicated. Furthermore, the dielectric sample in general has a complex shape and can be heterogeneous with regions of different dielectric constants. Even in the simple case of a uniform sample of a given dielectric constant, the probe-sample capacitance strongly depends on the actual tip geometry, which needs to be calibrated in situ, as described further below, and on the sample geometry, which needs to be accurately measured before/after the dielectric measurement. In general, except few exceptions, it is hard to find analytical formulas that accurately describe the probe-sample electrostatic interaction, and one needs to numerically solve the full electrostatic problem as described below in order to obtain accurate dielectric measurements.

13.3.1 Numerical Modelling

To obtain a capacitance model of the probe-sample system, in the case of an ideal dielectric material considered here for which capacitance is frequency-independent, one can simply implement and solve the corresponding electrostatic problem with a static applied potential. To this aim, the system is modelled as composed of homogeneous, linear and isotropic parts with no net fixed charges, no free charges, hence no conductivity and space-charge formation, and no losses (no frequency response of the material). The Laplace's equation is solved for the electric potential, ϕ_α,

$$\nabla^2 \phi_\alpha = 0 \tag{13.7}$$

in each homogeneous part of the system (α = air, sample, substrate) and imposing the discontinuity boundary conditions at the interfaces between different parts. Once the electric potential distribution has been obtained, one can compute capacitance (in current-sensing mode) and the capacitance first derivative (in force-sensing mode). The capacitance is calculated as

$$C \equiv \frac{Q_{\text{probe}}}{V}, \tag{13.8}$$

where V is the applied voltage and Q_{probe} is the total charge on the probe surface (or bottom electrode). This is obtained by integrating the electric potential gradient of the surface charge on the probe

$$Q_{probe} = \int_{probe} \sigma_{probe} dS = \int_{probe} (\varepsilon_0 \vec{\nabla} \phi_{probe}) \cdot \hat{n} dS \qquad (13.9)$$

The total electrostatic force acting on the probe can be calculated by integrating the Maxwell stress tensor on the probe surface as

$$\vec{F}_{probe} = \varepsilon_0 \int_{probe} \left[(\vec{E} \cdot \hat{n})\vec{E} - \frac{1}{2}\vec{E}^2\hat{n} \right] dS, \qquad (13.10)$$

where the electric field distribution is obtained as usual from $\vec{E} = -\vec{\nabla}\phi$. The capacitance first derivative is then given by using the relation

$$\frac{dC}{dz} = \frac{2F_{probe,z}}{V^2}, \qquad (13.11)$$

where a factor 2 appears instead of the factor 4 that appears in Eq. (13.2) because static (not dynamic) numerical calculations are carried out.

Figure 13.2 shows an example of numerically calculated electric potential distribution for an AFM probe on a dielectric layer (Fig. 13.2b,c) on a metallic substrate. The numerical calculations were done using the finite-element package COMSOL Multiphysics (AC/DC electrostatic module) and the direct axisymmetric 2D solver PARDISO.

13.3.2 Analytical Modelling

Numerical calculations are very effective and can be applied to virtually any system, but have the drawback of being time-consuming to implement. Ideally one would prefer to use analytical formulae as they are easier to handle and fit the experimental data. However, deriving analytical models that quantitatively predict the electrostatic force between the AFM probe and the sample is not straightforward even in the simplest cases such as laterally finite layers or spherical samples, as further discussed here below. The only analytical formulas available, given here below, apply to ultrathin/thin and wide layers, typically with thickness < 10–50 nm and width larger than a critical diameter (laterally infinite regime). Outside this range of thickness and width which depend on the probe size and geometry [4, 22, 25, 27], these formulas are inaccurate, and one has to restore to numerical calculations described above to extract the correct dielectric constant.

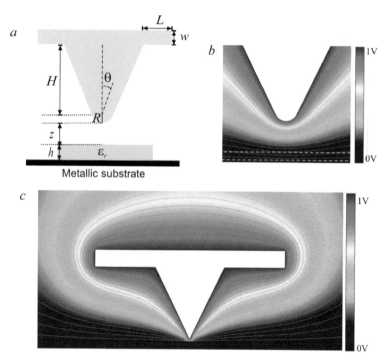

Figure 13.2 (a) Schematic representation of the probe on a dielectric layer on a conductive substrate that we implemented with numerical calculations, showing the main parameters used to model the tip–sample electrostatic interaction and extract the dielectric constant of the layer. (b,c) Example of calculated electric potential distribution for the case of dielectric layer (indicated with white dashed line) obtained by solving the full electrostatic problem (tip radius R = 20 nm, cone half-angle θ = 25°, cone height H = 12 μm, cantilever length L = 10 μm, layer dielectric constant ε_r = 2 and thickness h = 20 nm).

In the case of very thin, wide layers and under restricted conditions—as a rule of thumb, the layer thickness should be smaller than the tip apex radius and the layer width two-three times larger than that—the following analytical formulas apply. The capacitive contribution of the probe components (apex, cone and cantilever) can be derived as shown by Hudlet et al. for the case of a conductive substrate [31], substituting the tip–substrate distance z with the term $z + h/\varepsilon_r$, being h the layer thickness and ε_r the layer dielectric constant [4, 22, 25]. According to this model, one obtains

$$C_{apex}(z,\varepsilon_r) = 2\pi\varepsilon_0 R \cdot \ln\left(1 + \frac{R \cdot (1 - \sin\theta)}{z + \dfrac{h}{\varepsilon_r}}\right) + c_0 \tag{13.12}$$

$$C_{cone}(z,\varepsilon_r) = \frac{-2\pi\varepsilon_0}{\ln\left[\tan(\theta/2)\right]^2}\left[\left(z + \frac{h}{\varepsilon_r} + R(1-\sin\theta)\right)\ln\left(\frac{H}{z + \dfrac{h}{\varepsilon_r} + R(1-\sin\theta)}\right)\right.$$
$$\left. + R(1-\sin\theta) + \frac{R\cos^2\theta}{\sin\theta}\ln\left[z + \frac{h}{\varepsilon_r} + R(1-\sin\theta)\right]\right] + c_1 \tag{13.13}$$

$$C_{cant}(z,\varepsilon_r) = \varepsilon_0\alpha\frac{\pi\left[(L + H\tan\theta)^2 - (H\tan\theta)^2\right]}{z + \dfrac{h}{\varepsilon_r} + H} + c_2, \tag{13.14}$$

where θ is the cone angle, H the tip cone height and R the tip apex radius and c_0, c_1 and c_2 are constants independent from distance. The cantilever is modelled here as a disk of radius $L + H\tan\theta$ located on top of the tip cone, where L is the part of the disk radius not covered by the cone base, as indicated in Fig. 13.2a, and α is a numerical parameter close to unity that accounts for fringing effects. From these expressions, one obtains the corresponding capacitance first derivatives to be applied in force-sensing:

$$C'_{apex}(z,\varepsilon_r) = 2\pi\varepsilon_0\frac{R^2(1-\sin(\theta))}{\left(z + \dfrac{h}{\varepsilon_r}\right)\left(z + \dfrac{h}{\varepsilon_r} + R(1-\sin(\theta))\right)} \tag{13.15}$$

$$C'_{cone}(z,\varepsilon_r) = \frac{2\pi\varepsilon_0}{\ln[\tan(\theta/2)]^2}\left[\ln\left(\frac{H}{z+\dfrac{h}{\varepsilon_r}+R(1-\sin\theta)}\right) - 1 + \frac{R\cos^2\theta/\sin\theta}{z+\dfrac{h}{\varepsilon_r}+R(1-\sin\theta)}\right]$$

(13.16)

$$C'_{cant}(z,\varepsilon_r) = \varepsilon_0\alpha\frac{\pi\left[(L+H\tan\theta)^2-(H\tan\theta)^2\right]}{\left(z+\dfrac{h}{\varepsilon_r}+H\right)^2}$$

(13.17)

Equations (13.12)–(13.17) clearly show that capacitance and its derivatives not only depend on the sample dielectric constant ε_r and its geometry (here represented by the layer thickness h) but also and more strongly on the geometry/size of the probe. This is modelled here through four parameters, namely the tip apex radius, R, cone angle θ and the cone height, H, and the cantilever length L. Each of these parameters influences the interaction with different impact which depends on the tip-surface distance z and the specific size of the probe used, e.g. tip radius R of few nanometres or few hundreds nanometres. In general, the tip apex radius is the main parameter that impacts only in close proximity to the sample (few tens of nanometres) while the cone and cantilever geometry/size dominates the electrostatic interaction at hundreds of nanometre distances. This can be appreciated by plotting approach curves as a function of the tip-surface distance, z, as in Figs. 13.3 and 13.4. Importantly, while the cantilever has an impact on capacitance, it is negligible in the case of the capacitance derivatives. We remark that this holds only for thin/ultrathin layers treated here, as it can be shown that on thick dielectric substrates the cantilever length has an impact on the dielectric measurement and therefore has to be properly modelled [6, 25, 27].

We note that the modelling and the geometric parameters presented here are those that describe the majority of commercially available conductive probes (typically metal-coated or doped diamond-coated silicon probes) with a pyramidal tip attached at the end of a rectangular cantilever (with tip apex radii between

30 and 300 nm). For probes with a different shape, other geometric parameters may apply. As an example, in the case of commercial doped-silicon probes with no conductive coating, the end of the cone near the tip apex is normally much sharper than the rest of the cone; therefore a double-angle geometry needs to be adopted to correctly extract the tip geometry and the dielectric constant of the sample from the measurements. The use of a single-cone geometry in this case would lead to wrong dielectric constants, as we explained in details in Refs. [23, 24].

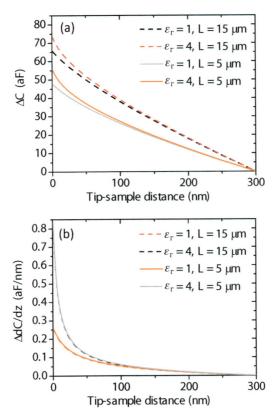

Figure 13.3 Total capacitance (a) and capacitance first derivative (b) approach curves calculated using the formulas for a dielectric layer of different dielectric constants (ε_r = 1 and 4) and different cantilever lengths (L = 5 and 15 μm). Other data: layer thickness h = 25 nm, tip radius R = 100 nm, cone height H = 12.5 μm.

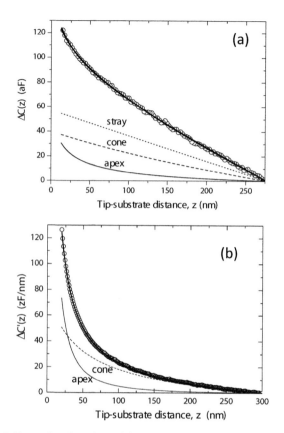

Figure 13.4 Example of probe calibration curves and fittings taken on a metallic surface. (a) Measured capacitance (circles) obtained applying the current-sensing approach and (b) measured capacitance first derivative (circles) obtained using electrostatic force approach and corresponding fittings to the theoretical models (thick solid lines). The apex (thin solid line), cone (dashed line) and stray contributions (dotted line) are also given.

13.4 Probe Geometry Calibration

As clearly shown in Fig. 13.4, the geometry of the probe has to be known with high accuracy in order to quantitatively extract the local dielectric constant of the sample from the tip–sample capacitance and its derivatives. Under some circumstances, such as when using large tips and measuring large samples (micrometer range), one can obtain the geometric parameters by taking electron microscopy images of the probe ex situ. In general, however, this approach is not sufficiently accurate and can lead to unacceptable errors in dielectric constant. In particular, when commercial silicon probes with tip radius of few tens of nanometres are used, few-nanometre accuracy in tip radius is needed which cannot be realistically achieved ex situ. Furthermore, the uncertainty in the probe geometry has increasingly higher impact with decreasing the size of the tip and the object. An accuracy below the nanometre was required in Ref. [23]. Therefore, calibrating the geometric parameters in situ by taking approach curves on the conductive substrate near the nanoparticles and fitting them to the theoretical model was key to quantitatively obtain the dielectric constants of the nanoparticles. This procedure was originally proposed in the context of EFM in Ref. [34] for calibrating the tip apex radius at short distances. We subsequently extended it to the larger distances [4–7, 22–27], which allow us to calibrate the cone aperture of the tip in addition to the tip radius, and systematically used it as probe calibration technique for dielectric measurements.

We first introduced and demonstrated this procedure for dielectric measurements in Ref. [4] applied to capacitance approach curves, $C(z)$, and then extended it to capacitance first-derivative curves, $dC(z)/dz$, taken on a conductive region of the substrate near the dielectric sample. We note that the same procedure was also used in Ref. [11]. The experimental curves are fitted to the tip–substrate theoretical models for the metallic surface given above, obtaining the *effective* geometrical parameters that allow extracting the dielectric constant of the sample with high accuracy. The theoretical model can be either the analytical formula presented above or a set of numerically calculated capacitance (or its derivatives) values. In the latter case, it can be systematically calculated as a function of all geometric parameters of the probe, covering the range of values of interest (R = 1–120 nm, θ = 5°–41°, H = 5–20 μm, L = 0–35 μm, z = 6–800 nm, w = 2 μm) and then fitted to the experimental data. Examples of probe geometry calibrations curves and fittings are reported in Figs. 13.4a,b, showing capacitance and capacitance derivative curves, respectively.

We note that in the case of capacitance, one would need to model and fit all the long-range stray contributions of the setup in addition to the cantilever. In practice, this can be avoided by noticing that all these stray components and the cantilever component sum up into a linear function of the tip displacement with respect to a reference distance, z_0, and the total capacitance variation can be approximated to

$$\Delta C \approx \Delta C_{apex} + \Delta C_{cone} + k_{stray} \cdot (z - z_0), \tag{13.18}$$

where k_{stray} is the slope of the linear function. Therefore, one can simply subtract the linear contribution and fit the remaining apex and cone contributions using the tip apex radius, R, and cone angle, θ, as adjustable parameters and keeping all the other geometric parameters at the nominal values. The apex radius and cone angle obtained in this way typically fall in the expected range specified by the probe manufacturers. More important, they show to be extremely accurate, allowing us to extract the dielectric constants of nanostructures as small as few tens of nanometres with uncertainty below 10% [21, 23].

13.5 Dielectric Constant Quantification

The procedure to extract and quantify the dielectric constant from local capacitance or capacitance derivatives measured with a scanning probe can therefore be summarized in the following main steps:

(1) Calibrate the tip geometry in situ by taking a capacitance (or capacitance derivative) approach curve on a metallic region of the substrate as described above and fitting it to the theoretical (analytical or numerical) model, thus obtaining all the unknown geometric parameters of the probe with high accuracy. Alternatively, if a bare metallic region is not available on the sample, e.g. in the case of using insulating substrates, use a different metallic substrate to calibrate the tip before and after the dielectric measurement.

(2) Measure the size/geometry of the dielectric sample. This is typically done using the AFM topography image taken before the capacitive images. Note however that if the sample is a small object of few tens of nanometres or smaller, tip–sample convolution effects will affect the width that is obtained from the topography image. These effects have to be properly taken into account to obtain the real width of the sample. Tip–sample convolution effects can lead to inaccurate or even wrong dielectric constants.

(3) Measure and quantify the capacitance or capacitance derivative of the sample using the current- or force-sensing approach as described further below. Measurements can be either approach curves taken in a single location of the sample or alternatively images taken using the double-pass scan procedure at a scan height, z, with respect to the metallic back substrate, which needs to be known with good accuracy.

(4) Fit the capacitive data to the theoretical (analytical or numerical) models as a function of the dielectric constant of the sample as the only fitting parameter. The analytical formula or numerical calculations need to be implemented using the geometric parameters of the sample and of the probe previously obtained by calibrating the probe.

This procedure was first successfully applied to capacitance approach curves and images in Refs. [4, 5, 22] using the low-frequency current-sensing approach (frequency < MHz) and then extended to the case of capacitance first-derivative images using electrostatic force microscopy [6, 7, 23–27]. Recently, the same procedure has been successfully adopted to obtain the dielectric constant at gigahertz from capacitance data measured using scanning microwave microscopy [13, 35].

13.6 Dielectric Measurement via Current-Sensing

The local capacitance can be measured at the nanoscale by using an AFM and standard alternating current (ac) detection. Under this measurement scheme, the AFM conductive probe is used as a nanometric electrode that is laterally displaced with nanometric spatial resolution over the sample, either in contact or non-contact mode, and which can be vertically controlled using the force feedback. A conductive substrate is needed to act as bottom electrode. The tip–substrate capacitance is thus obtained by applying an ac voltage, typically few Volts, and measuring the ac current flowing through the tip, as depicted in Fig. 13.5.

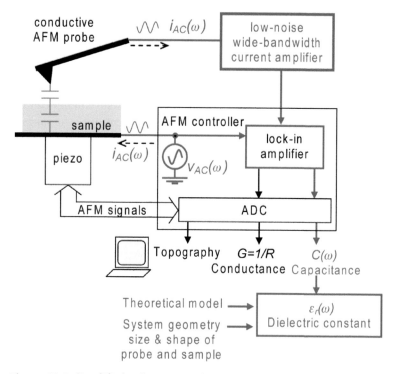

Figure 13.5 Simplified schematics of the setup for dielectric constant measurement using the current-sensing approach at low frequencies (< 1 MHz). A low-noise wide-bandwidth current-to-voltage amplifier is coupled to a lock-in amplifier, allowing simultaneous measurement (image) of tip-surface forces (topography) and sample admittance, thus obtaining the capacitance (admittance imaginary part). From local capacitance measurements, the local dielectric constant of the sample is extracted, provided that the probe and sample geometry/size are accurately known and the theoretical model reproduces them.

The total measured capacitance is typically around hundreds of femtoFarads given by the macroscopic parts of the AFM probe and the parasitics of the setup. The local capacitance probed by the tip apex is orders of magnitude smaller, in the attoFarad range [32]. Therefore only variations in local capacitance are measurable (not the absolute values), provided that they can be distinguished from non-local variations. In particular, we note that non-local contributions may also change while scanning the probe over the sample. Extreme care is therefore needed to determine the local variations from stray variations, as we described in the theoretical section. To this aim, a capacitance detector with high resolution, well below the attoFarad level, is required to detect local capacitance variations, which are typically a few attoFarads or smaller, over dominant non-local variations.

We note that measurements have been attempted using commercial impedance analyzers [36], which, however, are typically limited to femtoFarad resolution and do not allow to probe local capacitance, only non-local capacitances or the local resistive component. Capacitance bridge sensors are a better solution, and have been employed coupled to AFM as well as to scanning tunnelling microscope. They are the best benchtop equipment for capacitance measurement which can achieve attoFarad resolution, but at the expense of slow acquisition times, typically on the order of 1 s. This is far too long for the scanning probe approach, as it would require a few hours to take an image. The best strategy for dielectric measurements using a scanning probe consists of fully customizing the detector to the AFM probe, as we implemented in Refs. [4, 5, 22, 32] based on a low-noise transimpedance amplifier coupled to a lock-in amplifier. Using this approach, we showed that sub-attoFarad resolution is achieved in reasonable imaging times (5–10 s per line) [5]. By proceeding in this way, the local capacitance can be detected in a single point using approach curves (Fig. 13.6) as well as in imaging mode (Fig. 13.7) and from that the local dielectric properties of the sample area located under the tip apex can be precisely quantified by extracting the dielectric constant. Dielectric constant quantification is possible, however, under certain circumstances. In particular, the tip should be as near as possible from the surface (nanometre range). Non-local capacitance variations should be accurately analyzed, subtracted as detailed in the theoretical section and minimized. This can be done by taking images in **constant-height mode** instead of using lift mode typically used in electrical-SPM techniques [2]. In lift mode, the tip follows the topographic profile, introducing major variations in stray capacitance. Relative large tip apex (> 50 nm) and large samples (hundreds of nanometre width) also facilitate the measurement, as they both contribute to increase the signal. Under these experimental conditions, we first achieved to measure the dielectric constant of thin insulating layers (20–30 nm thickness) in a single-point mode by analyzing approach curves taken on a single location of the dielectric layer as shown in Fig. 13.6 [4, 22]. We then showed that dielectric *images* of molecularly thin monolayers (~5 nm thickness) (Purple Membrane) can be obtained in constant-height mode, as shown in Fig. 13.7b [5]. We remark that the measured capacitance value on the monolayer and double layer of Purple Membrane given in Fig. 13.7b are, respectively ~0.5 aF and ~1 aF, a level that we were able to detect by minimizing the capacitance noise level down to ~ 100 zeptoFarad using our fully customized setup based on a low-noise transimpedance amplifier [5, 32].

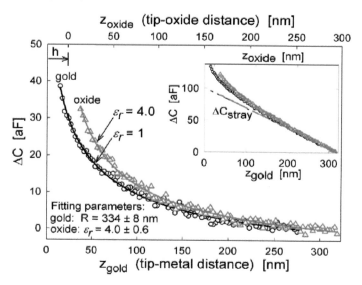

Figure 13.6 Dielectric constant measurement on a 25 nm thin oxide layer (triangles) in a single point. The probe geometry calibration curve taken on a gold surface (circles) is also given. Lines are the fitting to the analytical model after subtracting the stray linear contribution. Inset: raw data including the stray linear contribution (applied voltage of 1 V amplitude at 110 kHz). Reprinted from *Appl. Phys. Lett.*, 91, 243110 (2007), with the permission of AIP Publishing.

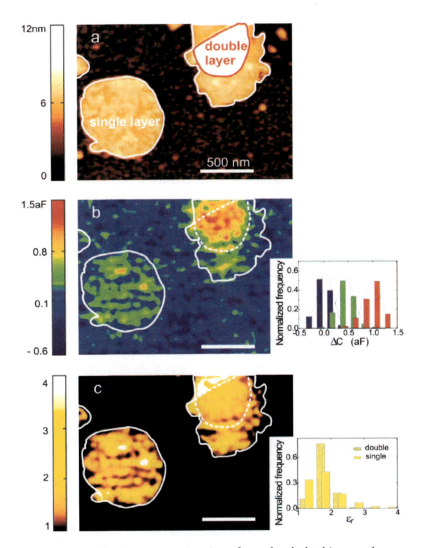

Figure 13.7 Dielectric constant imaging of a molecularly thin monolayers of Purple membrane (~ 5 nm thick) on a graphite substrate. Measured (a) topography, (b) capacitance and (c) extracted dielectric constant $\varepsilon_r \sim 2$ (scan speed 12 s/line applied ac voltage of 3 V amplitude, 54 kHz frequency). Reprinted with permission from *Nano Lett.*, 9, 1605 (2009). Copyright 2009 American Chemical Society.

We remark that the capacitance noise level also sets the spatial resolution that can be achieved in the dielectric constant measurement using the current-sensing approach. This is because the noise level imposes to use rather large tip radius (in the 50–100 nm range using our setup) to be able to detect the local capacitance signal. The results reported in Fig. 13.7 showed a lateral resolution of around 70 nm, which is set by the noise level of the equipment of ~150 zF/√Hz with 1 V ac applied signal. Yet, this is not the ultimate limit of this approach, which is set by the cantilever thermal vibrations. This is below 0.1 zF/√Hz in terms of capacitance noise—three orders of magnitude smaller than the noise of our capacitance detector. The spatial resolution of this approach could therefore be further improved by improving the noise level of the detection system. In particular, by minimizing the input capacitance of the transimpedance amplifier down to ~1 pF, one may reasonably improve the performance by a factor of 10 and reach ~10 zF/√Hz noise level with 1 V applied signal, achieving a few tens of nanometres in spatial resolution [5].

13.7 Dielectric Measurement via Force-Sensing

An alternative way to measure dielectric constant is to detect the electrostatic force between the tip and the substrate using EFM. With respect to the current-sensing approach, the EFM approach has the advantage of being easier to implement because it does not require low-noise current detectors. Furthermore, thanks to the high sensitivity of the cantilever to force variations, it allows to achieve higher spatial resolution. EFM can be implemented in different ways [3, 6–12], which detect the capacitive interaction by applying a dc or an ac voltage signal and by detecting the corresponding dc or ac variation in the mechanical deflection of the cantilever and, from this, the first (dC/dz) or the second (d^2C/dz^2) derivative of the tip–substrate capacitance. The ability of EFM to detect the conductivity/insulating nature of samples at the nanoscale has been demonstrated and routinely applied since the late 1980s soon after the invention of AFM. Yet, extracting the dielectric constant of insulating samples from EFM data had remained challenging for long, as it requires accurate measurement and quantification of the tip–sample capacitive interaction. We showed that dielectric constant measurement can be achieved by EFM in static [7] and dynamic mode [6, 23] by applying the quantification procedure and modelling that we developed for the current-sensing mode. Dynamic EFM in amplitude modulation showed higher electrical and spatial resolution, enabling us to measure the dielectric constants of objects as small as 10 nm in diameter [21, 23].

The dynamic detection scheme based on second-harmonic amplitude-modulation EFM is given in Fig. 13.8. An ac voltage of amplitude, v_{ac}, and angular frequency, ω, is applied between the AFM tip and the conductive back substrate. This induces the mechanical oscillation of the cantilever at double the frequency, 2ω, which is detected using a lock-in amplifier. Force images, $F_{2\omega}(x,y,z,\varepsilon_r)$, are then measured at *constant height* with respect to the substrate, and from this the first capacitance derivative in the z-direction (dielectric signal), $dC(x,y,z,\varepsilon_r)/dz$ is obtained as $dC(x,y,z,\varepsilon_r)/dz = 4 \cdot F_{2\omega}(x,y,z,\varepsilon_r)/v_{ac}^2$. To avoid non-local contributions from the probe, the dielectric signal is measured at constant height with respect to a reference position (x_0,y_0) on the bare substrate, and it is indicated as dC/dz, where $dC/dz = dC(x,y,z,\varepsilon_r)/dz - dC(x_0,y_0,z,\varepsilon_r)/dz$. The scan height, z, can be precisely determined by recording both the dc deflection and the capacitance gradient, dC/dz, as a function of the tip substrate distance at the image edges. The dc deflection approach curve provides the distance from the substrate, while the capacitance gradient curve allows to determine and compensate any vertical piezo drift by comparing it with the probe calibration curve.

Figures 13.9d–f show dielectric images of three dielectric nanoparticles of similar size (diameter around 30 nm) and different materials—polystyrene (PS, $\varepsilon_r = 2.6$), silicon dioxide (SiO$_2$, $\varepsilon_r = 3.9$) and aluminium oxide (Al$_2$O$_3$, $\varepsilon_r = 9.4$)—which were measured at the same distance of $z = 42 \pm 0.5$ nm from the substrate but in different experiments, that is, on different substrates and using different tips. Figure 13.9g gives the profiles across the centre of the nanoparticles, showing similar dielectric contrast of few zeptoFarad per nanometre, which however does not directly correspond to the dielectric constant of the material, as the signal of SiO_2 is larger than the one obtained for PS and Al_2O_3 (Fig. 13.9g). This clearly shows that for small objects the dielectric signal that is measured strongly depends on geometrical effects such as the size of the tip apex and of the nanoparticle, which slightly differ in the three experiments showing few-nanometre differences in tip or sample size.

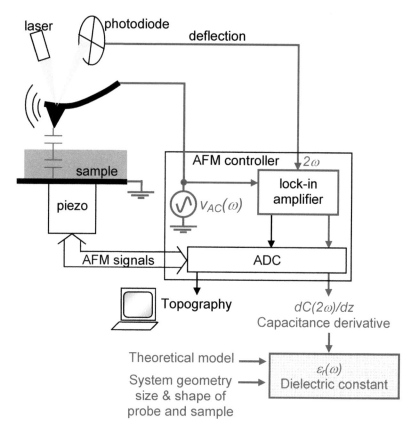

Figure 13.8 Simplified schematic of the setup for dielectric constant measurement using the second harmonic electrostatic force-sensing approach. An ac voltage signal at angular frequency ω is applied between the tip and substrate and the first derivative of capacitance is obtained by detecting the mechanical oscillation of the cantilever at double the frequency, 2ω. From this, the local dielectric constant of the sample can be extracted, provided that the geometry of the probe and of the sample are accurately known and the theoretical model precisely reproduces them.

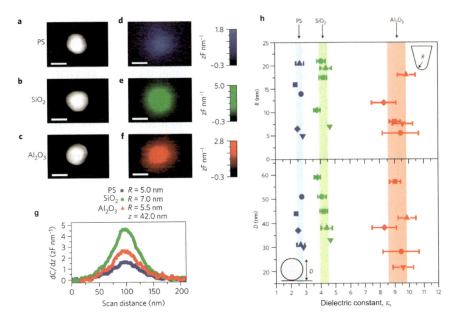

Figure 13.9 Dielectric constant measurement of nanoparticles of ~ 30 nm diameter using EFM. (a–c) Topography and (d–f) dielectric images (capacitance gradient, dC/dz) of PS (blue), SiO_2 (green) and Al_2O_3 (red) measured in constant-height, $z \sim 42$ nm, from the graphite substrate. (g) Measured dielectric profiles across the centre of the nanoparticles, taken with different probes of radius $R = 5.0$, 7.0 and 5.5 nm, respectively. The magnitude of the dielectric signal depends on both the dielectric constant of the nanoparticle and the tip radius. (h) Extracted dielectric constants in multiple experiments, carried out with different tips (radii) and on nanoparticles of different size (20–60 nm diameter) and materials. The dielectric constants of PS, SiO_2 and Al_2O_3 nanoparticles are obtained independent from geometrical artifacts. Reprinted from *Nat. Mater.*, 11, 808–816 (2012).

To correctly determine the dielectric constant of the nanoparticles, therefore, the dielectric signal (dC/dz) needs to be fitted to numerical calculations that take into account the exact geometry and size of the tip and of the nanoparticle. By doing that, we correctly obtained the expected dielectric constant of PS, SiO_2 and Al_2O_3. This means that despite the dielectric signal is strongly dependent on the tip and nanoparticle size, our approach precisely overcomes their influence. To further demonstrate it, we systematically repeated the measurement on nanoparticle of different diameter (D = 20–60 nm) and using different tips, covering the tip radius range (R = 1–25 nm). Figure 13.9h shows that the obtained dielectric constants are precisely separated in three groups corresponding to their nominal dielectric constants of PS, SiO_2 and Al_2O_3, with average values ε_r = 2.61 ± 0.20 (PS), 4.22 ± 0.32 (SiO_2) and 9.24 ± 0.60 (Al_2O_3). This shows the high precision of the technique, which enables to measure the dielectric constant of individual nanoparticles with uncertainty less than 10%. Furthermore, they are not required to be measured in a single experiment, as they can be measured separately in multiple experiments and on different substrates, with no need of a reference nanoparticle of known material on the substrate, because we obtain the absolute value (not variations) of ε_r.

Importantly, these results show that the technique allows **label-free material identification** of nanometric objects using a commercial AFM without the need of other instrumentation, by simply measuring their dielectric constants. The dielectric constant can be used as the fingerprint of the materials to discriminate nano-objects of identical shape but different composition that would be impossible to recognize from topography [23].

Another advantage of the technique is in that it enables to measure and distinguish not only the surface composition, but also the inner composition of the objects, allowing **sub-surface material identification**. This is because of the long-range nature of polarization forces, which are able to penetrate non-screening layers and reach the material inside. We demonstrated it by measuring and discriminating single viruses—the bacteriophage T7—from their empty capsids. These viruses are 60 nm-diameter nanoparticles, which consist of a 2.5 nm-thick protein shell (capsid) that encloses double-stranded DNA (dsDNA) well-organized in concentric layers. We found that the virus gives a dielectric response significantly larger than the empty capsid, obtained from the virus after intentional release of the DNA [23]. We obtained a dielectric constant of ~3 for the capsid shell in agreement with expected values for dry proteins (~2–5), while we obtained a dielectric constant of ~8 for the DNA inside the virus [21]. This was the first direct measurement of dielectric constant of DNA, which had remained unknown owing to the lack of experimental tools able to probe it. This is because quantitative dielectric measurements of DNA have been hampered at larger scale by the complex structure and chemical composition of DNA and its interaction with the solvent. We supported the experimental data by theoretically determining the DNA dielectric constant using atomistic molecular dynamic simulations. They were remarkably in agreement with the experimental value, yielding a dielectric constant of DNA around 8, sensibly higher than the dielectric response of proteins, arising from the sugar–phosphate backbone of DNA [21].

13.8 Dielectric Measurement in Liquid Environment

The experimental methods described above allow local dielectric measurements in dry or humid environment. Yet, for most of biological applications and for electrochemical studies, measurements need to be carried out in electrolytic environment. This has traditionally been a challenge because the presence of ions in the liquid medium makes it electrically conductive, and the methods developed for pure dielectric systems need to be modified. Furthermore, the formation of electric double layers on surfaces needs to be taken into account.

A complete and rigorous theoretical treatment of the problem is still to be presented. We have worked out an approximated treatment in which double-layer effects are included as an effective distributed surface capacitance and the liquid environment modelled as a medium with a given conductivity and dielectric constant. This approximated model can be solved by numerical calculations using a dynamic (or harmonic) solver to take into account the frequency-dependent response of the system due to the presence of different media with different conductivity and permittivity (Maxwell–Wagner relaxations). In particular, we have demonstrated that EFM measurements in electrolytic solutions can be local for high excitation frequencies (>MHz). This is because a significant voltage drop is obtained between the tip apex and the sample in the MHz range, while it vanishes at lower frequencies that are normally used in electrical-SPM techniques [2].

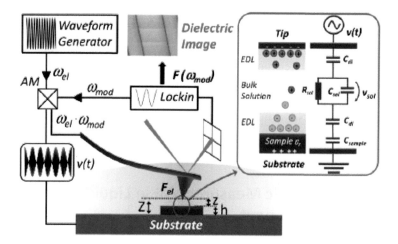

Figure 13.10 EFM measurement setup in electrolyte solution. An AM ac-potential with frequency ω_{el} > MHz and modulation frequency ω_{mod} < 10 kHz is applied between a conductive tip and a substrate in an electrolyte solution. An external lock-in amplifier detects the modulated bending of the cantilever, which depends, among other parameters, on the dielectric constant of the dielectric layer. Inset: Electrochemical model and equivalent circuit of a surface element of the tip-sample interaction region in solution. Reprinted from *Appl. Phys. Lett.*, 101, 213108 (2012), with the permission of AIP Publishing.

We experimentally verified it by taking dielectric images a SiO_2 layers as a function of the applied frequency using the setup given in Fig. 13.10. This setup allows high-frequency detection in the MHz range well beyond the mechanical resonance of the standard cantilevers. The ac voltage excitation at MHz frequencies is amplitude modulated at angular frequency in the kHz range and applied between the tip and the conductive substrate. The high-frequency excitation signal bends the cantilever in a static way due to the non-linear dependence of the actuation force on the applied voltage. The applied low-frequency modulation enables ac detection of this interaction, which depends on the dielectric constant of the sample, among other parameters, thus enhancing the signal-to-noise ratio of the measurement with respect to static detection. Figure 13.11 clearly demonstrates that only at high frequencies local dielectric contrast is achieved on SiO_2 layers. The critical frequency at which measurement locality is achieved increases with the ionic concentration of the solution, in agreement with a simple theoretical model [28, 29]. The dielectric contrast reaches a saturation value in the limit of high frequencies, in which conductive contributions are negligible and the system behaves like a pure dielectric system. Dielectric measurements can be performed in a very quantitative way by applying the same procedure that we developed for the air environment based on the calibration of the probe radius and quantitative analysis of approach curves or constant-height dielectric images. In this way, we demonstrated that the dielectric constant of a SiO_2 thin films [28] and of thin lipid bilayer [30] can be precisely measured in electrolyte solutions.

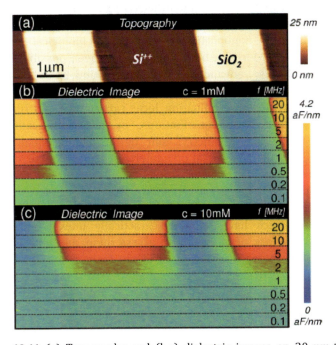

Figure 13.11 (a) Topography and (b,c) dielectric images on 20 nm-thin SiO_2 microstripes on a doped silicon obtained at constant height in electrolyte solutions with ion concentrations of 1 mM (b) and 10 mM (c) and applied voltage of 0.5 V. The applied excitation frequency was systematically decreased from 20 MHz to 0.1 MHz, showing that the dielectric contrast depends on the applied frequency and that it vanishes at low frequencies. The critical frequency above which local dielectric contrast is achieved increases with the ionic concentration. Reprinted from *Appl. Phys. Lett.*, 101, 213108 (2012), with the permission of AIP Publishing.

13.9 Conclusions

Measurement of dielectric properties at the nanoscale by using a scanning probe has recently shown great progress. In particular, two approaches have been established based on current sensing and electrostatic force sensing. They enable to precisely quantify the dielectric constant of materials and biological samples with high precision (less than 10% error) and with nanoscale spatial resolution—below 10–50 nm, depending on the approach used and the sample size. Here we have summarized our main results, through which we showed in the last years that the dielectric constant can be accurately quantified using these two approaches on a variety of inorganic samples (thin and thick dielectric films, dielectric nanoparticles) as well as biological samples (biomembranes, viruses, bacteria). This has been possible thanks to a number of advances, i.e. by using low-noise detection approaches; by fully understanding the probe-sample electrostatic interactions, thus discriminating local capacitance contributions from non-local stray contributions; by accurately calibrating in situ the probe geometry; and by developing quantitative theoretical (either analytical or numerical) models to correctly fit the experimental data. Our results show the broad range of applications and the great potential of dielectric microscopy. It allows the study of dielectric properties of nanostructures and molecules which have remained practically unexplored so far owing to the lack of tools able to access them, and which play a crucial role in many areas of research, from materials science and technology to chemistry and biology.

References

1. Hofer, W. A., Foster, A. S., and Shlugere, A. L. (2003). Theories of scanning probe microscopes at the atomic scale, *Rev. Mod. Phys.*, **75**, pp. 1287–1331.

2. Kalinin, S. V., and Gruverman, A (2007). *Scanning Probe Microscopy, Electrical and Electromechanical Phenomena at the Nanoscale* (New York, Springer).

3. Crider, P. S., Majewski, M. R., Jingyun, Z., Oukris, H., and Israeloff, N. E. (2007). Local dielectric spectroscopy of polymer films, *Appl. Phys. Lett.*, **91**, pp. 013102-1-3.

4. Fumagalli, L., Ferrari, G., Sampietro, M., and Gomila, G. (2007). Dielectric-constant measurement of thin insulating films at low frequency by nanoscale capacitance microscopy, *Appl. Phys. Lett.*, **91**, pp. 243110-1-3.

5. Fumagalli, L., Ferrari, G., Sampietro, M., and Gomila, G. (2009). Quantitative nanoscale dielectric microscopy of single-layer supported biomembranes, *Nano Lett.*, **9**, pp. 1604–1608.

6. Fumagalli, L., Gramse, G., Esteban-Ferrer, D., Edwards, M. A., and Gomila, G. (2010). Quantifying the dielectric constant of thick insulators using electrostatic force microscopy, *Appl. Phys. Lett.*, **96**, pp. 183107-1-3.

7. Gramse, G., Casuso, I., Toset, J., Fumagalli, L., and Gomila, G. (2009). Quantitative dielectric constant measurement of thin films by DC electrostatic force microscopy, *Nanotechnology*, **20**, pp. 395702-1-8.

8. Krayev, A. V., and Talroze, R. V. (2004). Electric force microscopy of dielectric heterogeneous polymer blends, *Polymer*, **45**, pp. 8195–8200.

9. Kumar, B., Bonvallet, J. C., and Crittenden, S. R. (2012). Dielectric constants by multifrequency non-contact atomic force microscopy, *Nanotechnology*, **23**, 025707-1-6.

10. Riedel, C., Arinero, R., Tordjeman, P., Lévêque, G., Schwartz, G. A., Alegria, A., and Colmenero, J. (2010). Nanodielectric mapping of a model polystyrene-poly (vinyl acetate) blend by electrostatic force microscopy, *Phys. Rev. E*, **81**, pp. 010801-1-4.

11. Lu, W., Wang, D., and Chen, L. W. (2007). Near-static dielectric polarization of individual carbon nanotubes, *Nano Lett.*, **7**, pp. 2729–2733.

12. Tevaarwerk, E., Keppel, D. G., Rugheimer, P., Lagally, M. G., and Eriksson, M. A. (2005). Quantitative analysis of electric force microscopy: The role of sample geometry, *Rev. Sci. Instrum.*, **76**, pp. 053707-1-5.

13. Gramse, G., Kasper, M., Fumagalli, L., Gomila, G., Hinterdorfer, P., and Kienberger, F. (2014). Calibrated complex impedance and permittivity measurements with scanning microwave microscopy. *Nanotechnology*, **25**, pp. 145703-1-8.

14. Lai, K., Ji, M. B., Leindecker, N., Kelly, M. A., and Shen, Z.-X. (2007). Atomic-force-microscope-compatible near-field scanning microwave microscope with separated excitation and sensing probes, *Rev. Sci. Instrum.*, **78**, pp. 063702-1-5.

15. Lai, K., Kundhikanjana, W., Kelly, M. A., and Shen Z.-X. (2011). Nanoscale microwave microscopy using shielded cantilever probes, *Appl. Nanosci.*, **1**, pp. 13–18.

16. Govyadinov, A. A., Amenabar, I., Huth, F., Scott Carney, P., and Hillenbrand, R. (2013). Quantitative measurement of local infrared absorption and dielectric function with tip-enhanced near-field microscopy, *J. Phys. Chem. Lett.*, **4**, pp. 1526–1531.

17. Tanaka, K., Kurihashi, Y., Uda, T., Daimon, Y., Odagawa, N., Hirose, R., Hiranaga, Y., and Cho, Y. (2008). Scanning nonlinear dielectric microscopy nano-science and technology for next generation high density ferroelectric data storage, *J. J. Appl. Phys.* **47**, pp. 3311–3325.

18. Esteban-Ferrer, D., Edwards, M. A., Fumagalli, L., Juárez, A., and Gomila, G. (2014). Electric polarization properties of single bacteria measured with electrostatic force microscopy, *ACS Nano*, **8**, pp. 9843–9849.

19. Van Der Hofstadt, M., Fabregas, R., Millan-Solsona, R., Juarez, A., Fumagalli, L., and Gomila, G. (2016). Internal hydration properties of single bacterial endospores probed by electrostatic force microscopy, *ACS Nano*, **10**, pp. 11327–11336.

20. Dols-Perez, A., Gramse, G., Calò, A., Gomila, G., and Fumagalli, L. (2015). Nanoscale electric polarizability of ultrathin biolayers on insulating substrates by electrostatic force microscopy, *Nanoscale*, **7**, pp. 18327–18336.

21. Cuervo, A., Dans, P. D., Carrascosa, J. L., Orozco, M., Gomila, G., and Fumagalli, L. (2012). Direct measurement of the dielectric polarization properties of DNA, *Proc. Natl. Acad. Sci. U. S. A.*, **111**, pp. E3624–E3630.

22. Gomila, G., Toset, J., and Fumagalli, L. (2008). Nanoscale capacitance microscopy of thin dielectric films, *J. Appl. Phys.*, **104**, pp. 024315-1-8.

23. Fumagalli, L., Esteban-Ferrer, D., Cuervo, A., Carrascosa, J. L., and Gomila, G. (2012). Label-free identification of single dielectric nanoparticles and viruses with ultraweak polarization forces, *Nat. Mater.*, **11**, pp. 808–816.

24. Fumagalli, L., Edwards, M. A., and Gomila, G. (2014). Quantitative electrostatic force microscopy with sharp silicon tips, *Nanotechnology*, **25**, pp. 495701-1-8.

25. Gomila, G., Gramse, G., and Fumagalli, L. (2012). Finite-size effects and analytical modeling of electrostatic force microscopy applied to dielectric films, *Nanotechnology*, **25**, pp. 255702-1-8.

26. Gomila, G., Esteban-Ferrer, D., and Fumagalli, L. (2013). Quantification of the dielectric constant of single non-spherical nanoparticles from polarization forces: eccentricity effects, *Nanotechnology*, **24**, pp. 505713-1-8.

27. Gramse, G., Gomila, G., and Fumagalli, L. (2012). Quantifying the dielectric constant of thick insulators by electrostatic force microscopy: Effects of the microscopic parts of the probe, *Nanotechnology*, **23**, p. 205703.

28. Gramse, G., Edwards, M. A., Fumagalli, L., and Gomila, G. (2012). Dynamic electrostatic force microscopy in liquid media, *Appl. Phys. Lett.*, **101**, pp. 213108-1-3.

29. Gramse, G., Edwards, M. A., Fumagalli, L., and Gomila, G. (2013). Theory of amplitude modulated electrostatic force microscopy for dielectric measurements in liquids at MHz frequencies, *Nanotechnology*, **24**, pp. 415709-1-10.

30. Gramse, G., Dols-Perez, A., Edwards, M. A., Fumagalli, L., and Gomila, G. (2013). Nanoscale measurement of the dielectric constant of supported lipid bilayers in aqueous solutions with electrostatic force microscopy, *Biophys. J.*, **104**, pp. 1257–1262.

31. Hudlet, A., Saint Jean, M., Guthmann, C., and Berger, J. (1998). Evaluation of the capacitive force between an atomic force microscopy tip and a metallic surface, *Eur. Phys. J. B*, **2**, pp. 5–10.

32. Fumagalli, L., Ferrari, G., Sampietro, M., Casuso, I., Martinez, E., Samitier, J., and Gomila, G. (2006). Nanoscale capacitance imaging with attoFarad resolution using ac current sensing atomic force microscopy, *Nanotechnology*, **17**, pp. 4581–4587.

33. Williams, C. C. (1999). Two-dimensional dopant profiling by scanning capacitance microscopy, *Annu. Rev. Mater. Sci.* **29**, pp. 471–504.

34. Sacha, G. M., Verdaguer, A., Martínez, J., Sáenz, J. J., Ogletree, D. F., and Salmeron, M. (2005). Effective tip radius in electrostatic force microscopy, *Appl. Phys. Lett.*, **86**, pp. 123101-1-3.

35. Biagi, M. C., Fabregas, R., Gramse, G., Van Der Hofstadt, M., Juárez, A., Kienberger, F., Fumagalli, L., and Gomila, G. (2015). Nanoscale electric permittivity of single bacterial cells at gigahertz frequencies by scanning microwave microscopy, *ACS Nano* **10**, pp. 280–288.

36. Shao, R., Kalinin, S. V., and Bonnell, D. A. (2003). Local impedance imaging and spectroscopy of polycrystalline ZnO using contact atomic force microscopy, *Appl. Phys. Lett.*, **82**, pp. 1869–1871.

Chapter 14

SPM-Based Capacitance Spectroscopy

Jian V. Li,[a] Giorgio Ferrari,[b] and Chun-Sheng Jiang[c]

[a]*Department of Aeronautics and Astronautics,*
National Cheng Kung University, Tainan, Taiwan
[b]*Dipartimento di Elettronica, Informazione e Bioingegneria,*
Politecnico di Milano, Milano, 20133, Italy
[c]*National Renewable Energy Laboratory, Golden, Colorado 80401, USA*

jianvli@mail.ncku.edu.tw, giorgio.ferrari@polimi.it, chun.sheng.jiang@nrel.gov

This chapter reviews SPM-based capacitance spectroscopy for scaling down the contact area of the traditionally macro-scale capacitance spectroscopy techniques—capacitance–voltage, admittance spectroscopy, and deep-level transient spectroscopy—to nanoscale dimensions. First, we present the scientific motivations behind the development of this instrument. After discussing the technical challenges anticipated in the implementation of such an instrument on an SPM platform, we describe the experimental setup and its associated electronic circuits. Last, we present preliminary results on SPM-based capacitance spectroscopy.

Capacitance Spectroscopy of Semiconductors
Edited by Jian V. Li and Giorgio Ferrari
Copyright © 2018 Pan Stanford Publishing Pte. Ltd.
ISBN 978-981-4774-54-3 (Hardcover), 978-1-315-15013-0 (eBook)
www.panstanford.com

14.1   Introduction

The motivation for the SPM-based capacitance spectroscopy lies in a growing need to better understand the electrical properties of inhomogeneous materials at microscopic scale, especially the technologically important polycrystalline semiconductors. The scientific community is currently hampered by a disconnection between the microscopic insights of materials and the macroscopic functional properties. The conventional capacitance–voltage (CV), admittance spectroscopy (AS), and deep-level transient spectroscopy (DLTS) techniques inspect the relevant functional properties at the macroscopic scale but do not have sufficient spatial resolution. Existing scanning electron microscope (SEM) and SPM-based microscopic instruments have sufficient spatial resolution but do not directly inspect the functional properties relevant to the defects problem of interest.

14.1.1   Motivation

The electronic properties of defects—their energetic locations in the bandgap E_t, concentrations N_t, the former two also jointly referred to as density-of-states, and capture cross sections σ—directly govern the performance of semiconductor materials for many applications. Conventional techniques such as DLTS [1], whose operating principles are described in Chapters 1–4, are routinely used to characterize the electronic properties of defects at macroscopic scale, yielding results consistent with microscopic understanding of first-principle prediction, structural, and compositional data [2–5]. While it has been successful largely in single-crystalline materials, the same methodology fails in more disordered materials, in particular polycrystalline materials, where the electronic signatures of defects obtained at the macroscopic scale exhibit weak if not zero correlation to microscopic data available from the atomic, structural, and theoretical perspectives.

For example, there have been many first-principle studies of point defects in the polycrystalline materials [6]. The CdTe source material for production-grade solar modules is generally of five to six 9 purity, which means that there are certainly impurities with concentration 10^{17} cm^{-3} or higher.

By intuitive expectation, the bandgap of CdTe should be littered by energy levels of extrinsic impurities or intrinsic defects. Yet, both electrical and optical experiments have not shown numerous defects signatures, e.g., see Fig. 14.1 [7]. The Landolt–Bornstein handbook [8] shows only a dozen or two experimentally observed defect signatures in polycrystalline CdTe materials. Consequently, the functional properties of polycrystalline materials—doping, defects, interface, recombination, lifetime—are poorly understood because the macroscopic-level observations are not built on top of the microscopic level knowledge (e.g., [9]). There are obvious questions on the roles played by the grain boundaries. Does the electric potential distribution at the grain boundaries affect the distribution of dopants [10]? How do microscopically measured carrier densities relate to the macroscopically measured values?

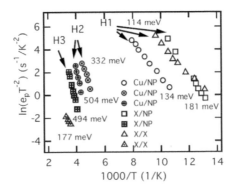

Figure 14.1 Disconnect between the microscopic prediction [6] and macroscopic observation [7] of defects.

Therefore, there is an acute need to bridge the microscopic and macroscopic worlds by experimentally extracting the localized functional properties of defects, doping, interface, and recombination from 100 nm to 1 μm. To achieve this goal, the scientific community needs the proper instrument.

14.1.2 Assessment of Existing Instruments

The general assessment of available instruments relevant to the scientific problem stated earlier is that: the techniques with sufficient spatial resolution do not inspect the relevant functionalities, whereas techniques inspecting the relevant functionalities do not yield sufficient spatial resolution. To directly extract the electrical functionalities of the polycrystalline materials, it is best for the instruments not be based on high-energy interaction of electrons or photons with nuclei or inner-shell electrons. SEM and SPM-based instrumentations are suitable because of the interaction between the low-energy stimuli and the electrically active electrons.

SPM is ideal for the task at hand because of the adequate spatial resolution stemming from a localized contact between the tip and the sample that also provides a conductive path [11]. However, existing SPM measurements of semiconductors are largely limited to the carrier transport in the DC regime, which are not useful for extracting the dynamic carrier capture/emission processes to and from defects. Scanning tunneling microscopy (STM) offers information on electronic structures but its working principle decides that its application is mainly in the atomic scale thus not suitable for scalable spatial resolution near 1 μm. SPM-based scanning Kelvin probe force microscopy (SKPFM) [12] is only able to extract electrical potential but offers no information to charge dynamics or electronic phenomenon in the bandgap. SPM-based scanning capacitance microscopy (SCM) [13] (described in Chapter 12) is successful for determining doping in Si-based metal–insulator–semiconductor (MIS) structures. However, it is an indirect measurement of doping requiring an auxiliary calibration procedure, operates at a fixed frequency of near 1 GHz that is higher than the dielectric relaxation frequency of many polycrystalline materials, and requires a high-quality insulator layer to form the MIS structure. Above all, it does not yield information besides doping and shallow defects. SEM-based cathodoluminescence [14] and SPM-based near-field scanning optical microscopy (NSOM) [15] methods have sufficient spatial resolution to tackle the particular scale of interest but do not offer enough physics information. Deep levels tend to be traps or recombination centers therefore exchange of carriers between them and the band edges are often non-radiative, i.e., not accessible by optical methods. Electron-beam or light-beam induced current (EBIC/LBIC) [16, 17] are good at visualizing the transport mechanism but they do not yield direct information on the defect properties.

Capacitance spectroscopy [18], as described in Chapters 1–4, 9, and 10, is a suite of techniques that inspect the charge response to experimentally manipulated electric field and/or potential. They yield a wealth of information on free carriers, the dynamics of carrier exchange between band edge and defect states, and electrostatic potential distribution. The CV method is one of the most relied-upon electrical characterization techniques [19] at device level to extract the depth profile of the free carrier density, depletion region width, and built-in voltage of a junction. For an MIS junction [20], CV is also used to extract the electrical properties of the insulator–semiconductor interface, namely the density of defect states D_{it} and fixed charge Q_t. AS inspects [21] the steady-state small-signal response of the carrier exchange process, i.e., capture and emission, between the defect states and band edges by varying the frequency and temperature. DLTS is the default method [22, 23] to characterize defects in semiconductors. It measures of the large-perturbation transient response of the carrier exchange process between the defect states and band edges by varying the emission rate and temperature. Both AS and DLTS produce three key electrical properties of the defects: their energetic locations in the bandgap, densities, and capture cross-sections.

Almost all capacitance spectroscopy methods have only been applied to macroscopic devices to extract the material functional properties, with a few exceptions discussed below [24, 25]. The attempts of SEM-based DLTS experiment in 1980s [24] were hampered by low signal-to-noise ratio, hence poor spatial resolution (Fig. 14.2). This is because the signal is generated from the small volume where the electron beam and material interact but the noise is generated from the entire sample. Therefore, SEM-based DLTS has not been adopted for material research since 1991. A recent work of AFM-DLTS [25] is limited in its spatial extent to within the depletion region, and limited to the cross-section instead of the planar view.

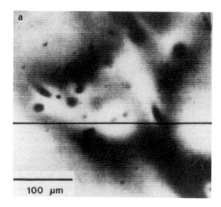

Figure 14.2 SEM-DLTS [24] with insufficient spatial resolutions.

14.2 Instrumentation

This section describes the considerations and construction of an instrument (Fig. 14.3) to implement the conventional capacitance spectroscopy methods in a SPM to access both functional properties and sufficient spatial resolution. This instrument serves the scientific goals (Fig. 14.3) of understanding the doping, interfaces, defects, and recombination of polycrystalline materials by resolving the electronic properties of defects using multiple spectroscopic modes and bridging the micro- and macroscopic worlds via SPM-based scanning.

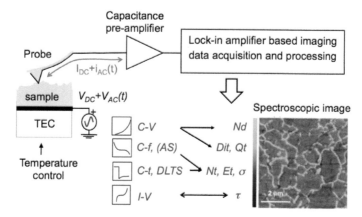

Figure 14.3 Schematic diagram of the concept of instrumentation and science outcome.

The focus of operational functionalities is on realizing the conventionally macroscopic-scale capacitance spectroscopy techniques of CV, AS, and DLTS on an SPM platform to enable extraction of localized functional material properties with 1 μm to 100 nm spatial resolution. The order of the presentation below is aimed at progressive understanding of the technical issues and development of their solutions evolving from the easiest task of voltage-dependence of capacitance at fixed frequencies (CV), to varied frequency and temperature (AS), and finally to transient response (DLTS).

14.2.1 The Technical Challenges

One challenging aspect of the instrumentation is the exceedingly small capacitance between the SPM tip and the sample, C_S. The depletion width of the metal–semiconductor Schottky contact formed between the SPM tip and the sample is ~100 nm, assuming a doping of 10^{16} cm^{-3} and a dielectric constant of 12. Assuming no 3D fringe effects, the associated depletion capacitance is ~10^{-7} F/cm^2 and C_S is 1 fF (10^{-15} F) for a 1 × 1 μm^2 area. For CV, the variation of capacitance can be ~50% of C_S. Therefore, the capacitance measurement circuit for a CV experiment with 8-bit digital resolution needs to resolve 2 aF at a few fixed frequencies. The best custom electronics reported [26, 27] is sufficient for CV experiment. For AS and DLTS, the smallest variation of capacitance may be ~1% of C_S, i.e. 10 aF for a 1 × 1 μm^2 area. Measuring a capacitance of 10 aF with 8-bit resolution (i.e., ~0.04 aF per bit) at sufficiently fast frequency is pushing the limit of custom circuit. It is worth noting that the coupling capacitor C_C between the cantilever body and the sample may reach 10–100 fF [28]. This capacitor C_C can be 10–100 times higher that the tip-sample capacitance C_S, if not removed.

The other challenging aspect of the instrumentation is the stability of electrical contact. The CV/AS/DLTS experiments are based on the depletion capacitance. The SPM tip, which is practically metallic, induces a depletion region in the semiconductor samples. Any instability of the contact directly translates into experimental error and may overwhelm the depletion capacitance signal. The situation is more severe considering that the coupling capacitor C_C is 100~1000 times higher than the depletion capacitance.

For both the AS and DLTS experiments, the temperature of the sample has to be varied within a range. This temperature variation also adds to the challenge of maintaining stable electrical contact. To realize the conventionally macroscopic-scale techniques of capacitance spectroscopy on an SPM platform, one has to devise schemes for orthogonal time/frequency–voltage–temperature scanning and real-space scanning. The first axis of scanning is temperature. At each temperature point, a real-space scanning will be conducted. At each pixel of the real-space scanning, the next scanning loop is voltage (CV), frequency (AS), or time (DLTS) depending on the mode of operation. Fiducial marks on the samples for spatial registration purpose may be implemented so that small drifts due to temperature variation are not a problem.

14.2.2   Instrumentation Development

The objective is to develop an apparatus for multimodal imaging spectroscopy of the electronic properties of defects using scanning probe microscopy (SPM) (Fig. 14.3). With this instrument, each pixel of an *X–Y* image in the real space contains the density-of-states and capture cross-section spectroscopy of defects in the energy space. The measurement electronics couples the *X–Y* plane spatial scanning with AC capacitance measurement scanning over voltage, time, frequency, and temperature at each pixel to achieve multimodal operation of CV, AS, and DLTS imaging. These functionalities significantly extend the boundaries of SPM instrumentation for characterizing defects in materials. Enabling this instrument (Fig. 14.4) are several novel concepts and components.

For the ultra high sensitivity pre-amplifiers required for the CV/AS/DLTS measurements, we use custom designed pre-amplifiers. We employ trans-impedance-amplifier-based custom circuits in the low-to-intermediate (100–10^7 Hz) frequency range to achieve the required capacitance sensitivity: <1 aF (10^{-18} F) out of a total of ~1 fF (10^{-15} F) for steady state CV/AS measurement, and sub-microsecond transient capacitance measurement with

8 aF resolution. According to estimates made earlier, this is sufficient for achieving 1 μm spatial resolution. Higher spatial resolution approaching 100 nm is possible by trading off capacitance resolution and measurement time (averaging).

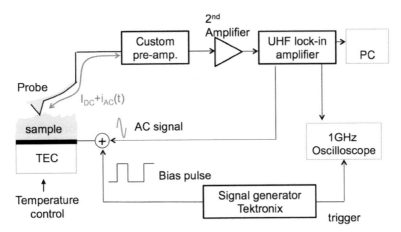

Figure 14.4 Schematic diagram of the instrument operating in the DLTS mode.

The SPM tip and sample interaction may be categorized into multiple types of contacts and junctions. For pristine semiconductor films, metallic SPM tips form Schottky contacts with the samples. It is also possible that the samples already have a P-N junction structure. In that case, the SPM tip only needs to form ohmic contacts with the top layer. Lastly, intentionally or not, one may form an insulator layer on top of the semiconductor. In that case, a metal–insulator–semiconductor junction is formed, which is similar to the SCM-mode operation but distinctive in that the operation is at low-to-medium frequency and based on absolute capacitance measurement. Alternatively, one can deposit arrays of nanoscale metal pads to form Schottky contacts with the sample while guaranteeing ohmic contact with the SPM tip.

The temperature control of the sample is achieved by a thermoelectric (Peltier) cooler/heater. It is better to enclose the entire SPM system together with the sample in a nitrogen-flown enclosure to avoid water condensation and increase temperature stability during operation. A wide and stable window of temperature control translates to a wide observation window for defect energy and spatial resolution, respectively. Considering that the temperature range in a thermoelectric controlled environment is more limited than that of a conventional cryostat, some newly developed tactics such as such as the temperature-rate duality [29] can be exploited to extract the same amount of defect information using less data. One strategy is to employ a larger range of rate measurement to compensate for a smaller range of temperature—essentially by stretching the rate axis while shrinking the temperature axis of an Arrhenius plot to preserve the same coverage of area, i.e., the detection window for defects. This can be implemented by expanding the sweeping range of frequency in AS and time in DLTS. Additionally, the 2-Dimensional Arrhenius method [29] is a novel approach to extract the energy and capture cross-section of defects solely based on the temperature-rate duality, i.e., without resorting to the conventional 1D Arrhenius plot. The 2D Arrhenius method more efficiently utilizes the defect response in the 2D temperature-rate plane (Fig. 14.5) thus requiring less data than the conventional 1D Arrhenius plot.

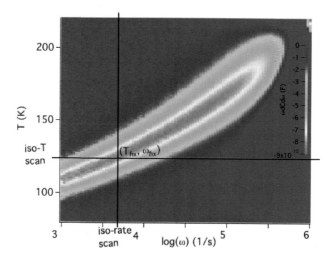

Figure 14.5 The defect response in the rate-temperature plane enables the 2-D Arrhenius method extraction.

14.3   Preliminary Results

This section describes the experimental results obtained by implementing the instrumentation described in the above section.

14.3.1   Capacitance

As shown in Fig. 14.4, the capacitance signal C_S, after being amplified by the pre-amplifier, is sent to a state-of-the-art lock-in amplifier UHFLI from Zurich Instruments capable of operating at 600 MHz. The frequency sweeping and DC voltage bias required for CV and AS are to be accomplished by the lock-in amplifier. A computer communicating with the UHFLI via a high-speed USB bus accomplishes additional digital acquisition and processing of signals. The preliminary data in Fig. 14.6 demonstrate feasibility of capacitance measurement using the experimental scheme described in the previous section.

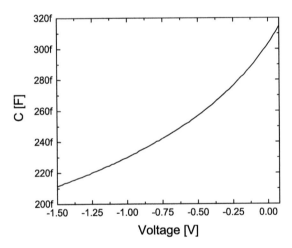

Figure 14.6 Preliminary capacitance data measured using the scheme shown in Fig. 14.3.

14.3.2   Capacitance–Voltage Scanning

We have acquired preliminary CV data using an earlier-generation single-input pre-amplifier, shown in Fig. 14.7. It demonstrates sufficient resolution of capacitance (aF). The spatial resolution (<1 μm) is excellent indicating good contact quality. The proof of concept of CV measurement of carrier density, obtained using the Mott–Schottky algorithm at each pixel, is also shown. The values match the range of experimentally observed carrier densities at macroscopic. There are not enough features in the "carrier density" image, possibly due to the rough morphology (~0.25 μm rms). In the future, we will remove the complexity due to morphology by polishing the sample.

Figure 14.7 Preliminary images obtained from a CIGS sample by an un-optimized pre-amplifier with simultaneous scans of morphology (a) with highest feature (lightest color) at 544 nm, capacitance output from the pre-amplifier (b) with arbitrary unit, and voltage-differential capacitance (c) which is proportional to carrier density. The scanning range of all three images is the same: 8 μm by 8 μm.

The CV technique is also used to extract the fixed charge and density of states at the interface of a metal–insulator–semiconductor structure. This is a function that is conspicuously missing in the SCM mode of operation. The reason is that the SCM resonant circuit measures only the derivative of capacitance (i.e., dC/dV) but not the capacitance itself, which is insufficient for recovering D_{it} and Q_t without additional information. Because the instrument described in this chapter measures capacitance directly, one can apply the commonly accepted algorithms to generate spatially resolved imaging of density of states D_{it} and fixed charge Q_t at the interface between an insulator and a polycrystalline material. This information will be very useful in understanding the interface phenomena and related passivation mechanisms of polycrystalline materials.

14.3.3   Capacitance Transients

Preliminary transient data (Fig. 14.8) acquired at a single-point contact ($\sim$25 μm^2) using a single-input pre-amplifier demonstrates successful measurement of capacitance transient measurement at microscopic scale. The transient capacitance signal taken from the demodulated auxiliary port of the UHFLI lock-in amplifier is connected to a high-speed (1 GHz bandwidth) and high-resolution (15-bit vertical digitization) oscilloscope to record the time-dependent waveform for further processing of DLTS spectra. Imaging operation has not been implemented. The response time is slow. Both limitations are not fundamental and can be improved in future. Note that we have observed some time constants of the capacitance transients are very long, ranging from seconds to tens of seconds (that shown in the figure is about 0.1 s) even at room temperature. This is possibly an indicator of metastability, which is prevalent in polycrystalline materials and actively being researched at the macroscopic scale. The capacitance transient imaging spectroscopy will provide new sights to unveil the microscopic origin of this phenomenon.

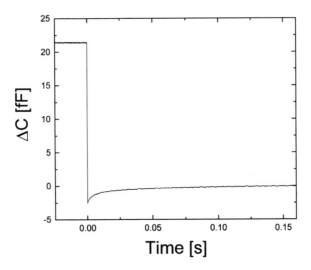

Figure 14.8 Preliminary capacitance transient data obtained from a Cu(In,Ga)Se$_2$ contact ($\sim$ 25 μm^2).

14.4 Summary

SPM-based CV, AS, and DLTS techniques can be developed through careful design and implementation. The end result is spatially resolved imaging of defect density, energy, and capture cross-section of polycrystalline materials. This instrument enables the acquisition of spatially resolved and energetically resolved electronic properties of defects.

References

1. D. V. Lang, deep-level transient spectroscopy–new method to characterize traps in semiconductors, *Journal of Applied Physics*, volume **45**, issue 7, 3023–3032 (1974).

2. H. G. Grimmeiss, Deep levels impurities in semiconductors, *Annual Review of Materials Sciences*, volume **7**, 341–376 (1977).

3. J. W. Chen and A. G. Milnes, Energy levels in silicon, *Annual Review of Materials Sciences*, volume **10**, 157–228 (1980).

4. S. T. Pantelides, The electronic structure of impurities and other point defects in semiconductors, *Review of Modern Physics*, volume **50**, issue 4, 797–858 (1978).

5. P. M. Mooney, Defect identification using capacitance spectroscopy, Chapter 2, in *Semiconductors and Semimetals*, volume 51B, Academic Press, New York (1999).

6. S. H. Wei and S. B. Zhang, Chemical trends of defect formation and doping limit in II-VI semiconductors: The case of CdTe, *Physics Review B*, volume **66**, issue 15, 155211 (2002).

7. J. V. Li, S. W. Johnsto, X. Li, D. S. Albin, T. A. Gessert, and D. H. Levi, Discussion of some trap signatures observed by admittance spectroscopy in CdTe thin-film solar cells, *Journal of Applied Physics*, volume **108**, Issue 6, 064501 (2010).

8. Landolt–Bornstein, New Series III/444B, Springer (2008).

9. O. Cojocaru-Miredin, P. Choi, D. Abou-Ras, S. S. Schmidt, R. Caballero, and D. Raabe, Characterization of grain boundaries in Cu(In,Ga)Se$_2$ films using atom-probe tomography, *IEEE Journal of Photovoltaics*, volume 1, issue 2, 207–212 (2011).

10. L. Kranz, C. Gretener, J. Perrenoud, R. Schmitt, F. Pianezzi, F. La Mattina, P. Blosch, E. Cheah, A. Chirila, C. M. Fella, H. Hagendorfer, T.

Jager, S. Nishiwaki, A. R. Uhl, S. Buecheler, and A. N. Tiwari, Doping of polycrystalline CdTe for high-efficiency solar cells on flexible metal foil, *Nature Communications*, volume 4 Article Number 2306 (2013).

11. T. Nakayama, O. Kubo, Y. Shingaya, S. Higuchi, T. Hasegawa, C. S. Jiang, T. Okuda, Y. Kuwahara, K. Takami, and M. Aono, Development and application of multiple-probe scanning probe microscope, *Advanced Materials,* volume 24, issue 13, 1675–1692 (2012).

12. C. S. Jiang, Microscopic electrical characterization of inorganic semiconductor-based solar cell materials and devices using AFM-based techniques, in *Scanning Probe Microscopy in Nanoscience And Nanotechnology*, vol 2, Book Series: Nanoscience and Technology (2011).

13. C. C. Willams, Two-dimensional dopant profiling by scanning capacitance microscopy, *Annual Review of Materials Science*, volume **29**, 471–504 (1999).

14. D. B. Holt and B. G. Yacobi, Cathodoluminescence characterization of semiconductors, Chapter 8, in *SEM Microcharacterization of Semiconductors*, ed., D. B. Holt and D. C. Joy, Published by Academic Press (1989).

15. P. F. Barbara, D. M. Adams, and D. B. O'Connor, Characterization of organic thin film materials with near-field scanning optical microscopy (NSOM), *Annual Review of Materials Science*, volume **29**, 433–469 (1999).

16. Holt D. B., The Conductive Mode, Chapter 6, in *SEM Micro-characterization of Semiconductors*, ed., D. B. Holt and D. C. Joy, Published by Academic Press (1989).

17. K. Durose, S. E. Asher, W. Jaegermann, D. Levi, B. E. McCandless, W. Metzger, H. Moutinho, P. D. Paulson, C. L. Perkins, J. R. Sites, G. Teeter, and M. Terheggen, Physical characterization of thin-film solar cells, *Progress In Photovoltaics*, volume **12**, issue 2–3, 177–217 (2004).

18. Chapters 1–4, 9, 10 in this book.

19. S. M. SZE, *Physics of Semiconductor Devices*, 2nd ed., Wiley, New York (1982).

20. E. H. Nicollian and J. R. Brews, *MOS (Metal Oxide Semiconductor) Physics and Technology*, Wiley, New York (2002).

21. D. L. Losee, Admittance spectroscopy of deep impurity levels: ZnTe Schottky barriers, *Applied Physics Letters*, volume **21**, issue 2, 54–56 (1972).

22. D. V. Lang, Recombination-enhanced reactions in semiconductors, *Annual Review of Materials Science*, volume **12**, 377–400 (1982).

23. D. K. Schroder, *Semiconductor Material and Device Characterization*, 3rd Edition, Wiley, New York (2006).

24. O. Breitenstein and J. Heydenreich, Scanning Deep Level Transient Spectroscopy, Chapter 7, in *SEM Microcharacterization of Semiconductors*, ed., D. B. Holt and D. C. Joy, Published by Academic Press (1989).

25. P. K. Paul, D. W. Cardwell, C. M. Jackson, K. Galiano, K. Aryal, J. P. Pelz, S. Marsillac, S. A. Ringel, T. J. Grassman, and A. R. Arehart, Direct nm-scale spatial mapping of traps in CIGS, *IEEE Journal of Photovoltaics*, volume 5, issue 5, 1482–1486 (2015).

26. G. Ferrari and M. Sampietro, Wide bandwidth transimpedance amplifier for extremely high sensitivity continuous measurements, *Review of Scientific Instruments*, volume **78**, 094703 (2007).

27. G. Ferrari, M. Farina, F. Guagliardo, M. Carminari, and M. Sampietro, Ultra-low-noise CMOS current preamplifier from DC to 1 MHz, *Electronics Letters*, volume **45**, issue 25, 44–45 (2009).

28. L. Fumagalli, G. Ferrari, M. Sampietro, and G. Gomila, Quantitative nanoscale dielectric microscopy of single-layer supported biomembranes, *Nano Letters*, volume **9**, issue 4, 1604–1608 (2006).

29. J. V. Li, S. W. Johnston, Y. Yan, and D. H. Li, Measuring temperature-dependent activation energy in thermally activated processes: A 2D Arrhenius plot method, *Review of Scientific Instruments*, volume **81**, 033910 (2010).

Chapter 15

Microwave Impedance Microscopy

Yong-Tao Cui[a] and Eric Yue Ma[b]

[a]*University of California, Riverside, USA*
[b]*Stanford University, USA*

yongtao.cui@ucr.edu, yuema@stanford.edu

15.1 Introduction

As demonstrated throughout this book, capacitance measurements provide essential information about electrical properties in semiconductors, which has powered the rapid development of nanoscience and technology in recent decades. As the feature size of semiconductor devices continues to scale down, it has become increasingly important to make capacitance characterization on the nanoscale. Scanning probe microscopy (SPM) is a class of techniques that use nanoscale probes, typically with sharp tips, to interact with the sample on length scales all the way down to individual atoms. Previous chapters have introduced several SPM techniques to measure the spatial profiles of capacitance in materials. In this chapter, we introduce a relatively

Capacitance Spectroscopy of Semiconductors
Edited by Jian V. Li and Giorgio Ferrari
Copyright © 2018 Pan Stanford Publishing Pte. Ltd.
ISBN 978-981-4774-54-3 (Hardcover), 978-1-315-15013-0 (eBook)
www.panstanford.com

new member of the SPM family, the microwave impedance microscopy (MIM) (Lai et al., 2008). As its name suggests, MIM measures the complex impedance (or its inverse, admittance, to be exact) between a sharp conductive tip and solid-state samples at microwave frequencies (~100 MHz to 10 GHz), through which it characterizes the local complex permittivity of the material.

The MIM technique is built upon the fundamentals of capacitance measurements. In general, capacitance characterizes the material's capability to screen an applied electrical field. Such screening capability is affected by both the dielectric and conductive properties of the material, which can be characterized by the complex permittivity $\hat{\varepsilon} = \varepsilon' + i(\varepsilon'' + \sigma/\omega)$, where ε' is the real part of the dielectric constant, ε'' the imaginary part that characterizes the dielectric loss, σ the conductivity, and ω the frequency. AC measurement of the impedance can provide information on both the real and imaginary parts of $\hat{\varepsilon}$, which represent the dielectric and conductive properties of the material. This is particularly useful for the study of semiconductors—Unlike in metals where conduction dominates its electrical response or in insulators where conduction is negligible, the conductivity in semiconductors, sensitive to many material parameters, can be comparable to dielectric screening at application-relevant frequencies and is of crucial importance to the functionality of many semiconductor devices. Therefore, the ability to characterize the conductivity in addition to the dielectric constant through nanoscale impedance measurements is highly valuable for semiconductor research.

The readers may wonder whether direct resistance measurement (current against voltage) would be more straightforward in measuring conductivity. While this is true for many cases and it has also been implemented in a version of SPM techniques known as conductive atomic force microscopy (c-AFM), it has certain limitations. In particular, the contact resistance at the tip-sample interface needs to be relatively low in order for the sample resistance to dominate. In practice, achieving a low tip-sample contact resistance is quite challenging. This could be due to a native oxidation layer forming on the surface of the sample or a Schottky contact that creates an insulating depletion layer especially in the case of a metal tip contacting semiconductors. This can be seen from analysis of the tip-sample contact and the equivalent circuit. As shown in Fig. 15.1a, the presence of the insulating layer acts as a capacitor between the tip and the sample, blocking direct low-frequency current flow while allowing high-frequency excitation used in MIM to go through. The screening-type impedance measurements are also more "local"—the sample's response only comes from a small region that the electric field reaches. In comparison, the c-AFM measurement requires the current to flow through the entire sample and be collected at a counter-electrode. This may not work for samples with both highly insulating and conductive regions, for example as shown in Fig. 15.1b.

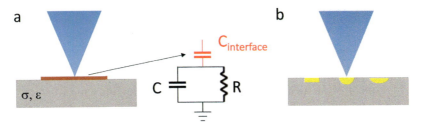

Figure 15.1 (a) A schematic of the tip-sample interface and the equivalent circuit. The red layer indicates the insulating interface layer. (b) Illustration of conductive paddles (yellow) embedded in an insulating environment (gray).

In cases of such insulating tip-sample interface there is an optimal range for the working frequency. As an empirical guidance, it can be estimated by analyzing the frequency dependence of the admittance associated with the equivalent circuit. In typical cases, the optimal frequency is achieved when the real and imaginary parts of the sample admittance are comparable, i.e., $\omega C \approx 1/R$. (The readers are encouraged to calculate this relationship.) C and $1/R$ are proportional to ε and σ, respectively, with the same geometrical dependence. So the above relationship can be converted to $\omega \varepsilon \approx \sigma$. We can then calculate this frequency value by choosing values for ε and σ that are most relevant for the materials of interest. For the case of semiconductors, the typical resistivity range is between

10^{-3} and 10^{7} $\Omega \cdot$ cm, so we choose the mid-point value, 10^{2} $\Omega \cdot$ cm and a typical value for ε is $\sim$10. This leads to an optimal frequency $\sim$2 GHz, which is in the microwave regime (0.1 to 10 GHz). In general, to probe a higher conductivity favors the use of a higher frequency.

To summarize our introduction so far, impedance measurement at microwave frequencies in a SPM configuration has the potential to provide valuable information about the nanoscale dielectric and conductive properties in semiconductors. In the remainder of this chapter, we will first describe the technical components of an MIM measurement, including the method for measuring impedance at microwave frequencies, the MIM probes, and the typical operation modes, and then provide a few examples to demonstrate how MIM can be used to obtain nanoscale electrical properties in semiconductors.

15.2 Technical Components of MIM

A typical MIM measurement includes the following components (Fig. 15.2):

(1) MIM Probe. It should allow conduction of a microwave signal to its sharp tip apex with minimum loss. We will describe several versions of MIM probes in Section 15.2.1.

(2) Scanning platform. Similar to other SPM techniques, the scanning platform should allow easy mounting of the MIM probe and raster scanning of either the probe or sample with high spatial resolution and stability. A topography feedback mechanism compatible with the probe is preferred. The control electronics should be able to record multiple external signals that are generated by the MIM electronics.

(3) MIM electronics. The MIM electronics measures both the imaginary and real parts of the tip-sample impedance with high sensitivity. We will describe a typical implementation of the MIM electronics in Section 15.2.2.

(4) MIM operation mode. To facilitate the MIM data acquisition and interpretation, we have developed several operation modes which will be described in Section 15.2.3.

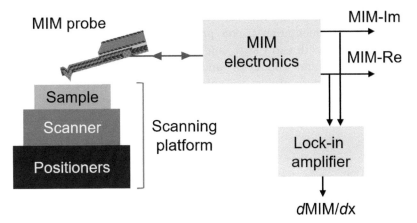

Figure 15.2 Components of MIM.

15.2.1 MIM Probes

The probe is an essential component of MIM measurement because it directly interacts with the sample: the microwave signal delivered to the probe generates a strong electric field near the tip apex which induces screening and dissipation in the sample, and the resulting tip-sample impedance is the quantity we aim to measure in MIM. The size of the tip apex primarily determines the spatial resolution of MIM, so it is important to have a robust sharp tip apex that can be maintained during scanning. To achieve these functionalities, we have developed several versions of MIM probes.

One class of probes are based on micro-fabricated cantilevers (Fig. 15.3). The structure consists of a cantilever attached to a probe body, similar to the probes commonly used in atomic force microscopy (AFM). Therefore they are fully compatible with major AFM platforms. Laser reflection off the back surface of the cantilever is used to detect the cantilever deflection and serves as the topography feedback mechanism. A metal tip is formed at the end of the cantilever, and a metal trace is deposited on the cantilever and probe body to conduct microwave signal to the tip. For such cantilever probe structure, it is important to fully shield the microwave conducting trace apart from the tip apex (Fig. 15.3a). Otherwise a large parasitic capacitance will arise and add a noisy, poorly controlled background to the total tip-sample impedance that can complicate the data interpretation. In our implementation (Gen5 probes, Fig. 15.3), the cantilever is deposited with extra Si_3N_4 and Al layers on both sides to fully shield the center conductor, and only the pyramid shaped tip is exposed. These probes are batch fabricated for small variation and consistent microwave performance among probes. Tip apex size of ~50 nm is routinely achieved.

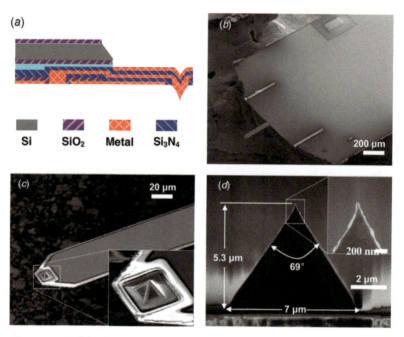

Figure 15.3 (a) Cross-sectional view of shielded cantilever probes. (b) SEM image of the handle chip. (c) Close-up view of the cantilever and metal tip. (d) Side view of the tip and its sharp apex. The diameter of the tip apex is less than 50 nm. Adapted from Yang et al. (2012).

A further development of cantilever probes uses heavily doped silicon to form the tip portion which can be made sharper than the metallic version (Gen6 probes, Fig. 15.4). A silicon piezo-resistive element is also built into the cantilever (Fig. 15.4d), which provides a mechanism to sense cantilever deflection without the need of laser reflection and can be used to provide topography feedback in situations where optical feedback is either difficult to implement (such as in cryogenic experiment) or unfavorable due to photosensitivity (such as in many semiconductors).

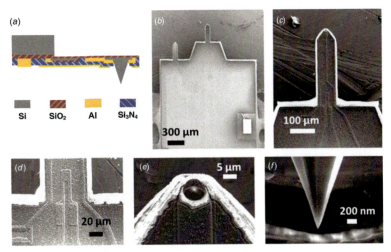

Figure 15.4 (a) Cross-sectional view of Gen6 shielded cantilever probes. (b) SEM image of the handle chip. (c) Close-up view of the cantilever. (d) Close-up view of the piezo-resistor at the root of the cantilever. (e) Close-up view of the heavily doped Si tip. (f) Side view of the tip and its sharp apex. The diameter of the tip apex is less than 50 nm. Adapted from Yang et al. (2014).

Another class of probes are based on quartz tuning forks. The quartz tuning fork (TF) is a high quality factor mechanical resonator that can be excited electrically. It has been widely exploited in the SPM community as a self-topography-sensing platform. In our implementation of the TF-based MIM sensor (Fig. 15.5), a electrochemically etched metal tip (either W or Pt/Ir), similar to those used in scanning tunneling microscope (STM), is attached to one prong of the TF. The metal wire is positioned roughly perpendicular to the sample surface. The other end of the wire is soldered to the center conductor of a coaxial cable to make microwave connection to the MIM electronics. In this configuration, although the wire is not shielded as in the cantilever probe case, the exposed wire is much further away from the sample (>100 μm) than the cantilever probes (<10 μm) such that the parasitic capacitance is much smaller, making it possible to achieve stable MIM measurement without shielding. Similar to the piezo-resistive cantilever probes, TF-based sensors are particularly useful in situations where optical feedback is not available or needs to be avoided. Its microwave performance is better than the piezo-resistive probes with silicon tips in cryogenic environments, especially at higher frequencies.

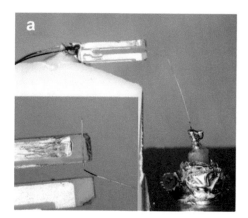
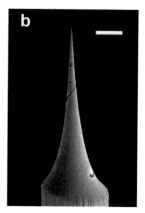

Figure 15.5 (a) Photograph of a TF-based MIM sensor. Inset shows how the etched W wire is glued to the tuning fork. (b) Scanning electron microscopy (SEM) micrograph of a typical electrochemically etched W tip. The scale bar is 10 µm. Adapted from Cui et al. (2016a).

15.2.2   MIM Electronics

A simplified equivalent circuit diagram of MIM is shown in Fig. 15.6. An incident microwave field is launched towards the probe through an impedance matching network, and the reflected signal is amplified and demodulated into two orthogonal voltage components, the MIM-Im and MIM-Re.

Because the relevant length scale of the probe and sample is much smaller than the wavelength of microwave, we are in the (electro-)quasi-static limit and can use lumped elements, namely impedance Z and its inverse admittance Y to describe the interaction between the MIM probe and the sample. We define the tip-sample admittance Y_{ts} as the admittance contribution that changes appreciably in a typical MIM scan, which is practically between the sample and the region up to few microns around the tip apex; the admittance between the rest of the probe and the ground is Y_{probe}, and can be regarded as a constant during a scan. Y_{ts} and Y_{probe} are in parallel, i.e., $Y_{total} = Y_{ts} + Y_{probe}$. The complex impedance matching network transforms Y_{ts} into a value $y(Y_{total})$ close to the characteristic admittance of the transmission line, $Y_0 = 1/50\ \Omega^{-1}$ to enhance the delivery of the microwave into the probe. The microwave reflection coefficient at the impedance matching network is thus

$$\Gamma = \frac{Y_0 - y(Y_{total})}{Y_0 + y(Y_{total})}.$$

The measured MIM signals MIM = (MIM-Re) + i(MIM-Im) is simply the reflection signal demodulated at a reference phase ϕ with respect to the incident signal:

$$\text{MIM} = V_{in}e^{i\phi}\Gamma = V_{in}e^{i\phi}\frac{Y_0 - y(Y_{total})}{Y_0 + y(Y_{total})},$$

where V_{in} is the incident microwave signal (Fig. 15.6). We can redefine it as

$$\text{MIM} = e^{i\phi}f(Y_{total}).$$

In typical experiments $|Y_{ts}| \ll |Y_{probe}|$, so we can treat Y_{ts} as a small perturbation to the total admittance and Taylor expand MIM around $Y_{total} = Y_{probe}$ to get

$$\text{MIM} \approx e^{i\phi}f(Y_{probe}) + e^{i\phi}f'(Y_{probe})Y_{ts}, \tag{15.1}$$

where $f'(Y)$ is the first derivative of $f(Y)$ with respect to Y. Therefore, $e^{i\phi}f(Y_{probe})$ and $e^{i\phi}f'(Y_{probe})$ are simply complex constants. We can further choose a demodulation phase ϕ to cancel the phase in $f'(Y_{probe})$ to make it a real number. (This can be done by, for example, scanning a purely capacitive feature while tuning ϕ.) So finally we obtain

$$\text{MIM} \approx rY_{ts} + c,$$

or more explicitly

$$(\text{MIM-Re}) + i(\text{MIM-Im}) \approx r[\text{Re}(Y_{ts}) + i\,\text{Im}(Y_{ts})] + c,$$

where r is a real constant and c is a complex constant.

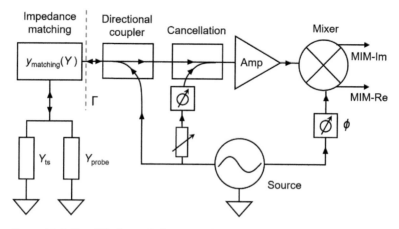

Figure 15.6 Simplified circuit diagram of MIM.

In the MIM experiment, we are more interested in the changes in MIM signals due to variations of sample's electrical properties, so a more useful form is

$$\Delta(\text{MIM-Im}) = r\text{Im}(\Delta Y_{ts}), \qquad (15.2a)$$

$$\Delta(\text{MIM-Re}) = r\text{Re}(\Delta Y_{ts}), \, r \in \mathfrak{R}. \qquad (15.2b)$$

Another trick worth mentioning is the implementation of cancellation signal. As we can see from Eq. 15.1, the background term in the MIM signal is related to the reflection determined by the total probe admittance $y(Y_{total})$. Although $y(Y_{total})$ can be very close to Y_0, the matching is never perfect so there is always a finite reflection background (typically –10 dB or more relative to the incident power). However, the signal change due to Y_{ts} is even smaller than the background. So to increase the sensitivity, we use a directional coupler to add another signal to the reflected microwave, and by carefully tuning the phase and amplitude of the added signal the background signal can be further reduced to the same level as the signal associated with Y_{ts}. After such background cancellation, further amplification can be used without saturating the amplifier. A capacitance sensitivity better than 1 aF can be routinely achieved.

In a nutshell, the MIM electronics can be regarded as a lock-in amplifier operating at microwave frequencies. The demodulated

MIM-Im and MIM-Re signals are proportional to the imaginary and real components of the tip-sample admittance. Realistic MIM electronics are more complex, but such principle holds in most cases. Next we describe how MIM signals can be manipulated and coupled with scanning operation.

15.2.3 MIM Operation Modes

15.2.3.1 Direct MIM

MIM signals are recorded as the probe is scanned on the sample surface. Typical scan rate is around 1 Hz per line with up to ~500 pixels. So we low-pass filter the MIM signals with a bandwidth of 1 kHz. This is the most straightforward method to record the MIM signal, which we refer to as the direct MIM detection. This mode can be used in almost any AFM mode, such as contact mode or tapping mode. As discussed previously, the raw MIM signals have a compensated DC background which is susceptible to long-term drift of the electronics. Line-leveling to a known reference level (e.g. the substrate, or when the tip is reasonably far away from the sample) is usually required. As a result, the MIM signals obtained are only relative values, so care is needed for the data interpretation.

Another complication of direct MIM detection is the coupling to surface topography. Even when the tip follows the surface topography during scanning, the stray coupling between the exposed portion of the microwave path and the sample still contributes a significant capacitance that varies as the tip moves up and down to follow the surface topography. This may result in artifacts that need to be addressed carefully.

15.2.3.2 Differential MIM

MIM signals can also be recorded in differential mode. This requires a modulation to Y_{ts} at a certain frequency (typically 10–100 kHz) and demodulation of the resulting oscillation amplitude of MIM signals at the same frequency. A natural way to generate the modulation is to use tapping mode AFM or TF-based sensor. In this case, the tip-sample distance is modulated by the oscillation of the cantilever/TF at the mechanical resonance frequency. The differential MIM signals demodulated at this frequency would then represent dMIM$/dz$, which is linearly proportional to dY_{ts}/dz with no undetermined offset. This can be seen by taking the derivative of Eq. 1.2:

$$\frac{d(\text{MIM} - \text{Im})}{dz} = r\frac{\text{Im}(dY_{ts})}{dz} \tag{15.3a}$$

$$\frac{d(\text{MIM} - \text{Im})}{dz} = r\frac{\text{Re}(dY_{ts})}{dz}, \ r \in \mathfrak{R}. \tag{15.3a}$$

This is a very powerful result because we now have an absolute measure of a physical quantity dY_{ts}/dz up to a scaling constant, without having to worry about any background or offset. Thanks to the noise-rejection nature of the lock-in measurements, dMIM$/dz$ images are largely drift-free, and the absolute signal at each pixel has concrete physical meaning (Fig. 15.7).

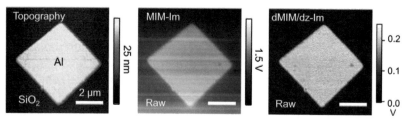

Figure 15.7 Topography, unprocessed MIM-Im and dMIM$/dz$-Im image of an Al dot on SiO$_2$ taken with a tuning fork–based MIM sensor at room temperature. Adapted from Cui et al. (2016a).

The dMIM$/dz$ mode can also reject most stray coupling from surface topography. This is because the major contribution to dY_{ts}/dz comes from near the tip apex where z_{ts} is small. Such rejection is enhanced for smaller tip vibration amplitudes and smaller average z_{ts}. Due to the same reason, the spatial resolution can also be improved by limiting the contribution further away from the tip apex.

The response curves for dMIM$/dz$ are almost identical to that for raw MIM signals, because the raw MIM response curves mostly only scale with increasing tip-sample distance.

The *d*MIM/*dz* mode is not without its disadvantages. For example, by definition it requires non-contact or tapping mode operation, and therefore would not work for samples that require hard contact to maximize sensitivity. There is also often a compromise between choosing the excitation parameters best for topography sensing and those best for maximizing the signal-to-noise ratio and spatial resolution of *d*MIM/*dz* measurements.

In addition to modulation on the tip-sample distance, other types of modulation of Y_{ts} can also be generated but usually specific to sample/experiment. For example, an oscillating voltage bias can be applied to the tip to generate an electric field (typically at 10–100 kHz) to modulate the samples' electrical properties. The resulting differential MIM signals would represent *d*MIM/*dV* or *d*MIM/*dE*. For example, it can modulate the doping level thus the conductivity in a semiconductor, which is similar to the modulation method used in scanning capacitance microscopy (SCM). In ferroelectrics, the dielectric constant can also be modulated in this way, and the differential MIM signals can be used to detect ferroelectric domains. The right choice of modulation is useful not only for removing backgrounds and drifts, but also for exploring new physics.

15.2.3.3 Multi-Frequency MIM

As we discussed briefly in the introduction, the working frequency of the impedance measurement determines the active window of the conductivity range that MIM is mostly sensitive to. So in order to make the measurements more quantitative, it is desirable to operate MIM at multiple microwave frequencies over the same region of the sample and preferably with the same tip. Practically, the working frequency is set by the impedance matching network, and we have implemented different methods to achieve multi-frequency matching for multi-frequency detection.

The first approach is to make the frequency of matched impedance tunable. For cantilever probes, Y_{probe} in the GHz frequency range are on similar orders of magnitude as Y_0, so we typically use single-stub matching or lumped-element matching (Fig. 15.8a,b). The matching frequency of such impedance matching network can be tuned with electronically tunable lumped elements. The most common example is a varactor— voltage-controlled capacitor. This allows continuous tuning of the MIM working frequency by about an octave (a factor of 2).

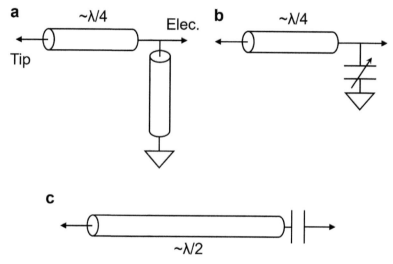

Figure 15.8 (a,b) Single-stub and lumped-element impedance matching networks for shielded cantilever probes. (c) Half-wavelength resonator matching network for TF-based sensors.

The second approach is to use a fixed matching network with multiple discrete matching frequencies. This is most easily achieved with the half-wavelength resonator matching network for a probe with a smaller Y_{probe}, such as the tuning fork–based MIM sensors (Fig. 15.8c). Figure 15.9a shows the S11 (reflection coefficient, dips means matched impedance) of such a sensor—it contains multiple resonances which corresponds to multiple available frequencies for MIM. Figures 15.9b,c show MIM-Im and dMIM/dz-Im images of a square Al dot on SiO_2 taken at 1 and 9.3 GHz respectively. The images look similar because neither Al nor SiO_2 has appreciable frequency-dependent σ or ε' in this frequency range. This method allows a frequency range up to a decade, but only at discrete values.

The added frequency degree of freedom can provide valuable information, not only for more quantitative determination of local complex permittivity, but also to identify anomalous contributions with unexpected frequency dependence.

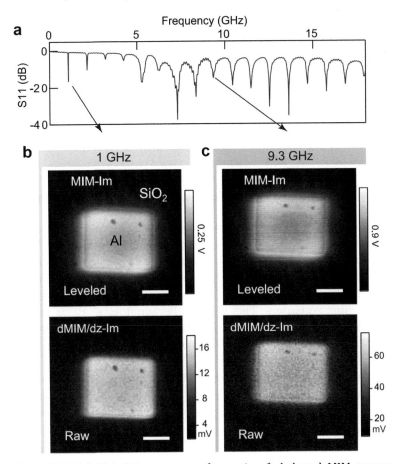

Figure 15.9 (a) The S11 spectrum of a tuning fork–based MIM sensor with half-wavelength resonator type impedance matching network. MIM-Im and dMIM/dz-Im images of an Al dot on SiO$_2$, taken at the 1 (b) and 9.3 GHz (c) resonance of the sensor at room temperature. The scale bars are 2 μm. Adapted from Cui et al. (2016a).

15.3 MIM Measurements of Semiconductors

At ambient conditions, MIM can be performed simultaneously with topography measurements on most AFM platforms. Since it measures the tip-sample admittance by analyzing the reflected microwave signal at the tip-sample interface, MIM does not require any special preparation of the sample such as a counter-electrode. The measured MIM signal is dominated by the properties from a sample volume in the vicinity of the tip-sample interface, typically on the order of the tip apex size, ~100 nm, regardless of the inhomogeneity in the sample. These features make MIM a versatile technique to study a wide variety of systems with meso-/microscopic conductivity variations. In this section, we present a few representative MIM studies on semiconductors to demonstrate its capability. We start by introducing the concept of MIM response curves.

15.3.1   MIM Response Curve with Varying Conductivity

The primary quantity MIM measures is the complex tip-sample admittance. The use of such lumped element is justified because at microwave frequencies the length scale of the tip-sample interaction region is much smaller than the microwave wavelength $(\sim 10^{-2}$ m$)$, so we can use the electro-quasi-static approximation. The tip-sample admittance Y_{ts} is solely determined by the spatial profile of the complex dielectric constant $\hat{\varepsilon}(r) = \varepsilon'(r) + i[\varepsilon''(r) + \sigma(r)/\omega]$ for the entire region around the tip-sample interface.

We can see that the $\hat{\varepsilon}(r)$ profile contains more degrees of freedom than spatial profile of $Y_{ts}(r)$ that are obtained through MIM. Therefore, it is straightforward to solve the forward problem, i.e., calculating $Y_{ts}(r)$ for any given $\hat{\varepsilon}(r)$, which can be done using finite-element method. On the other hand, it is much more difficult to solve the inverse problem, i.e., extracting $\hat{\varepsilon}(r)$ from a measured $Y_{ts}(r)$. If certain constraints of $\hat{\varepsilon}(r)$ are known, for example, the tip and sample geometry, partially known structure of $\hat{\varepsilon}$ (such as uniformity along a certain direction), etc., it is possible to simulate the forward problem (sometimes iteration is necessary) to match the measured $Y_{ts}(r)$ and determine the profile of $\hat{\varepsilon}(r)$. Therefore it is instructive to examine the response of $Y_{ts}(r)$ when a certain parameter such as ε' or σ is tuned for a well-defined geometry, which we refer to as MIM response curves.

Figure 15.10 shows an example response curve simulated for the geometry shown in panel (a). The sample has a uniform resistivity, ρ, which is the tuning parameter. Figure 15.10b plots the simulated $Im(\Delta Y_{ts})$ and $Re(\Delta Y_{ts})$ as the resistivity ρ is tuned over many orders of magnitude. $Im(\Delta Y_{ts})$ is a monotonic function of ρ that varies within a finite "sensitivity window" and saturates outside, whereas $Re(\Delta Y_{ts})$ peaks at the center of the sensitivity window and gradually decreases in both directions. This is a very general feature for MIM response curves, independent of the tip geometry: the $Im(\Delta Y_{ts})$ characterizes the screening property of the sample and therefore saturates to different levels in the conductive or insulating limit; $Re(\Delta Y_{ts})$ characterizes the dissipation and is only finite for "lossy" materials—very insulating materials are lossless and the field is so well screened in very conductive materials that the dissipation also becomes negligible. The center and width of the sensitivity window are determined by the real dielectric constant of the sample and the microwave frequency—in particular, a higher frequency will shift the sensitivity window to a higher conductivity.

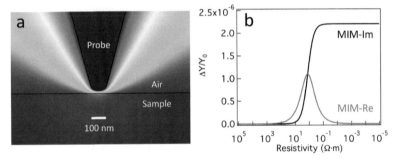

Figure 15.10 (a) The quasi-static potential profile of the simulation, showing the tip-sample geometry. (b) Simulated imaginary and real parts of the admittance contrast scaled by Y_0 as a function of resistivity.

15.3.2 Example 1: SRAM Device

The MIM response curve can then be used as a guide to interpret the MIM images qualitatively, without the need of point-by-point simulations. In particular, the MIM-Im channel, being monotonic as a function of conductivity, can be straightforwardly interpreted. Figure 15.11 shows an example study on a SRAM sample provided by Bruker Corporation, CA. The sample (Fig. 15.11a) contains regions with different doping types and concentrations, and has been used as standard test samples for scanning capacitance microscopy (SCM). In this study, in addition to AFM and MIM measurement, a low frequency bias voltage (typical amplitude of 0.5 V at 5 kHz) is applied to the tip to modulate the doping concentrations in the sample, and the resulting dMIM-Im/dV is equivalent to the SCM measurement.

Comparing the MIM (Fig. 15.11c) and SCM (Figs. 15.11d,e) results, we can see that the major features agree very well—the highly doped regions (bright color in the amplitude image of dMIM-Im/dV) also shows higher conductivity (bright color in the MIM-Im image). The MIM-Im image shows more details in the spatial profile. Interestingly, the p-type epi-layer (p-epi) region which is supposed to have the lowest doping levels shows a higher conductivity than the lightly doped regions (n-LDD and n-channel). This can be explained by analyzing the difference between MIM and SCM. In SCM, the large low-frequency voltage modulation drives carriers in and out of the sample region underneath the tip, which creates a modulation in the depth of the depletion layer that gives rise to the capacitance change. In MIM, the applied microwave voltage is on the order of millivolts which does not significantly modify the carrier densities. The probing depth of MIM is ~100 nm or so, which corresponds to only a thin layer near the surface of the epi-layer region. The pattern of the n-epi region with higher conductivity correlates with the pattern of the thin protection oxide used during the implantation step. Therefore, it is likely the dopants leaked through the thin oxide modify the doping levels in the surface layer. SCM is not sensitive to this doping variation in the thin surface layer because it probes the much thicker depletion layer modulated by the large voltage bias. On the other hand, MIM only probes the conductivity of a relatively thin layer near surface, therefore can detect such changes. (More detailed discussion can be found in Kundhikanjana et al. (2013).)

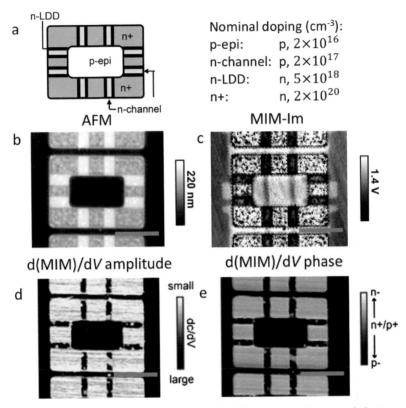

Figure 15.11 (a) The structure of the SRAM sample and nominal doping info. (b) AFM (c) MIM-Im (d) dMIM-Im/dV amplitude (e) dMIM-Im/dV phase. Adapted from Kundhikanjana et al. (2013).

15.3.3 Example 2: Standard Calibration Si Sample

To obtain more quantitative information from MIM measurement, it is necessary to perform calibration on known samples and/ or detailed simulation. Figure 15.12a shows the MIM-Im image taken on an IMEC standard n-type doped silicon sample. It has several layers with well-defined doping levels spanning a few orders of magnitude. The MIM measurement is performed using a commercial MIM electronics, ScanWave, from PrimeNano, Inc. The microwave frequency is 3 GHz. The data is taken line-by-line in a two-pass mode: for each line, the tip is first scanned across the sample in contact mode, then the tip is lifted from surface by 100 nm and a second pass is scanned by following the topography measured during the first scan so that the tip-sample distance is maintained at 100 nm during the second pass. The difference in MIM data taken during the first and second passes are then calculated. This mode is similar to the dMIM/dz mode but with a larger height differential, i.e., 100 nm vs. ~10 nm in dMIM/dz mode, but nonetheless removes the uncontrolled DC background signal.

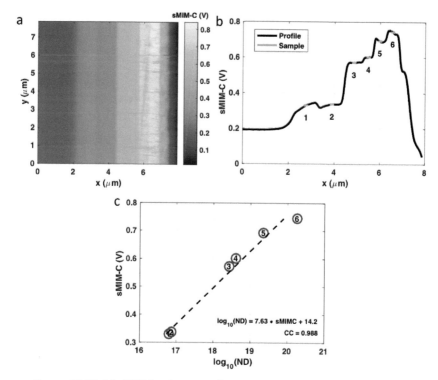

Figure 15.12 (a) MIM-Im images of an n-type IMEC standard sample. (b) The average profile with calibration samples highlighted in green. (c) Plot of the MIM-Im calibration values vs. the published values of log doping. The linear fit is a calibration that can be applied to subsequent unknown doped samples. Adapted from Friedman et al. (2016).

The obtained MIM image clearly shows the varying MIM-Im signals in layers with different doping levels. We could then extract the MIM-Im value in each region from the average line profile (Fig. 15.12b) and plot it against the known values of the doping concentration (Fig. 15.12c). It shows a largely monotonic behavior as expected—a higher doping level corresponds to a higher conductivity and in turn a higher MIM-Im signal. The dependence is approximately linear so that it can further be used to determine the doping levels in an unknown sample. It should be noted that the MIM signal is sensitive to tip and sample geometry, so such calibration procedure needs to be performed for each probe and the sample to be measured should have a similar structure as the calibration sample. In this case, the calibration sample has a uniform doping level in the depth direction, so the calibration data only applies to silicon samples (having the same dielectric constant) with uniform doping in a volume larger than the probing volume. (More details can be found in Friedman et al. (2016).)

15.3.4   Example 3: Two-Dimensional Semiconductors

Simulation can also be used to provide quantitative data interpretation, especially for samples with conductivity values in the sensitivity window of the MIM response curve. A recent study demonstrates such capability in a two-dimensional (2D) semiconductor material, MoS_2 (Wu et al., 2016). A few-layer MoS_2 flake is exfoliated on Si substrate with a 300 nm SiO_2 layer. Gold electrical contacts are made to the MoS_2 flake and a thin (15 nm) Al_2O_3 layer is then deposited to cover the entire device (Fig. 15.13a).

MIM and transport measurements are performed simultaneously as the carrier density is varied such that the Fermi level is tuned from inside the band gap to the conduction band. In this process, the source-drain conductance shows a turn-on behavior. (Fig. 15.13b) Correspondingly, the MIM images also show transitions in both MIM-Im and MIM-Re channels (Fig. 15.13c). This MoS_2 flake contains both three monolayer (3ML) and four monolayer (4ML) regions. The MIM signal in each region is rather uniform throughout the transition, and the behaviors in two regions are also similar only with a slight shift in gate values.

The overall source-drain conductance can be used to calculate the average 2D conductivity of the sample (G_{DS}). Figure 15.13d plots the MIM signals as a function of G_{DS} which show excellent agreement with the simulated result for the same geometry (Fig. 15.13e). Such agreement nicely confirms the validity of the finite element modeling of the tip-sample interaction. In this sample, there is also an interesting feature at the sample edge which is different from the bulk behavior. This shows the power of spatially resolved measurement. More detailed discussion regarding this feature can be found in Wu et al. (2016).

It should be noted that good agreement with simulation depends on the knowledge of the sample properties, including the geometry, dielectric constant, and expected uniformity of the conductivity, etc. For certain materials, the tip in contact with the sample surface can induce a thin depletion layer due to Schottky contact. Such interface layer can be tricky to simulate. So it is preferred to create a well-defined behavior at the tip-sample contact. The use of Al_2O_3 here is an example of such practice. Non-contact scans can also be used to avoid the interface issue.

15.4 Remarks and Future Directions

To summarize our discussion, microwave impedance microscopy measures the admittance associated with the tip-sample interaction through microwave electric field. By implementing sensitive microwave electronics, it resolves both the imaginary and real parts of the admittance, providing information on materials' complex permittivity on sub-100 nm length scale. The shielded cantilever probes allow stable MIM measurement and are fully compatible with commercial AFM platforms. Minimum sample preparation is required to perform MIM measurement. The more recent developments of tuning fork–based probes, differential and multi-frequency operation modes have strengthened and further expanded the capability of MIM, enabling studies on an even larger variety of systems, including semiconductors as well as other exotic materials (Lai et al., 2011; Ma et al., 2015a,b,c; Cui et al., 2016b).

On the other hand, MIM is still a relatively new technique and is constantly evolving. Many aspects are actively being improved and developed. New generations of MIM probes aim at obtaining a higher spatial resolution. This requires not only a sharper tip apex, but also improved probe designs to minimize unwanted

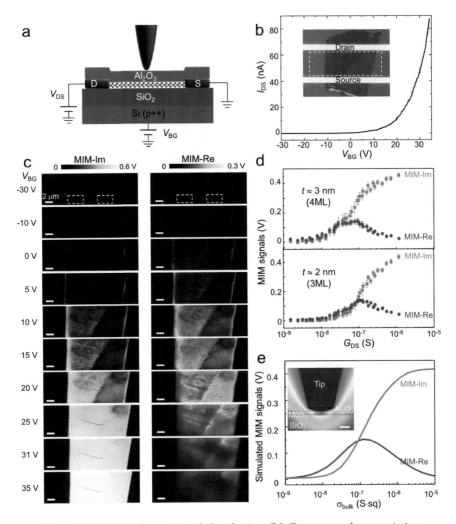

Figure 15.13 (a) Schematics of the device. (b) Transport characteristics of the device. The inset shows an optical image of the device. The white dotted box marks the scan region of the MIM measurement. (c) MIM-Im and MIM-Re images of the MoS_2 region at selected gate voltages. (d) Average MIM signals inside the white dashed boxes in (c) as a function of the source-drain conductance G_{DS}. (e) Simulated MIM signals as a function of the bulk sheet conductance σ_{bulk}. The inset shows the modeling geometry and the quasi-static potential distribution when the MoS_2 layer is insulating. The scale bar is 50 nm. Adapted from Wu et al. (2016).

stray coupling between tip and sample, such as a higher aspect ratio, fully shielded tips, probes containing non-linear material in order to create a more concentrated electric field profile, etc. New electronics are being developed to enable continuous broadband impedance spectroscopy measurements, which would improve the quantitative measurement of conductivity and enable MIM to differentiate dielectric loss from conductivity. Time-resolved MIM is also being developed to study dynamics of electronic and optoelectronic materials. Advanced algorithms will provide guidance toward solving the inverse problem more efficiently. Integrating multiple SPM capabilities into a single probe can help disentangle problems involving not only the electrical, but also magnetic degrees of freedom. With these further developments in sight, MIM is on the way to become a more powerful and standard technique that will help uncover rich science as well as provide guidance for applications.

References

Cui, Y.-T., Ma, E. Y. and Shen, Z.-X. (2016a). Quartz tuning fork based-microwave impedance microscopy, *Rev. Sci. Instrum.* **87**, p. 063711.

Cui, Y., Wen, B., Ma, E. Y., Diankov, G., Han, Z., Amet, F., Taniguchi, T., Watanabe, K., Goldhaber-Gordon, D., Dean, C. R., and Shen, Z. (2016). Unconventional correlation between quantum Hall transport quantization and bulk state filling in gated graphene devices. *Physical Review Letters*, **117**, 18, p. 186601. doi:10.1103/PhysRevLett.117.186601, URL: http://link.aps.org/doi/10.1103/PhysRevLett.117.186601.

Friedman, S., Amster O., Yang, Y. and Stanke, F. (2016). Nanoscale capacitance and capacitance-voltage curves for advanced characterization of electrical properties of Si and GaN structures using scanning microwave impedance microscopy (sMIM), ISTFA 2016: Proceedings from the 42nd International Symposium for Testing and Failure Analysis (November 6–10, 2016, Fort Worth, Texas, USA), URL https://www.asminternational.org/home/-/journal_content/56/10192/ZCP2016ISTFA080/PUBLICATION.

Kundhikanjana, W., Yang, Y., Tanga, Q., Zhang, K., Lai, K., Ma, Y., Kelly, M. A., Li, X. X. and Shen, Z.-X. (2013). Unexpected surface implanted layer in static random access memory devices observed by microwave impedance microscope, *Semicon. Sci. Tech.* **28**, p. 025010.

Lai, K., Kundhikanjana, W., Kelly, M. and Shen, Z. X. (2008). Modeling and characterization of a cantilever-based near-field scanning microwave impedance microscope. *Rev. Sci. Instrum.* **79**, p. 063703.

Lai, K., Kundhikanjana, W., Kelly, M. A., Shen, Z. X., Shabani, J. and Shayegan, M. (2011). Imaging of Coulomb-driven quantum Hall edge states, *Phys. Rev. Lett.* **107**, p. 176809.

Ma, E. Y., Bryant, B., Tokunaga, Y., Aeppli, G., Tokura, Y. and Shen, Z.-X. (2015a). Charge-order domain walls with enhanced conductivity in a layered manganite. *Nat. Commun.* **6**, p. 7595.

Ma, E. Y., Calvo, M. R., Wang, J., Lian, B., Mühlbauer, M., Brüne, C., Cui, Y.-T., Lai, K., Kundhikanjana, W., Yang, Y., Baenninger, M., König, M., Ames, C., Buhmann, H., Leubner, P., Molenkamp, L. W., Zhang, S.-C., Goldhaber-Gordon, D., Kelly, M. A. and Shen, Z.-X. (2015b). Unexpected edge conduction in mercury telluride quantum wells under broken time-reversal symmetry. *Nat. Commun.* **6**, p. 7252.

Ma, E. Y., Cui, Y.-T., Ueda, K., Tang, S., Chen, K., Tamura, N., Wu, P. M., Fujioka, J., Tokura, Y. and Shen, Z.-X. (2015c). Mobile metallic domain walls in an all-in-all-out magnetic insulator. *Science* **350**, pp. 538–541.

Wu, D., Li, X., Luan, L., Wu, X., Li, W., Yogeesh, M. N., Ghosh, R., Chu, Z., Akinwande, D., Niu, Q. and Lai, K. (2016). Uncovering edge states and electrical inhomogeneity in MoS$_2$ field-effect transistors, *Proc. Natl. Acad. Sci. U.S.A.* **113**, pp. 8583–8588.

Yang, Y., Lai, K., Tang, Q., Kundhikanjana, W., Kelly, M. A., Zhang, K., Shen, Z.-X. and Li, X. X. (2012). Batch-fabricated cantilever probes with electrical shielding for nanoscale dielectric and conductivity imaging, *J. Micromech. Microeng.* **22**, p. 115040.

Yang, Y., Ma, E. Y., Cui, Y.-T., Haemmerli, A., Lai, K., Kundhikanjana, W., Harjee, N., Pruitt, B. L., Kelly, M. and Shen, Z.-X. (2014). Shielded piezoresistivecantilever probes for nanoscale topography and electrical imaging, *J. Micromech. Microeng.* **24**, p. 045026.

Index